SCIENTIA · IMPERII · DECVS · ET · TVTAMEN
SCIENCIA

Springer Proceedings in Physics 69

Springer Proceedings in Physics

Managing Editor: H. K. V. Lotsch

44 *Optical Fiber Sensors*
Editors: H. J. Arditty, J. P. Dakin, and R. Th. Kersten

45 *Computer Simulation Studies in Condensed Matter Physics II: New Directions*
Editors: D. P. Landau, K. K. Mon, and H.-B. Schüttler

46 *Cellular Automata and Modeling of Complex Physical Systems*
Editors: P. Manneville, N. Boccara, G. Y. Vichniac, and R. Bidaux

47 *Number Theory and Physics*
Editors: J.-M. Luck, P. Moussa, and M. Waldschmidt

48 *Many-Atom Interactions in Solids*
Editors: R .M. Nieminen, M. J. Puska, and M. J. Manninen

49 *Ultrafast Phenomena in Spectroscopy*
Editors: E. Klose and B. Wilhelmi

50 *Magnetic Properties of Low-Dimensional Systems II: New Developments*
Editors: L. M. Falicov, F. Mejía-Lira, and J. L. Morán-López

51 *The Physics and Chemistry of Organic Superconductors*
Editors: G. Saito and S. Kagoshima

52 *Dynamics and Patterns in Complex Fluids: New Aspects of the Physics–Chemistry Interface*
Editors: A. Onuki and K. Kawasaki

53 *Computer Simulation Studies in Condensed Matter Physics III*
Editors: D. P. Landau, K. K. Mon, and H.-B. Schüttler

54 *Polycrystalline Semiconductors II*
Editors: J. H. Werner and H. P. Strunk

55 *Nonlinear Dynamics and Quantum Phenomena in Optical Systems*
Editors: R. Vilaseca and R. Corbalán

56 *Amorphous and Crystalline Silicon Carbide III, and Other Group IV-IV Materials*
Editors: G. L. Harris, M. G. Spencer, and C. Y.-W. Yang

57 *Evolutionary Trends in the Physical Sciences*
Editors: M. Suzuki and R. Kubo

58 *New Trends in Nuclear Collective Dynamics*
Editors: Y. Abe, H. Horiuchi, and K. Matsuyanagi

59 *Exotic Atoms in Condensed Matter*
Editors: G. Benedek and H. Schneuwly

60 *The Physics and Chemistry of Oxide Superconductors*
Editors: Y. Iye and H. Yasuoka

61 *Surface X-Ray and Neutron Scattering*
Editors: H. Zabel and I. K. Robinson

62 *Surface Science: Lectures on Basic Concepts and Applications*
Editors: F. A. Ponce and M. Cardona

63 *Coherent Raman Spectroscopy: Recent Advances*
Editors: G. Marowsky and V. V. Smirnov

64 *Superconducting Devices and Their Applications*
Editors: H. Koch and H. Lübbig

65 *Present and Future of High-Energy Physics*
Editors. K.-I. Aoki and M. Kobayashi

66 *The Structure and Conformation of Amphiphilic Membranes*
Editors: R. Lipowsky, D. Richter, and K. Kremer

67 *Nonlinearity with Disorder*
Editors: F. Abdullaev, A. R. Bishop, and S. Pnevmatikos

68 *Time-Resolved Vibrational Spectroscopy V*
Editor: H. Takahashi

69 *Evolution of Dynamical Structures in Complex Systems*
Editors: R. Friedrich and A. Wunderlin

Volumes 1–43 are listed on the back inside cover

R. Friedrich A. Wunderlin (Eds.)

Evolution of Dynamical Structures in Complex Systems

Proceedings
of the International Symposium
Stuttgart, July 16–17, 1992

With 137 Figures

Springer-Verlag
Berlin Heidelberg New York
London Paris Tokyo
Hong Kong Barcelona
Budapest

Dr. Rudolf Friedrich
Dr. Arne Wunderlin

Institut für Theoretische Physik, Universität
Pfaffenwaldring 57, W-7000 Stuttgart 80, Fed. Rep. of Germany

ISBN 3-540-55506-4 Springer-Verlag Berlin Heidelberg New York
ISBN 0-387-55506-4 Springer-Verlag New York Berlin Heidelberg

Library of Congress Cataloging-in-Publication Data
Evolution of dynamical structures in complex systems : proceedings of the international symposium, Stuttgart, July 16–17, 1992 / R. Friedrich, A. Wunderlin, eds. p. cm. – (Springer proceedings in physics ; 69) Includes bibliographical references and index. ISBN 3-540-55506-4 (alk. paper). – ISBN 0-387-55506-4 (New York : alk. paper)
1. System theory–Congresses. I. Friedrich, R. (Rudolf). II. Wunderlin, A. (Arne), 1947– . III. Series : Springer proceedings in physics ; v. 69.
Q295.E86 1992 003'.7–dc20 92-15954 CIP

Printed in Germany

Typesetting: Camera ready by authors/editors
54/3140 - 5 4 3 2 1 0 - Printed on acid-free paper

Preface

“Evolution of Dynamical Structures in Complex Systems” is dedicated to the founder of synergetics, Hermann Haken, on the occasion of his 65th birthday. This volume is an attempt to gather together and review the new results and developments achieved by researchers from various fields during the last few years. The contents bear witness to the great success in the development of general approaches to synergetic systems as well as remarkable progress in the more traditional fields of synergetics such as lasers and nonlinear optics, hydrodynamics, condensed matter physics, biology, and sociology.

Since their inception, the concepts of synergetics and rigorous mathematical theories have been extended to other scientific disciplines such as medicine, artificial intelligence and synergetic computers, and psychology. Here too, these ideas have yielded new insights, raised unexpected questions and produced innovations in both theoretical and experimental projects. The conception of self-organization, the central theme of Hermann Haken's scientific work, has stimulated epistemological studies that draw relations between synergetics and the German romantic “Naturphilosophie”. It is fascinating to observe how these intuitive notions of self-organization, etc., have now evolved into a precise and holistic scientific comprehension of synergetic systems.

We express our deep gratitude to Dr. Angela Lahee from Springer-Verlag for her valuable help during the preparation of this book.

Stuttgart
March 1992

R. Friedrich
A. Wunderlin

Contents

Part IV Biology and Medicine

Part V Artificial Intelligence and Synergetic Computers

Part VI Psychology and Social Sciences

Part VII Epistemology

Part I

General Approaches

On the Principles of Synergetics

A. Wunderlin

Institute of Theoretical Physics and Synergetics,
University of Stuttgart, Pfaffenwaldring 57/4,
W-7000 Stuttgart 80, Fed. Rep. of Germany

Abstract:

This article presents basic methods of synergetics in an elementary way and aims to collect basic notions and principles developed in the realm of synergetics as a common language for the various contributions to this book from quite different scientific disciplines.

1. Introduction

Synergetics, the science of cooperation and self-organization, was founded by Hermann Haken in the year 1969, on the occasion of a lecture course at Stuttgart University (see especially [1] for a comprehensive appreciation of his complete work). This far-reaching decision of his, which has had an impact on many different scientific fields, was strongly influenced by the following fundamental observation: Nature confronts us with a large diversity of systems including seemingly simple ones but also extremely complex systems, whose striking behaviour, remarkable features, and even existence cannot be explained and understood from a pure thermodynamic reasoning. The common characteristic of all these systems is the noteworthy phenomenon that they exhibit the surprising and distinctive property of spontaneous self-organization on macroscopic scales which results for example in a spatial, temporal, spatio-temporal ordering, and/or special functioning. Consequently, to approach these systems theoretically one has to ask - among other questions - about various limitations of one of the most fundamental classical disciplines of physics, namely thermodynamics. Pioneering ideas and suggestions became necessary to create a completely new theory consisting of fresh notions and novel principles in order to overcome the difficulties that arose.

Springer Proceedings in Physics, Vol. 69
Evolution of Dynamical Structures in Complex Systems
Editors: R. Friedrich · A. Wunderlin

Systems of the above mentioned kind are encountered in the animate as well as the inanimate world, and the number of examples is enormous [2]. Here we shall only list some of them. The most prominent representatives in the field of physics are the laser and the devices of non-linear optics, phenomena of spontaneous pattern formation in hydrodynamic arrangements, magneto-hydrodynamics, and plasma physics [3,4]. But these systems can be equally well observed in other scientific disciplines. The generation of spatial, temporal, and spatio-temporal patterns in autocatalytic chemical reactions has meantime become a field in its own right [3,5-7]. Macroscopic ordering phenomena have their reside in various domains of biology. Examples are provided by phenomena in morphogenesis [3,8,9] and the coordination of animal or human body movements [10,11]. Processes of self-organization can be identified in the blood circulation, the heartbeat as well as the still mysterious time and space dependent macroscopic behaviour of electric potentials measured in the human brain (EEG) through its different stages [13,14]. Finally the observation of spontaneous self-organization even in the human sciences should at least be mentioned [15,16].

The foundation of synergetics initiated the search for a unified theory of these important phenomena. And it should be noted that most of the fundamental questions which Hermann Haken raised at the outset were later answered exhaustively by himself (eg. [3,4,12]).

In this introduction we shall concentrate ourselves to a profound problem which has been discussed in the context of the foundation of synergetics. It reveals important relations to the question of whether there are different laws and principles separately valid in the animate and inanimate world (compare e.g. [17]). This possibility has not been excluded but rather suggested by two different principles which were formulated at the end of the nineteenth century, i.e. Darwin's principle and Boltzmann's principle. Because the dichotomy expressed through these principles was of great and fruitful importance for the development of synergetics, we shall take the opportunity to portray them to some extent (see also [18]).

1.1 Boltzmann's Principle

We shall begin with the discussion of Boltzmann's principle [19] which introduces a statistical interpretation of entropy. This principle relates quantities of macroscopic phenomenological thermodynamics such as temperature, pressure etc. to statistical properties (averages) of the microscopic motion of the particles which constitute the thermodynamic system. The correspondence between the microscopic and the macroscopic world is manifested in a universal way; that is, all general relations are formulated independently of any special system under consideration. The subject of thermodynamics, however, is exclusively confined to equilibrium states of many-body systems. In isolated systems these equilibrium states are characterized by a maximum value of the entropy S which is attributed to the various possible particular macroscopic states the system may admit:

$$S = S_{max} \quad . \tag{1.1}$$

An external disturbance acting on such a system under suitably chosen conditions will therefore reduce its entropy: $S < S_{max}$. If we isolate this system again after the disturbance, the second law of thermodynamics guarantees that only such processes can occur within the system which are consistent with the demand

$$S \rightarrow S_{max} \quad \text{(monotonically)}. \tag{1.2}$$

In all cases the final macroscopic equilibrium state corresponding to $S = S_{max}$ is assumed to be unique. Relations between distinct equilibrium states of a system, which are adopted in cases of different experimental preparations (constraints), can be accomplished by Carnot cycles where heat and work is exchanged with the external world in a reversible manner.

Clausius introduced the notion of entropy in a thermostatic and more formal manner [20]. The statistical interpretation of this fundamental quantity is due to Boltzmann. Following Boltzmann, the entropy is related to the number of microscopic states W which are connected to the realization of a given macroscopic state characterized by S through:

$$S = k_B \ln W \quad , \tag{1.3}$$

where k_B denotes Boltzmann's constant. From (1.1) we conclude that the equilibrium state of a system is distinguished by

$$W = W_{max} \quad . \tag{1.4}$$

We can therefore propose that the uncertainty of our knowledge about the actual microscopic state which realizes a given macroscopic state becomes maximal in the case of equilibrium states. It is in this sense that we can attribute to equilibrium states the notion of maximal microscopic disorder (under given external conditions). From (1.2) we immediately extract the consequence that an isolated thermodynamic system in the course of time proceeds monotonically to a state of maximal disorder.

1.2 Darwin's Principle

From the above considerations it becomes evident that the equilibrium states of thermodynamics exclude the existence of a living biological system. In fact, approximately at the same time a completely different principle has been formulated for living systems, i.e. Darwin's principle [21]. In a more popular version it states that the development of species to more and more complex entities has to be understood as a result of defective reproduction. The resulting mutants start to compete with the original species and eventually the fittest version of the different copies will survive. The result of this competition might yield qualitative changes and provide a path to more and more complex and higher forms of species. Clearly, this mechanism should be at work at each stage of development.

1.3 Beyond Thermodynamics

Not all of the problems which result from confronting the consequences of these principles have yet been solved completely. However, it became evident from incompatibilities that one had to follow new roads, to develop original and unusual ideas to resolve inconsistencies. A solution had to go far beyond the presuppositions of thermodynamics. This attitude gains still more weight when one is not prepared to accept different laws and principles for the animate and inanimate world. With this in mind new notions have been developed in synergetics and previously known ones have been given new interpretations during the last twenty years and more. The novel principles which have been discovered seem to be equally valid in the animate and inanimate world. Their contents are the main subject of this contribution which is a tutorial introduction to synergetic notions and principles and aims to provide an understanding of the more specialist contributions to this book from various fields.

The article is organized as follows. Section 2 presents a characterization of the systems of synergetics by summarizing the main ingredients. Sufficient conditions are presented which guarantee the ability of a complex system to exhibit processes of spontaneous self-organization on macroscopic scales. In section 3 the single mode laser is treated as a special example in a rather pedagogical way, following a route which has been suggested by Hermann Haken [3]. Many of the novel fundamental steps and interpretations introduced in synergetics can be exhibited in a simple way. They are executed in some detail to gain insight into how the notion of self-organization can be formulated concisely in mathematical terms. The general theory of synergetic systems is outlined in section 4 starting from a mesoscopic level. In section 5 a collection of other systems is listed which can be handled from a point of view which is suggested by the general method of section 4. There are situations where this general theory is no longer applicable, i.e. when the subsystems become so complicated that they cannot be properly handeled in mathematical terms. This yields the need for the development of phenomenological (section 6) and macroscopic synergetics (section 7). In the latter formulation the special role of fluctuations, which can even become constitutive, is taken into account and included into the macroscopic considerations. This is performed through the discussion of the relation between the order parameter concept and the principle of maximizing information entropy. Section 8 presents an important application of synergetics, the synergetic computer, and section 9 draws some general conclusions.

2. Ingredients of the Systems of Synergetics

Several important common features can be identified in the systems treated in synergetics. Taking another point of view they can be considered as necessary conditions for the occurrence of self-organization [3,4,12]. All of these systems are open systems, i.e. there is a continuous flow of energy, matter, information etc. through them. In the following we shall measure these flows by a set of control parameters $\{\sigma\}$:

$$\{\sigma\} = \sigma_1, \sigma_2, \ldots \tag{2.1}$$

Processes of self-organization may also occur in closed systems which then have to be initially prepared in an extreme non-equilibrium state. Self-organization is then observed through transients. Furthermore, the control parameters can become functions of time which in effect yields the sweeping through an instability. In situations where these parameters cannot be controlled perfectly from outside they have to be considered as fluctuating quantities. Here, however, we shall merely assume the control parameters to be quasi-static quantities.

The second ingredient of the systems we shall deal with in the following is that they are built from many subsystems. These can be of a single type as will be the case in a model of the laser, or of different kinds such as in chemical reactions, for example. The subsystems define the mesoscopic scale of our analysis in space and time. When these subsystems are simple enough we can define a state vector $\underset{\sim}{U}$ which is constructed from all the variables which characterize states of the subsystems and their interactions.

Examples are provided by the following systems: In a solid state laser the variables are the amplitudes of the different modes of the light field, the polarization of the atoms, and their inversion when they are considered as two level atoms. In hydrodynamics the subsystems are fluid elements; they are described by their density, position, and momentum. In chemistry the concentrations of different reactants form the set of subsystems, etc.

The set of state vectors form the state space Γ. Roughly we may say that when we know the state vector of a system under consideration, we possess exhaustive knowledge about its mesoscopic state. (This statement has to be modified correspondingly when we allow for fluctuations within the system or external fluctuations.) For these reasons we expect the time evolution of the state vector to be of great importance for our knowledge about a system when we become interested in predictions about its behaviour in the future. In mathematical terms this is condensed in an evolution equation for the particular system. For a large class of systems this equation can be written in the (symbolic) form

$$\dot{\underset{\sim}{U}} = \underset{\sim}{N}(\underset{\sim}{U},\nabla,\{\sigma\},t) + \underset{\sim}{F} \quad , \tag{2.2}$$

where $\dot{} = \partial_t$, the derivative with respect to time t. $\underset{\sim}{N}$ denotes a vector field which depends nonlinearly on $\underset{\sim}{U}$. Spatial inhomogeneities are represented through the ∇-symbol. Because we are considering open systems the vector field can be influenced by changes in the control parameters. This is taken into account through the dependence of $\underset{\sim}{N}$ from $\{\sigma\}$. Finally, in an externally forced system, the vector field can depend explicitly on time t. $\underset{\sim}{N}$ fixes that contribution to the systems dynamics which is referred to as the systematic or deterministic part of its motion. The second term on the right side of (2.2) accounts (symbolically) for the fluctuations. When we have

$$d\underset{\sim}{F} = f(\underset{\sim}{U}, \ldots)d\underset{\sim}{\eta}(t) \quad , \tag{2.3}$$

where the differentiable matrix f is also allowed to depend on the control parameters, and $\underset{\sim}{\eta}(t)$ corresponds to gaussian white noise with zero mean, i.e.

$$\langle\underset{\sim}{\eta}(t)\rangle = 0 \quad ,$$

$$\langle\underset{\sim}{\eta}(t)\underset{\sim}{\eta}(t')\rangle = I\,\delta(t-t') \tag{2.4}$$

(I is the unit matrix $I_{ik} = \delta_{ik}$), one has to consider (2.2) as an Ito or Stratonovich stochastic differential equation. From a mathematical point of view we can say that (2.2) defines a system on a mesoscopic level.

If equations (2.2) model a physical system or a chemical reaction we expect that for a certain well-defined parameter range σ, they have to represent thermodynamic equilibrium as a universal attractor of the deterministic part as well as appropriately model the behaviour of the system in its neighborhood (for example in the region of linear irreversible thermodynamics). Clearly this also means that dissipation is included in the equations.

Various generalizations of the evolution equations (2.2) exist. On a mesoscopic scale there can be important non-local effects in time and space which are the result of memory and interactions over long distances in the system, respectively. Time delay effects might also be included. Furthermore, on microscopic scales which are distingished in that we have to take quantum mechanics into account to describe the system properly, these equations have to be interpreted as Heisenberg equations of motion of operators which are attributed to the observables [22]. In the following we shall confine ourselves to evolution equations of the type (2.2). Before examining these equations generally, we shall discuss the single mode laser as an example.

3. Determination of States of Self-Organization from the Evolution Equations. The Single Mode Laser as a Paradigm of Synergetics

In exploiting the evolution equations (2.2) one should be

tacitly aware of the fact that we are confronted with a complicated set of non-linear stochastic differential equations in a high dimensional phase space Γ. There exists no general recipe for solving such a difficult set of equations. To give an impression of the difficulties which had to be overcome, we shall discuss the historically most prominent example, the solid state laser, guided by the method introduced by Hermann Haken [3,22].

3.1 Model of a Laser

A solid state laser consists of a host crystal, where the laser active atoms are embedded. The crystal is placed into an optical cavity. For simplicity these atoms are treated here as two level atoms. They are assumed to be located far enough from each other in a way that their wave functions do not overlap.

Obviously we have a system which fulfills all criteria of a synergetic system. It is an open system (the laser is pumped from outside), furthermore it is composed of many (identical) subsystems, the laser active atoms, and the system has the ability of spontaneous self-organization on macroscopic scales. Self-organization will be manifested in the coherent action of the laser active atoms over the volume of the whole crystal.

The light field of the laser within the cavity can be described by its electric field strength $\underset{\sim}{E}(\underset{\sim}{x},t)$, where $\underset{\sim}{x}$ denotes the space coordinate. The electric field is decomposed into the eigenmodes $\underset{\sim}{u}_\lambda(\underset{\sim}{x})$ of the cavity according to

$$\underset{\sim}{E}(\underset{\sim}{x},t) = \sum_\lambda N_\lambda [b_\lambda(t) - b_\lambda^*(t)] \underset{\sim}{u}_\lambda(\underset{\sim}{x}) \quad , \tag{3.1}$$

where N_λ is a normalization factor and the set $b_\lambda(t)$ denotes the complex amplitudes which measure the strength of excitation and the phase of the different modes λ.

The state of the two-level atoms is characterized through their polarization $\underset{\sim}{P}(\underset{\sim}{x},t)$ and inversion density $D(\underset{\sim}{x},t)$. Both of these mesoscopic densities are divisable into the contributions from the single atoms. When we enumerate the atoms by the index μ and describe their location by the vector $\underset{\sim}{x}_\mu$ we obtain for the polarization density

$$\underset{\sim}{P}(\underset{\sim}{x},t) = \sum_\mu \underset{\sim}{p}_\mu(t)\ \delta(\underset{\sim}{x} - \underset{\sim}{x}_\mu) \tag{3.2}$$

where $\underset{\sim}{p}_\mu(t)$ is the polarization of the atom μ. Expressing as usual $\underset{\sim}{p}_\mu(t)$ through the atomic dipole matrix element of the two-level atoms $\underset{\sim}{\Theta}_{12}$

$$\underset{\sim}{p}_\mu = -\ \alpha_\mu(t)\underset{\sim}{\Theta}_{12} - \text{c.c.}\ , \tag{3.3}$$

$\alpha_\mu(t)$ becomes a complex measure of the polarization of the single atom μ. Correspondingly $d_\mu(t)$, the inversion of the single atoms, constitutes the inversion density

$$D(\underset{\sim}{x},t) = \sum_\mu d_\mu\ \delta(\underset{\sim}{x} - \underset{\sim}{x}_\mu)\ . \tag{3.4}$$

In a semiclassical theory, where the light field is assumed to behave classically and the atoms are treated quantum mechanically, one derives by the application of well established approximations the following closed set of equations for the interaction of light and matter in a given optical resonator

$$\dot{b}_\lambda = -\ (i\omega_\lambda + \kappa_\lambda)b_\lambda - i\sum_\mu g^*_{\mu\lambda}\alpha_\mu + F_\lambda \tag{3.5}$$

$$\dot{\alpha}_\mu = -\ (i\nu + \gamma)\alpha_\mu + i\sum_\lambda g_{\mu\lambda}b_\lambda d_\mu + \Gamma_\mu \tag{3.6}$$

and

$$\dot{d}_\mu = \gamma_{11}(d_0 - d_\mu) + 2i\sum_\lambda(g^*_{\mu\lambda}\alpha_\mu b^*_\lambda - \text{c.c.}) + \Gamma_{d\mu} \tag{3.7}$$

These equations have been written in their spatially homogeneous version, where all the atoms are treated equally. ω_λ denotes the frequencies which are associated with the cavity modes, ν the transition frequency of the single atom. Damping terms have been added: κ_λ describes the losses in the cavity, γ is attributed to the finite atomic life time and γ_{11} to the damping of the atomic inversion. The quantity $g_{\mu\lambda}$ characterizes the interaction of the field and the atoms. For the sake of completeness we have added fluctuating forces F_λ, Γ_μ, and $\Gamma_{d\mu}$. Their detailed properties can be derived from a complete microscopic quantum field theoretical treatment of the system, when it is additionally coupled to different heat baths which provide damping and fluctuations in a way consistent with quantum mechanics [22]. The so-called unsaturated inversion d_0 takes the role of the control parameter which measures the external pumping of the system. We do not consider changes in the cavity dimension. We note that we have a set of basically non-linear equations which defines the mesoscopic level of consideration and can be seen as a special realization of the general form (2.2).

The number of equations (3.5-7) is enormous: about 10^{18}. Clearly, the remark made at the beginning of this section applies: It is impossible to present a complete solution for this set of equations. On the other hand it appears useless even to know the complete solutions because it is impossible to manage the huge amount of information needed to specify initial conditions and their transformation in time along the solution curves. Accepting this point of view the situation turns out to be rather similar to problems which are related to the foundations of statistical mechanics (e.g.[23]). As we shall see, however, the solution here will be quite different.

3.2 Collective Variables, Order Parameters, and Slaving

In order to demonstrate how to cope with these equations successfully, we shall choose a heuristic way which will be exploited in the next section when we discuss the general formalism. We shall assume that the inversion of the atoms is homogeneous over the whole crystal

$$d_\mu = d \quad . \tag{3.8}$$

Under this circumstance it becomes possible to prove that only one mode λ of the light field can be macroscopically excited. The system prefers the mode which is in resonance with the atomic frequency ν. Taking this for granted we can suppress the index λ and introduce the collective variables P and D in a straightforward way. We use

$$P = \sum_{\mu} g_{\mu}^{*} \alpha_{\mu} \tag{3.9}$$

and

$$D = \sum_{\mu} d_{\mu} = Nd \quad , \quad D_0 = \sum_{\mu} d_0 \quad , \tag{3.10}$$

where N denotes the total number of laser active atoms. Transforming the variables into a rotating frame by setting

$$P = \tilde{P}\, e^{-i\nu t} \quad \text{and} \quad b = \tilde{b}\, e^{-i\nu t} \tag{3.11}$$

we drop the tilde and choose the coupling constant g independent of the atom μ to obtain

$$\dot{b} = -\kappa b - iP + F \quad , \tag{3.12}$$

$$\dot{P} = -\gamma P + i|g|^2 Db + \Gamma \quad , \tag{3.13}$$

$$\dot{D} = \gamma_{11}(D_0 - D) + 2i(Pb^* - c.c.) + \Gamma_D \quad . \tag{3.14}$$

Here the fluctuations are defined corresponding to the transformations (3.9-11). For the time being we shall neglect these fluctuations and concentrate on the systematic motion inherent to these equations. The following observation is central to a further simplification of the still complicated equations (3.12-14): For a good cavity laser there exists a pronounced hierarchy in the damping constants

$$\kappa \ll \gamma_{11} \ll \gamma \ . \tag{3.15}$$

It is important to note that these damping constants can be attributed to characteristic time scales on which the different variables change.

At this stage the most fundamental principle of synergetics, the slaving principle, comes into play. On the level of our heuristic considerations we are led to the conclusion that on macroscopic scales the slowest variable will rule the behaviour of the complex many body system. Indeed we can use the observation expressed through (3.15) to approximately solve the equations for the fast variables. In a first step we formally integrate (3.13) to obtain

$$P(t) = i|g|^2 \int_{-\infty}^{t} \exp[-(\gamma(t-\tau)]D(\tau)b(\tau)d\tau \quad . \tag{3.16}$$

Here we have confined ourselves to the long term behaviour to avoid transient solutions resulting from the initial conditions. Now we make use of the hierarchy in time scales (3.15). The characteristic time scales corresponding to the changes of the inversion and the field mode are much larger than that of the polarization. As a consequence we shall expect that the polarization has already reached an equilibrium value which, however, is prescribed by the slow variation of the inversion and the field mode. Mathematically this fact can be expressed through the following approximation

$$P(t) \approx i|g|^2 D(t)b(t) \int_{-\infty}^{t} \exp[-(\gamma(t-\tau)]d\tau \quad . \tag{3.17}$$

Equation (3.17) reflects the fact that the polarization follows the inversion and the field instantaneously (adiabatic principle). In other words, the long lived variables enslave the behaviour of the short lived ones. Performing the elementary integration remaining in (3.17) we obtain

$$P(t) \approx i|g|^2 D(t)b(t)/\gamma \quad . \tag{3.18}$$

Relation (3.18) yields a considerable simplification of the set of differential equations (3.12-14):

$$\dot{b} = -\kappa b + (|g|^2/\gamma)D(t)b(t) \tag{3.19}$$

and

$$\dot{D} = \gamma_{11}(D_0 - D) - 4(|g|^2/\gamma)D(t)|b(t)|^2 \quad . \tag{3.20}$$

In a similar fashion we can again apply the idea of slaving by taking into account the first pair in the hierarchy (3.15). The formal integration of (3.20) yields

$$D = D_0 - (4|g|^2/\gamma) \int_{-\infty}^{t} \exp[-\gamma_{11}(t-\tau)]D(\tau)|b(\tau)|^2 d\tau \quad . \tag{3.21}$$

This equation can be solved iteratively by a standard procedure for the solution of Volterra integral equations of the second kind. We put

$$D \approx D^{(1)} + D^{(2)} + \dots , \qquad (3.22)$$

assuming the contribution of the integral in (3.21) to be a small quantity. By comparing different orders of magnitude we then obtain

$$D^{(1)} = D_0 \qquad (3.23)$$

and

$$D^{(2)} = -(4|g|^2/\gamma) \int_{-\infty}^{t} e^{-\gamma_{11}(t-\tau)} D_0|b(\tau|^2 d\tau \approx -\frac{4|g|^2}{\gamma\gamma_{11}} D_0|b(t)|^2 . \qquad (3.24)$$

In evaluating (3.24) we have again taken into account consequences of the different roles of slow and fast variables. Inserting the result (3.24) in the equation for the field mode (3.19) we finally obtain

$$\dot{b} = -(\kappa - |g|^2 D_0/\gamma)b - 4|g|^4 D_0(\gamma^2\gamma_{11})^{-1}|b|^2 b . \qquad (3.25)$$

Equation (3.25) represents the basic result which we have obtained from a heuristic application of the slaving principle: The complete macroscopic time dependent action of the complex laser system can be exhaustively understood from an effective equation of motion of one variable, namely the complex field amplitude b(t) of the surviving mode. In synergetics Hermann Haken has coined the notion "order parameter" for this most important variable. Its value determines the inversion D and the polarization P through (3.24) and (3.18), respectively. They play the role of enslaved variables. The instantaneous value of the order parameter can be considered as a qualitative and quantitative measure of spontaneous macroscopic ordering of the system. In the case of a single mode laser the newly evolving stable state is observed above the so-called laser threshold which is marked by the condition

$$D_0 = \kappa\gamma|g|^{-2}. \tag{3.26}$$

3.3 A Discussion of the Results

The idea of slaving in connection with the occurrence of collective modes has considerably reduced the complicated set of original equations (3.5-7) to an analytically manageable equation of one complex order parameter b(t). It therefore turns out that the slaving principle not only becomes of fundamental importance from a theoretical point of view but also provides us with a practically applicable tool to handle quite complex systems with comparatively simple non-linear equations of motion: The spontaneous formation of macroscopic ordering far from thermal equilibrium is systematically mapped by time scale arguments onto the motion of - in general - few order parameters.

There are further general aspects which can be abstracted from the laser example. Here we especially mention the notion of non-equilibrium potentials - a concept introduced by Graham and Haken [24] and extended by Graham and his coworkers (e.g. [25]). The introduction of non-equilibrium potentials has been achieved through the observation that we can write (3.25) in the form

$$\dot{b} = -\frac{\partial V}{\partial b^*}, \tag{3.27}$$

if we choose V (up to an arbitrary constant)

$$V = (\kappa - |g|^2 D_0/\gamma)|b|^2 + 2|g|^4 D_0(\gamma^2\gamma_{11})^{-1}|b|^4 . \tag{3.28}$$

Equation (3.27) can now be interpreted as representing the overdamped motion of a particle in the potential $V(b,b^*)$. The minima of the potential become the stable stationary points of the system the maxima correspond to unstable ones. V therefore has all the properties of a Lyapunov function of the system. For this reason $V(b,b^*)$ rules the stability (and also the fluctuations) of the system. The shape of the potential V is changed characteristically by the variation of the control parameter D_0. The new ordered state above the laser threshold is reached through an instability and the minimum finally assumed is selected purely by chance (when for simplicity b is considered as a real variable), which reflects the role of the fluctuations. The transition can be classified as a symmetry breaking transition where symmetry is restored through the action of the fluctuations. Following Haken we can give these observations the rank of principles: Ordered states are created through <u>instabilities</u> far from thermal equilibrium and <u>fluctuations</u> are responsible for which of the possible ordered states of a system will eventually be acquired in a particular realization.

There is an important remark which lays the basis for the notion of self-organization. Quite unspecific changes in a control parameter, the external pump of the laser, generate a macroscopically highly ordered state. The evolving macroscopic order is not prescribed by external influences but has to be considered as the result of the internal dynamics of the system itself.

Another comment concerns the laser as a system of considerable importance for the foundation of synergetics. The laser model as described above is the best known system of synergetics: The general theory starts from first principles of physics, i.e. quantum mechanics and quantum electrodynamics. Each step in the theory can be controlled precisely up to the prediction of the macroscopically evolving state. It therefore becomes possible in the case of the laser to derive the macroscopic ordered states from a complete microscopic theory.

3.4 A Remark About Chaos in the Laser System

We have taken the single mode laser instability, which is connected with the transition from a lamp to laser action, as the simplest representative of an instability leading from microscopic chaos (in the sense of Boltzmann) to the spontaneous macroscopic ordering of a system. However, there exist higher order instabilities [3,22,26] which will also be the concern of several articles in this book. Furthermore there is a possibility to observe chaotic states of the single mode laser. In fact, when we choose the field mode b(t) in (3.12-14) as a real variable and P(t) as purely imaginary the equations can be mapped onto the Lorenz equations [27] by scaling and shifting the variables. When the cavity damping becomes large the conditions for the Lorenz instability can be met and a strange behaviour of the laser light is yielded. Taking into account that b, P, and D are collective variables of the system, we may interpret this behaviour as macroscopically chaotic, generated by the three collective variables of the system. They can again be considered as order parameters: All other degrees of freedom are enslaved by these few macroscopic variables. We conclude that the concept of order parameters and slaving can be extended and proves to be of importance even when states of macroscopic chaos occur in a system.

4. Outline of the Systematic Theory

A systematic theory has to start from equations (2.2) on a mesoscopic level. The concrete interpretation of the variables is fixed by the context in which the equations are developed and analyzed. An important step consists in the construction of appropriate collective modes for the system. We can perform this task systematically by confining our attention to a reference state in the state space which we shall denote by $\underset{\sim}{U}_0$. We shall consider the reference state to be a solution of the purely deterministic version of equations (2.2)

$$\dot{\underset{\sim}{U}}_0 = \underset{\sim}{N}(\underset{\sim}{U}_0, \nabla, \{\sigma\}) \quad . \tag{4.1}$$

$\underset{\sim}{U}_0$ is assumed to be a comparatively simple object in phase space. Such objects correspond to by low dimensional attractors. In the simplest case $\underset{\sim}{U}_0$ is a stationary state, i.e.

$$\dot{\underset{\sim}{U}}_0 = 0 \quad . \tag{4.2}$$

$\underset{\sim}{U}_0$ is then taken as a solution of the equation

$$\underset{\sim}{N}(\underset{\sim}{U}_0,\nabla,\{\sigma\}) = 0 \quad . \tag{4.3}$$

Time dependent reference solutions are also allowed, e.g. limit cycles, where

$$\underset{\sim}{U}_0(t) = \underset{\sim}{U}_0(t + T) \quad , \tag{4.4}$$

and where T measures the time for the smallest period. Still more complicated situations are accessible, such as quasiperiodic states, etc., see [4,28].

A considerable simplification results from the assumption that we can confine ourselves to some limited region around the reference state. This means that we do not aim to explore the whole state space of the system, but only the reference state itself and a certain piece of its neighbourhood. As a consequence we expect that the collective modes will also depend strongly on the chosen reference state. Thus our theory will be nonlinear but in general only locally valid in phase space.

For the time being we shall assume that we have chosen the control parameters in such a way that the reference state is asymptotically stable in the sense of Lyapunov. This can be examined satisfactorily by linear stability analysis. The reason is, loosely speaking, that when the system is linearly stable there always exists a neighborhood of the reference state in which the complete non-linear deterministic system shows the same stability behaviour. Furthermore, there is an orientation preserving homomorphism between the corresponding trajectories. (These facts are rigorously formulated in the Hartman-Grobman theorem, e.g.[28]).

The linearized deterministic version of (2.2) in the neighbourhood of the reference state can be constructed via the ansatz

$$\underset{\sim}{U} = \underset{\sim}{U}_0 + \underset{\sim}{q} \quad , \tag{4.5}$$

where $\underset{\sim}{q}$ measures the deviation from the stationary state $\underset{\sim}{U}_0$. Using (2.2) and (4.1) we obtain the equation for $\underset{\sim}{q}$ in the form

$$\dot{\underset{\sim}{q}} = \underset{\sim}{N}(\underset{\sim}{U}_0 + \underset{\sim}{q},\nabla,\{\sigma\}) - \underset{\sim}{N}(\underset{\sim}{U}_0,\nabla,\{\sigma\}) + \underset{\sim}{F} \quad , \tag{4.6}$$

and its linearized deterministic version reads

$$\dot{\underset{\sim}{q}} = L(\underset{\sim}{U}_0, \nabla, \{\sigma\})\underset{\sim}{q} \quad , \tag{4.7}$$

where the elements of the linear matrix $L = (L_{ik})$ are given by

$$L_{ik} = \partial N_i/\partial U_k|_{\underset{\sim}{U}_0} \quad . \tag{4.8}$$

Note that the linearized matrix depends on the reference state $\underset{\sim}{U}_0$ and the set of control parameters $\{\sigma\}$. L may also depend on gradients (compare (4.7)). The linear problem is solved by transforming (4.7) to an algebraic eigenvalue problem (which becomes especially simple when equations (2.2) are independent of gradients). This is achieved by putting

$$\underset{\sim}{q} = \underset{\sim}{q}_0 \exp(\lambda t) \quad , \tag{4.9}$$

from which we get

$$\lambda \underset{\sim}{q}_0 = L \underset{\sim}{q}_0 \quad . \tag{4.10}$$

The solution of (4.10) gives the eigenvalues λ_j and the corresponding right-hand and left-hand eigenvectors $\underset{\sim}{O}_j$ and $\bar{\underset{\sim}{O}}_j$. These eigenvectors form a biorthonormal set and span the state space. The eigenvectors $\underset{\sim}{O}_j$ are identified with the collective modes of the system. In more general cases, where additional space dependence is allowed, (4.10) defines a linear partial differential equation in real space which must be solved using appropriate boundary conditions [3,4]. Additional problems connected with infinitely extended systems then arise (see the article of M. Bestehorn).

When $\underset{\sim}{U}_0$ represents a periodic motion, the corresponding linear problem consists of the determination of the Floquet exponents and eigenvectors which have to fulfill Floquet's theorem. In a situation where the reference state is quasiperiodic, Haken [4] has given general conditions which guarantee the existence of a complete set of eigenvectors. The general method which will be outlined below is therefore applicable.

If we choose $\underset{\sim}{U}_0$ as a stationary state it becomes evident from (4.9) that the reference state $\underset{\sim}{U}_0$ is stable as long as

$$\mathrm{Re}\ \lambda_j < 0 \text{ for all } j. \tag{4.11}$$

The eigenvalues can now be influenced from outside by changing the control parameters. The equation

$$\mathrm{Re}\ \lambda_j(\sigma) = 0 \tag{4.12}$$

for some j then marks the border between linear stability and instability of the given reference state. Geometrically (4.12) is interpreted as a (critical) surface of codimension 1 in the the parameter-space $\{\sigma\}$. A discussion of these different surfaces (enumerated by the index j) provides valuable insight into the possible - and observable - instabilities [28] which can occur along our path chosen through the space of control parameters. The simplest example is a situation where we move along a one-dimensional manifold through this space. We will then generically meet just one of the surfaces determined by (4.12). Other, more complicated versions of instabilities, cannot be observed in a real system because a small disturbance will push us away from such an instability. Indeed, a more complicated instability would arise when two eigenvalues change the sign of their real part at the same place. In the control parameter space this corresponds two of the manifolds (4.12) cutting each other, or, in other words, to a manifold of codimension 2, which generically cannot be met when we move along a one-dimensional manifold.

The instabilities (L is assumed to be a real matrix) which become possible in our elementary example are connected with a change in sign of one real eigenvalue or of the real part of two conjugate complex eigenvalues. (Clearly, when we move the control parameters along a manifold of higher dimension the number of possible instabilities correspondingly increases.)

Close to an instability, considerations based on a linear analysis are no longer adequate to predict the behaviour of a system. On the other hand we note that it is not possible to appropriately treat the full equations of the system. The problem therefore consists in filtering out the relevant nonlinear contributions which leads to an exhaustive description of the system on macroscopic scales. This goal can be achieved by classifying the collective variables by their distinct behaviour near an instability. Passing the instability surface in phase space, some of the collective variables become linearly unstable; most of them, however, will remain linearly stable. This observation can be used to split the following general ansatz for $\underset{\sim}{q}(t)$

$$\underset{\sim}{q}(t) = \sum_j \xi_j(t) \underset{\sim}{O}_j \quad , \tag{4.13}$$

where we divide the set $\{\xi(t)\}$ into stable modes which are collected into a vector $s(t)$ and unstable ones, $\underset{\sim}{u}(t)$. The unstable slow modes will dominate the macroscopic behaviour of the system beyond the instability. In other words, the system generates a hierarchy in time scales near an instability, similar to that observed in a laser. To profit from this perception we shall first transform the equation of motion (4.6) into an equation for the collective modes $\xi(t)$. This can be performed by using the biorthogonality of right-hand and left-hand eigenvectors. The result reads

$$\dot{\xi}_j = \Lambda_{jk}\xi_k + N_j(\{\xi\}) + F_j \quad . \tag{4.14}$$

Here Λ denotes the diagonalized version of the linear matrix L or its Jordan form. (Summation is taken over dummy indices.) N contains the remaining nonlinear contributions of the original vector field and $\{\xi\}$ denotes the set of collective mode amplitudes. F_j represents the correspondingly transformed fluctuating forces.

The already suggested division of variables into unstable and stable ones leads to two sets of equations

$$\dot{\underset{\sim}{u}} = \Lambda_u \underset{\sim}{u} + Q(\underset{\sim}{u},\underset{\sim}{s}) + \underset{\sim}{F}_u \quad , \tag{4.15}$$

and

$$\dot{\underset{\sim}{s}} = \Lambda_s \underset{\sim}{s} + \underset{\sim}{P}(\underset{\sim}{u},\underset{\sim}{s}) + \underset{\sim}{F}_s \quad , \tag{4.16}$$

where we have applied the above introduced vector

notation. $\tilde{Q}$ represents the part of the nonlinearities $N_i(\{\xi\})$, projected on the subspace spanned by the unstable variables, while $\tilde{P}$ represents the other part in the subspace of the stable variables. Obviously both terms take into account the coupling between the unstable and the stable modes.

The interpretation of $\tilde{F}_u$ and $\tilde{F}_s$ is similar. Up to this stage we have only appropriately transformed our original equations (2.2), and no information contained in the original equations has got lost.

Considerable simplification can be achieved by utilizing the different role of the variables in these two sets of collective variables. This can be performed in complete analogy to the discussion of the different variables in the case of the single mode laser: The stable variables still remain damped near the instability and adjust to the motion of the unstable modes. In the simplest case this adjustment is instantaneous (adiabatic principle). The presentation for the single mode laser, however, can be considerably generalized by seeking simplified solutions of the equation of motion of the stable modes in the form

$$\tilde{s} = \tilde{s}(\tilde{u}(t)) + \tilde{s}_1 \quad . \tag{4.17}$$

$\tilde{s}(u(t))$ is chosen under the condition that s_1 carries only small perturbations in addition to a free dynamic motion and these transformed stable modes include the influence of fluctuations. The important simplification is due to the first term which expresses the fact that the macroscopically relevant movement of the stable modes is completely determined by the unstable modes. Equation (4.17) summarizes the mathematical formulation of the slaving principle introduced by Hermann Haken in 1975 [29] in its general form. (We note, however, that the explicit construction of (4.17) may still be a formidable task [30].)

From (4.17) we can reduce the complicated set of equations (4.15) and (4.16) to a set of equations for the unstable modes alone. In accordance with our previous considerations we call the unstable modes the order parameters. Inserting (4.17) into (4.15) we obtain as final result the order parameter equations

$$\dot{\underset{\sim}{u}} = \Lambda_u \underset{\sim}{u} + \underset{\sim}{Q}(\underset{\sim}{u}, \underset{\sim}{s}(\underset{\sim}{u})) + \tilde{\underset{\sim}{F}}_u \tag{4.18}$$

with effective fluctuations $\tilde{\underset{\sim}{F}}_u$. In general (4.18) is a set of low - dimensional non-linear stochastic differential equations which now represents the motion of the complex system on macroscopic scales. Its solutions in the deterministic case for example, correspond to the macroscopically observed patterns

$$\underset{\sim}{U} = \underset{\sim}{U}_0 + \sum_\ell u_\ell(t)\, \underset{\sim}{O}_\ell + \sum_m s_m(\underset{\sim}{u}(t))\, \underset{\sim}{O}_m \quad . \tag{4.19}$$

Many examples of the application of the slaving principle are presented in this book, and many fields are mentioned in which it has found applications. Further consequences will arise when we discuss the behaviour of more difficult complex systems. They are built from subsystems which are themselves of complex nature, so that it becomes impossible to formulate their dynamics on a mesoscopic level. A central discovery consists in the following result: The macroscopic behaviour of a system becomes to a large extent independent of the microscopic or mesoscopic details of the dynamics of the subsystems as well as their interactions.

We note in passing that the slaving principle can also be formulated for discrete time processes. It has also been applied in connection with Master- and Fokker-Planck equations [30].

Consider a probability density of the original variables $\underset{\sim}{U}$, $P(\underset{\sim}{U},t)$. After the transformations outlined above $P(\underset{\sim}{U},\tilde{t})$ becomes a function of unstable and stable modes

$$P(\underset{\sim}{U},t) \rightarrow P_1(\{\xi\},t) = P_1(\underset{\sim}{u},\underset{\sim}{s};t) \quad . \tag{4.20}$$

Enslaving is manifested in the ansatz [12,30]

$$P_1(\underset{\sim}{u},\underset{\sim}{s};t) = p(\underset{\sim}{s}|\underset{\sim}{u};t) f(\underset{\sim}{u},t) \quad , \tag{4.21}$$

where p denotes the conditional probability for the enslaved variables to assume certain values under the condition that the order parameters have prescribed values. f is the probability density for the unstable modes. The problem then consists in finding approximate expressions for the conditional probability density p to construct the order parameter equation which becomes an effective equation for the probability density f.

5. Further Examples of Non-Equilibrium Systems with Known Mesoscopic Dynamics

In the foregoing sections we have elucidated important general aspects of a synergetic system. The general method always becomes applicable when the dynamics of the system on a mesoscopic scale is well established. Here we shall add results concerning systems in other fields of physics as well as other scientific disciplines, where these conditions are met.

Closely related to laser physics are phenomena of nonlinear optics (see the article by A. Schenzle). Nonequilibrium phenomena have been observed in various systems of hydrodynamics [3,31] e.g. in the convection instability. Here a fluid layer is heated from below. The macroscopically ordered states are regular patterns of the velocity field, for example highly ordered roll formations or the famous Benard cells [3], as well as various instabilities of higher order (see the article of M. Bestehorn for details). Applications to magnetohydrodynamics include the problem of the generation of a magnetic field in stars or planets (the article by F. Busse). Ordered states of a plasma have also been considered by utilizing the general method [3,4].

In chemical reactions such as the model reaction of the Brusselator [7] similar instabilities have been discovered; as a an example of spatio-temporal patterns the so-called spiral waves were observed in the Belousov-Zhabotinski reaction. Closely related to these attempts is the model of Gierer and Meinhardt [3,8,9], which describes the cell differentiation along concentration fields of chemical agents. Here, a prominent role is played by the principle of long-range inhibition and short-range activation [3,4].

The basic observation common to all of these systems can be summarized as follows. Several classes of instabilities can be identified. A typical order parameter equation can be constructed for each class, the same for all these different systems (cf. section 6). This result expresses the fact that the macroscopic behaviour of quite different systems becomes similar close to an instability: Each transition of the same class, being observed in quite different systems, can be mapped onto the same type of order parameter equations. This implies that synergetics may be considered as a theory of mathematical structures and forms which builds the common basis for an interdisciplinary approach. Furthermore these observations and their experimental examination justify the use of concepts of dynamic systems theory to analyze complex systems of synergetics.

6. Phenomenological Synergetics

As already noted, we are not able to construct the mesoscopic dynamics for many interesting complex systems. This is always the case when the subsystems are themselves complicated and detailed dynamics and interactions of the subsystems are not completely known or even unknown. Such situations occur in biological systems when the subsystems are cells constituting a plant, an organ, parts of an organism, etc. The same problem is even more pronounced in the study of systems within the human sciences: in the unexpected behaviour of a patient in psychology or in social sciences where the subsystems are humans or groups of humans, and little is known about the conditions of the single members. As we shall see, however, the methods of synergetics still remain applicable, but have to be used with special care.

The reason why the methods of synergetics still remain of considerable value is a result of the observation that the behaviour of the systems becomes universal near critical points. This is reflected in the mathematical theory: We are able to combine systems into equivalence classes. Each member of a special class behaves qualitatively the same near an instability. These considerations naturally lead to the notion of normal forms. A normal form can be considered as the simplest representative of such a class. These classes can be fixed from the purely macroscopic behaviour of the systems.

Macroscopically we change the control parameters $\{\sigma\}$ in an unspecific manner. In the discussion of the linear stability analysis we have observed that the dimension of the manifold in parameter space along which the parameters are changed, strongly (but not completely) determines the number of real parts of eigenvalues which can become unstable at the same place when we reach a critical region. Because our interest is devoted to the main principles we shall proceed as simply as possible and confine ourselves to a one-dimensional manifold along which our control parameters are varied. (compare with the discussion in section 4). More difficult situations have been analyzed, for example in [4,28].

We start from the linear problem (4.7) and assume that the matrix L is real. If the eigenvalue whose border is crossed is complex, a further eigenvalue exists which is the complex conjugate of the given one. When the corresponding critical surface is transversely crossed a Hopf bifurcation is observed. The representative normal form reads

$$\dot{u} = (\varepsilon - i\omega)u - c|u|^2 u \; . \qquad (6.1)$$

Here u is the complex order parameter, c denotes a generally complex constant and ε symbolizes the control parameter

$$\varepsilon = \varepsilon(\{\sigma\}) \quad . \qquad (6.2)$$

(We note that (6.1) contains the order parameter equation of the laser as a special case when we put $u = b$).

If the critical surface corresponds to a real eigenvalue we observe a saddle-node bifurcation. Its normal form is

$$\dot{u} = \varepsilon - u^2 \; . \qquad (6.3)$$

Here the order parameter is a real quantity. If we additionally require that the original state remains a stationary (but unstable) state we have a transcritical bifurcation with the corresponding normal form

$$\dot{u} = \varepsilon u - u^2 \quad . \qquad (6.4)$$

Finally, when we have a symmetry such as inversion symmetry as a condition for the order parameter we will typically observe a pitch-fork bifurcation

$$\dot{u} = \varepsilon u - u^3 \quad , \qquad (6.5)$$

which can be discussed in terms of a double well potential.

We have mentioned these simple examples just to give an idea of how a classification scheme of possible instabilities and their normal forms can be constructed. Clearly, when we change control parameters along a manifold of higher dimension the number of possible instabilities increases and they become more and more complicated. Eventually we reach a situation where no complete classification scheme is available [28].

A reasoning from purely macroscopic data becomes possible in the following way. When we know about our system that control parameters are changed along a one-dimensional manifold we conclude that a Hopf- or saddel-node bifurcation become typical possible instabilities. For a complex system we do not even know our precise location in parameter space when we start to move, so we additionally have the possibility that the system will remain in its original state. Similarly we may argue that when there is an inversion symmetry present we may observe a Hopf bifurcation or a pitch-fork instability.

We conclude that the theory of normal forms provides us with a tool to give predictions in the form of well-defined possibilities. We furthermore note that this method goes far beyond a simple linear extrapolation because it takes into account the nonlinear nature of a complex system.

An important example solved along this way of reasoning is given by Haken et al. [10] where human hand movements are analyzed. An important generalization including techniques developed for pattern recognition is presented in the article by Friedrich and Uhl.

The remaining problem for complex systems consists in a correct identification strategy: There is no a priori rule telling us how to identify and how to interpret the order parameters.

7. Macroscopic Synergetics

In developing the basis of a phenomenological approach of synergetics we have completely neglected fluctuations. These have, however, been shown to play an important and even constitutive role [3,4] in the understanding of complex systems. Applying normal forms to stochastic case would only be of approximative value. To give a macroscopic theory a rigorous foundation which includes both the deterministic aspect as well as the stochastic one, Hermann Haken introduced concepts of information theory into synergetics so as to give the macroscopic theory a rigorous foundation which includes both the deterministic as well as the stochastic aspect. He named his approach "The Second Foundation of Synergetics" [12]. Again the idea consists of the treatment of complex systems on the basis of purely macroscopic quantities. Being aware that there is no unique rule to identify these quantities appropriately, his new foundation is based on guesses. One tries to make an unbiased guess at the mesoscopic processes which give rise to the macroscopically observed structures. The mathematical method used is initiated by the maximum entropy principle which has been formulated in its most general form by Jaynes (e.g.[12]).

7.1 Survey of the Method

The procedure can be summerized as follows. We fix a set of macroscopic variables f_j and try to make an unbiased guess about the corresponding probability distribution P_k, where k labels the different mesoscopic states of the system under consideration. The rule with which to perform the unbiased guess is to maximize the information entropy

$$i = \sum_k P_k \ln P_k \quad , \tag{7.1}$$

under the condition that the actually measured mean values f_j obtain their observed value

$$\sum_k P_k f_j^{(k)} = f_j \quad . \tag{7.2}$$

$f_j^{(k)}$ is the value which f_j takes in the mesoscopic state k. Clearly, the P_k must be normalized

$$\sum_k P_k = 1 \quad . \tag{7.3}$$

Mathematically this problem can be solved by using the method of Lagrange multipliers. The fundamental question now consists in the determination and identification of the appropriate constraints (7.2). Gibbs has answered this question in the case of equilibrium thermodynamics (e.g.[3,12]) but his solution does not extend to systems far from thermal equilibrium or to non-physical systems. The central new idea is to measure a whole set of moments; if we call the variables U_j, we must measure quantities like

$$<U_j>, \quad <U_j U_k>, \quad \ldots \tag{7.4}$$

etc. (at least up to fourth order) which we identify with the f_j introduced above. In this way it becomes possible to construct stationary probability distributions in a quite natural way. However, time-dependent processes can also be analyzed if we interpret the index j in an appropriate manner.

7.2 On the Analysis of Time-Dependent Processes

When the moments represent conditional moments of the form

$$<U_j> \rightarrow <U_\ell(t+\tau)>_{\underset{\sim}{U}(t)} \tag{7.5}$$

one can construct a path integral by systematically applying Jaynes' principle. The corresponding short time propagator of the system turns out to be a Fokker-Planck operator, if the system is assumed to undergo a continuous Markov process [12]. This opens up the possibility to extract the underlying dynamics and statistics of the system from a time series [32]. The analysis starts from the N-time joint distribution function which can be written in the form

$$P_N = P(\underset{\sim}{U}_N, t_N; \underset{\sim}{U}_{N-1}, t_{N-1}; \; \ldots; \underset{\sim}{U}_1, t_1) \tag{7.6}$$

with the time intervalls τ correspondingly enumerated. The expression for the information is then

$$i = - \int DU \; P_N \; \ln P_N \; . \tag{7.7}$$

In the following we shall assume that we are dealing with a continuous Markov process so that (7.6) splits into a product of transition probabilities $P(\underset{\sim}{U}_{i+1}, t_{i+1} | \underset{\sim}{U}_i t_i)$

$$P_N = \Pi \; P(\underset{\sim}{U}_{i+1}, t_{i+1} | \; \underset{\sim}{U}_i t_i) \; P_o(\underset{\sim}{U}_1, t_1) \tag{7.8}$$

Where P_o denotes the initial probability density. The whole procedure can be performed for many-variable problems including non-stationary Markov processes. Here we shall exhibit the different steps by means of an example, namely that of a stationary Markov process and a single variable. The chosen constraints are then

$$f_1 = \langle U_{i+1} \rangle_{U_i} \tag{7.9}$$

and

$$f_2 = \langle U^2_{i+1} \rangle_{U_i} \quad , \tag{7.10}$$

i.e. the first two conditional moments. Here the state variable U is prescribed at time i and is then measured at a later time i + 1, whereby the measurement applies to an ensemble of possible outcomes and the average over this ensemble is taken. In the following we shall make the substitution

$$i + 1 \to i + \tau \tag{7.11}$$

to exhibit the fact that the time interval τ eventually becomes infinitesimally small. The maximum information entropy principle immediately yields

$$P(U_{i+\tau} | U_i) = \exp\{\lambda + \lambda_1 U_{i+\tau} + \lambda_2 U^2{}_{i+\tau}\} \tag{7.12}$$

with λ, λ_1, and λ_2 as Lagrange multipliers. There are several further conditions which have to be imposed. The transition probability (7.12) must be normalized (which fixes λ) and must fulfill in the limit $\tau \to 0$

$$P \to \delta(q_{i+\tau} - q_i) \quad . \tag{7.13}$$

from which we get

$$\lambda_2 = \frac{Q}{\tau} \quad , \tag{7.14}$$

where Q may still be a function of U_i. Furthermore we must have in the same limit

$$U_{i+\tau} - \frac{\lambda_1}{2\,\lambda_2} \to U_{i+\tau} - U_i \quad . \tag{7.15}$$

Assuming that λ_1 and λ_2 can be expanded into power series in τ we find

$$\frac{\lambda_1}{2\,\lambda_2} = U_i + \tau K(U_i) + \ldots \tag{7.16}$$

(Note that λ_1 and λ_2 may be functions of U_i because the conditional moments were moments under given U_i.) Putting these results together the final expression for the transition probability P (compare (7.12)) reads

$$P = N \exp\{-\frac{Q}{\tau}(U_{i+\tau}-U_i-\tau K(U_i))^2\} \quad (7.17)$$

with N as a normalization constant. This is the well-known short time propagator of the Fokker-Planck equation. It now becomes an obvious task to derive the explicit form of the corresponding Fokker-Planck equation. In conclusion we may state that the maximum information entropy principle in connection with the constraints of the first and second moment allow us to explicitly derive the path integral solution and furthermore to reconstruct the Fokker-Planck equation in which the drift and diffusion terms can be determined [32].

7.3 Relation Between the Slaving Principle and Jaynes' Principle

In the following we shall identify the index k in (7.1) with $\underset{\sim}{u}$, and $\underset{\sim}{s}$, i.e. to the indices of the unstable and stable modes used in the slaving principle [12,31]. We therefore make the identification

$$k \rightarrow \underset{\sim}{u}, \underset{\sim}{s} \quad ,$$

$$P_k \rightarrow P(\underset{\sim}{u},\underset{\sim}{s}) \quad . \quad (7.19)$$

The information then explicitly reads

$$i = - \sum_{\underset{\sim}{u},\underset{\sim}{s}} P(\underset{\sim}{u},\underset{\sim}{s}) \ln P(\underset{\sim}{u},\underset{\sim}{s}) \quad . \quad (7.20)$$

The slaving principle implies that the joint probability distribution for $\underset{\sim}{u}$ and $\underset{\sim}{s}$ can be split into the form

$$P(\underset{\sim}{u},\underset{\sim}{s}) = p(\underset{\sim}{s}|\underset{\sim}{u}) f(\underset{\sim}{u}) \quad (7.21)$$

where p can be split into a product over the individual indices $\underset{\sim}{s}$. Obviously

$$\sum_{\underset{\sim}{s}} p(\underset{\sim}{s}|\underset{\sim}{u}) = 1 \quad (7.22)$$

must hold. By means of (7.21) we may cast (7.20) into

$$i = -\sum_{\underset{\sim}{u}} f(\underset{\sim}{u}) \ln f(\underset{\sim}{u})$$
$$- \sum_{\underset{\sim}{u}} f(\underset{\sim}{u}) \sum_{\underset{\sim}{s}} p(\underset{\sim}{s}|\underset{\sim}{u}) \ln p(\underset{\sim}{s}|\underset{\sim}{u}) \quad (7.23)$$

or in short

$$i = i_f + \sum_{\underset{\sim}{u}} f(\underset{\sim}{u}) i_s(\underset{\sim}{u}) \quad , \quad (7.24)$$

where i_f and i_s refer to the information of the order parameters and slaved modes, respectively. To demonstrate the significance of this result, we quote a typical result for $p(\underset{\sim}{s}|\underset{\sim}{u})$, namely

$$p(\underset{\sim}{s}|\underset{\sim}{u}) = N \exp \{(\underset{\sim}{s}-\underset{\sim}{s}(\underset{\sim}{u}))^T Q (\underset{\sim}{s}-\underset{\sim}{s}(\underset{\sim}{u}))\} \; . \quad (7.25)$$

By a simple shift of coordinates $\underset{\sim}{s}$ one may readily convince oneself that

$$i_p(\underset{\sim}{u}) = -\sum_{\underset{\sim}{s}} p(\underset{\sim}{s}|\underset{\sim}{u}) \ln p(\underset{\sim}{s}|\underset{\sim}{u}) \quad (7.26)$$

becomes independent of $\underset{\sim}{u}$. In other words (7.26) does not change in the transition region. Therefore, the only relevant contribution to (7.24) stems from

$$i_f = -\sum_{\underset{\sim}{u}} f(\underset{\sim}{u}) \ln f(\underset{\sim}{u}) \; . \quad (7.27)$$

The macroscopic behaviour is completely determined by the order parameters $\underset{\sim}{u}$. Thus the slaving principle is in accordance with Jaynes' principle and even provides a microscopic derivation of Jaynes' principle in the region close to nonequilibrium phase transition points (see also the article by W. Ebeling).

7.4 Some Concluding Remarks

In the above discussion we have shown how a macroscopic approach to synergetics can be constructed. This approach now appears on a similar footing as traditional thermodynamics. That field may be based on adequate macroscopic quantities, such as energy, particle numbers etc. which serve as constraints for a maximum entropy principle from which the well-known relations of thermodynamics follow. The Lagrange parameters which thereby occur acquire a physical meaning, for instance, temperature or chemical potential.

Here, we were able to devise a similar approach at least for the class of non-equilibrium phase transitions where the macroscopic behaviour of a system is governed by a few order parameters. We have shown that we may either introduce these order parameters as macroscopic variables, or we may determine them from measured data on a set of suitably chosen variables, characterized by their moments up to the fourth order.

Of course, one may now study the physical meaning of the Lagrange multipliers in these transitions. There are several problems left for future research, for instance, even though we now know the adequate constraints for systems in thermal equilibrium and for those close to non-equilibrium phase transitions, the choice of constraints for the region in between remains open.

8. Application to Pattern Recognition and the Synergetic Computer

The successful macroscopic description of complex systems by means oforder parameters and the close relationship between the order parameter equations and normal forms have supported the development of phenomenological and macroscopic approaches [12]. These were developed for the treatment of complex systems which consist of subsystems that are themselves of a complex structure. To give an example of the application of a macroscopic approach based on the order parameter concept, we shall discuss aspects of a synergetic computer and its relation to brain functioning.

In trying to understand the functioning of the human brain, a powerful method consists in attempting to construct a machine that shows properties and abilities similar to those of the brain. As is well known nowadays, the sequential data processing of a normal computer cannot be a candidate for simulating the behaviour of the human brain. Indeed, the strategies of the human brain are quite different from any step-by-step method of recognizing for instance a picture or a melody. The strategy seems rather to be such that patterns are identified as a whole rather than by sequentiall combining elements of the picture.

In this situation the macroscopic description of complex systems and the order parameter concept open quite new perspectives. These have been applied in the so-called synergetic computer. The problem of pattern recognition was used to test the abilities and the functioning of a synergetic computer [34]. We shall, however, first explain the fundamental ideas of the functioning of a synergetic computer by using the simple example of the convection instability.

In the simplest (idealized) case of the convection instability in a rectangular geometry we have a pure roll pattern. There are two different realizations of this fluid streaming pattern which are both equally likely. This can be seen immediately from the corresponding order parameter equation that reads (cf. [3])

$$\dot{u} = \varepsilon u - u^3 \quad . \tag{8.1}$$

The result of the theory is perhaps best understood from the discussion of the overdamped motion of a particle in a symmetric double well potential. The valleys of the curve can be interpreted as the stable positions of a moving ball; it will come to rest at these points. Similarly a maximum of the curve characterizes an unstable situation. If we now attribute the valleys to the two available streaming patterns in a roll formation, we see that both are of equal rank but only one can actually be realized.

Following the ideas of pattern recognition, we can identify these two patterns of motion with so-called prototype patterns which the fluid can "recognize". Now the question is, how can this system recognize an offered pattern? The offered pattern is usually disturbed and can be incomplete. This situation is analogous to stirring the fluid in the layer. A number of the collective modes will then become excited. But they all die out and the final long term motion of the fluid is again represented by our ball in a double well potential. Our ball will move to the minimum which is closest to its initial position. During this movement the pattern is reconstructed, i.e. the offered pattern is recognized. Several comments are in order: We have presented the idea of a synergetic computer by means of a very simple example. But this proposal can be generalized (see below) and a complete theory has been developed and successfully applied to the problem of recognition of faces and other objects [34]. An important implication of this concept is that our system acts as an associative memory: incomplete and disturbed patterns that are offered are completed and corrected and thus brought into accordance with the originally stored patterns ("prototype patterns"). If the prototype pattern consists of a person's face and name, and the offered pattern is only the face, the synergetic computer acting as an associative memory provides us with the name of the person, i.e. the computer has "recognized" him or her.

It is important to mention that the dynamics of the recognition process is governed by the order parameters which characterize the ordered states of the system. These ordered states are, as we have seen, attributed to the prototype patterns.

We shall now explain the model of a synergetic computer applied to the problem of pattern recognition. Our patterns will consist of two-dimensional pictures, photographs etc. We digitize these objects by a grid and enumerate the cells of the grid as components of a multi-dimensional vector. The values of the components are taken proportional to the grey values in each cell. The form of this vector is given by

$$\underset{\sim}{v} = (v_1, v_2, \ldots, v_n) \tag{8.2}$$

which represents a prototype pattern. In applications we have to consider several prototye patterns which we distinguish by an additional index, say k

$$\underset{\sim}{v}^{(k)} \quad . \tag{8.3}$$

The basic idea now is to construct a complex system which is able to attribute the corresponding prototype pattern to a given pattern, represented by a vector $\underset{\sim}{q}(0)$. This can be achieved by subjecting the vector $\underset{\sim}{q}$ to the following dynamics

$$dq_j/dt = -\partial V/\partial q_j \quad , \tag{8.4}$$

where j enumerates the components of $\underset{\sim}{q}$. Here the potential V consists of three parts:

$$V = V_1 + V_2 + V_3 \quad . \tag{8.5}$$

The first part is connected to the so-called learning matrix [34]

$$V_1 = 1/2 \sum_k \lambda_k (\underset{\sim}{v}^{(k)} \underset{\sim}{q})^2 \tag{8.6}$$

and induces the offered pattern $\underset{\sim}{q}(0)$ to evolve towards the subspace that is spanned by the prototype patterns. (For the sake of simplicity we have here assumed that the different prototype patterns are orthogonal to each other. This restriction can be removed by using adjoint prototype patterns in the case where they are not orthogonal. In general one only needs linear independence between the prototype patterns.) The second part V_2

$$V_2 = B/4 \sum_{k \neq k'} (\underset{\sim}{v}^{(k)} \underset{\sim}{q})^2 (\underset{\sim}{v}^{(k')} \underset{\sim}{q})^2 \tag{8.7}$$

(B is a constant) discriminates between the different patterns. Finally V_3 guarantees global stabilization of the system:

$$V_3 = C/4 \left(\sum_k (\underset{\sim}{v}^{(k)} \underset{\sim}{q})^2\right)^2 . \tag{8.8}$$

Equations (8.3) have been implemented on a serial computer and applied to face-recognition [34]. It turned out that the computer is able to single out persons from groups, can recognize deformed and rotated patterns, has the ability to learn, and can model the oscillatory behaviour of the human brain while recognizing ambiguous patterns. The chosen dynamics models a massive parallel computing device.

9. Conclusions

We have demonstrated the usefulness of the order parameter concept in connection with the slaving principle while treating problems involving pattern formation and pattern recognition. An outline of different methods has been given starting from a mesoscopic and macroscopic level. The "top down" approach introduced by Hermann Haken in analyzing complex systems such as brain functioning, which is provided by the synergetic computer, seems to be a particularly promising ansatz for simulating strategies of human recognition from a completely new point of view.

References

[1] R. Graham: in Lasers and Synergetics, (eds. R.Graham and A. Wunderlin) Springer-Verlag (Berlin, 1987)

[2] Springer Series in Synergetics (Vols. 1-57), (ed. H. Haken) Springer-Verlag (Berlin 1977 - 1992)

[3] H. Haken: Synergetics. An Introduction (3rd edition) Springer-Verlag (Berlin 1983)

[4] H. Haken: Advanced Synergetics Springer-Verlag (Berlin 1983)

[5] C. Vidal and A. Pacault (eds.): Nonlinear Phenomena in Chemical Dynamics Springer-Verlag (Berlin 1978)
Y. Kuramoto: Chemical Oscillations, Waves, and Turbulence Springer-Verlag (Berlin 1984)

[6] A. Mikhailov: Foundations of Synergetics, Vols. 1,2 Springer-Verlag (Berlin 1991)

[7] P. Glansdorff, I. Prigogine: Thermodynamic Theory of Structure, Stability and Fluctuations Wiley, New York (1971)
G. Nicolis and I.Prigogine: Self-Organization in Non-Equilibrium Systems Wiley, New York (1977)

[8] H. Meinhardt, A. Gierer: J. Cell. Sci. 15, 321 (1974)

[9] H. Haken, H. Olbrich: Analytical Treatment of Pattern Formation in the Gierer-Meinhardt Model of Morphogenesis, J. Math. Biology 6, 317 (1978)

[10] H. Haken, J.A.S. Kelso, H. Bunz: Biological Cybernetics 51, 347 (1985)

[11] H. Haken, A. Wunderlin: In The Natural-Physical Approach to Movement Control (H.T.A. Whiting, O.G. Meijer, P.C.W. van Wieringen eds.) VU University Press, Amsterdam (1990)

[12] H.Haken: Information and Self-Organization Springer-Verlag Berlin (1988)

[13] A. Fuchs, R. Friedrich, H. Haken, D. Lehmann in Computational Systems - Natural and Artificial ed. H. Haken Springer-Verlag Berlin (1987)

[14] R. Friedrich, A. Fuchs, H.Haken in Synergetics of Rhythms in Biological Systems eds. H. Haken, H.P. Koeppchen, Springer-Verlag Berlin (1992)

[15] W. Weidlich, G. Haag: Concepts and Models of Quantitative Sociology Springer-Verlag Berlin (1982)

[16] A. Wunderlin, H. Haken: Some Applications of Basic Ideas and Models of Synergetics to Sociology. Springer Series in Synergetics, Vol. 22 (ed. E. Frehland) Springer-Verlag (1984).

[17] H. Bergson: Oeuvres Presses Universitaires de France, Paris (1963)
[18] H. Haken: Synergetik-Nichtgleichgewichts-Phasenuebergaenge und Selbstorganisation in Physik, Chemie, Biologie, Nove Acta Leopoldina NF 60, Nr. 265, 75 (1989)
[19] L. Boltzmann: Vorlesungen ueber Gastheorie, Leipzig (1896, 1898)
[20] S. Clausius: Ann. Phys., Lpz. 93, 481 (1854)
[21] C. Darwin: On the Origin of Species by Means of Natural Selection Murray, London (1859)
[22] H. Haken: Laser Theory, in Encyclopedia of Physics Vol. XXV/2C, ed. S. Fl{gge, Springer-Verlag Berlin (1970)
[23] Landau-Lifshitz: Course of Theoretical Physics Vol.V Pergamon Press London (1952)
[24] R. Graham, H. Haken: Z. Physik 243, 289 (1971)
R. Graham, H. Haken: Z. Physik 245, 141 (1971)
[25] R. Graham in Noise in nonlinear dynamical systems (F. Moss, P.V.E. Mc Clintock eds.) Cambridge University Press New York (1989)
[26] C.O. Weiss, R.Vilaseca: Dynamics of Lasers Weinheim (1991)
[27] H. Haken: Phys.Lett. 53A, 77 (1975)
[28] V.I. Arnol'd: Geometrical Methods in the Theory of Ordinary Differential Equations in Grundlehren der mathematischen Wissenschaften 250, Springer (Berlin 1983)
[29] H.Haken: Z. Physik B21, 105 (1975)
[30] A. Wunderlin, H. Haken: "Generalized Ginzburg-Landau Equations, Slaving Principle and Center Manifold Theorem
Z. Physik B44, 135 (1981).
H. Haken, A. Wunderlin: Slaving Principle for Stochastic Differential Equations with Additive and Multiplicative Noise and for Discrete Noisy Maps
Z. Physik B47, 179 (1982).
H. Haken, A. Wunderlin: Dynamical Systems Described by Discrete Maps with Noise. Springer Series in Synergetics Vol. 12 (eds. C. Vidal and A. Pacault), p. 15 (1982).
[31] M. Bestehorn, R. Friedrich, A. Fuchs, H. Haken, A. Kuhn, A. Wunderlin: Synergetics Applied to Pattern Formation and Pattern Recognition, in Optimal Structures in Heterogeneous Reaction Systems, Springer Series in Synergetics Vol. 44 (ed. P.J. Plath) Springer-Verlag Berlin (1989)

[32] L. Borland, H. Haken: To be published

[33] H. Haken, A. Wunderlin: A Macroscopic Approach to Synergetics in Structure, coherence and chaos in dynamical systems (eds. P.L Christiansen, R.D. Parmentier) Manchester University Press 1989

[34] H. Haken: Pattern Formation and Pattern Recognition An Attempt at a Synthesis in Pattern Formation by Dynamical Systems and Pattern Recognition (ed. H. Haken) Springer-Verlag Berlin, Heidelberg (1979)
H. Haken: Computers for Pattern Recognition and Associative Memory in Computational Systems - Natural and Artificial (ed. H. Haken, Springer-Verlag Berlin, Heidelberg (1987)
A. Fuchs and H. Haken: Biol. Cybern. 60, 17, (1988);
Biol. Cybern. 60, 107, (1988)
H. Haken: Synergetics in Pattern Formation and Associative Action in Neural and Synergetic Computers, (ed. H. Haken) Springer-Verlag Berlin,

Elements of a Synergetics of Evolutionary Processes

W. Ebeling

Institut für Theoretische Physik, Humboldt-Universität,
Invalidenstr. 42, O-1040 Berlin, Fed. Rep. of Germany

Abstract. Some of the basic principles of a synergetics of evolutionary processes are developed. Evolutionary processes are defined as (quasi-) infinite, branching chains of self- organization cycles. Typical properties as self-reproduction, criticality, valuation, stochasticity and historicity are discussed. A general (formal) stochastic model is developed and the main principles and stages of early evolution are discussed.

1. Introduction

Synergetics, the science developed by Hermann Haken is based on the term self-organization. Haken defines synergetics as "a rather new field of interdisciplinary research which studies the self-organized behavior of systems leading to the formation of structures and functionings" [1-3]. With the term self- organization we denote in the following the spontaneous creation of order in pumped systems, which are operating beyond a critical distance from equilibrium. Historically the developement of the concepts of selforganization is connected with the names of famous scientists like Boltzmann, Schrödinger, Bertalanffy, Turing, Prigogine, Haken and Eigen. The exploration of the relation between order and chaos belongs to the central themes of the theory of self-organization.

Haken's approach is interdisciplinary from the very beginning. He writes: "The central question in synergetics is whether there are general principles which govern the self-organized formation of structures and/or functions in both the animate and in the inanimate world" [4]. After more than twenty years of a successful developement of the science synergetics there is no more doubt that such general principles exist. Therefore the question left to workers in synergetics after Haken is to work out such principles in more detail. We are following this route here, concentrating on the general principles of evolution. In contrast to our earlier approach which was mainly

Springer Proceedings in Physics, Vol. 69
Evolution of Dynamical Structures in Complex Systems
Editors: R. Friedrich · A. Wunderlin

devoted to the "physics of evolution" [5-7] we are concentrating here on the more general principles, i.e., to find the elements of a "synergetics of evolution".

2. General Aspects of a Synergetics of Evolution

Evolution is, roughly speaking, the story how the partially ordered cosmos surrounding us was created out of the chaos after the big bang. More than 2000 years ago the greek philosophers considered CHAOS and COSMOS as the two basic aspects of the world. CHAOS was considered as its completely disordered original state from which the COSMOS, i.e. the partially ordered state was created. Our modern understanding of evolution is still in the same spirit. The main task of a synergetics of evolution is, to find out the governing principles of the creation of our recent COSMOS. Understanding these principles is the only reliable way to make predictions about its possible future. Among the many open problems is especially the story about the creation of life, which is connected with the generation of information and value [8-17].

Evolution in nature and in society may be considered as an infinite chain (or a spiral) consisting of self-organization cycles [5-7]. Each of the cycles consists of the following stages [7, 15]:

(i) A relatively stable stationary state starts to become unstable.

(ii) The instability triggers a process of self-organization which creates a new structure.

(iii) As a result of the self-organization process a new relatively stable stationary state is reached, which may turn into a new cycle.

These stages may be represented in the following scheme:

———> state n ————>—————>———— state n+1 ———>
instability selforganization

The most "elegant" way which evolution uses to make long chains of such basic steps is self-reproduction in combination with mutation and selection [5-10]. Mutation is basically error reproduction which is a stochastic phenomenon. The main element in evolution is the appearance of the NEW. The NEW is in our definition the first appearance of a new quality (elementary particles, nuclei, atoms, molecules, species etc.) in the historical process. In general a new quality will appear for the first time only in one or a few cases. This essential instance requires a stochastic formulation in terms of occupation numbers [17-19]. The idea of the occupation number formalism is just to count the number of existing cases and to express it by integers N = 0,1,2,.... Only this discrete description enables us to describe the essential difference between the state of no instances of certain quality on the one hand and one or a few instances of the new quality on the other hand.

Selection is a process which is connected with competition and valuation. Competition is a dynamical process which is observed in a large class of physical and non-physical dynamical systems consisting of many elements (particles, laser modes, individuals, technologies etc.). According to our earlier work [5] we define competition as a collective process in dynamical systems consisting of many elements of different species having a common goal (representing certain value), which cannot be reached by all members of the game. In ecology one speaks about competition if a factor which is necessary for the survival is used by two or more members of the community. As a rule, competition is connected with selection, a term coined by Charles Darwin introducing the slogan "survival of the fittest". We define selection as a special type of a coherent behaviour in a system consisting of many subsystems (elements, species, ...) which all have in principle the conditions for existence. But due to competition a coherent process leads to the extinction of one ore some of them; this is selection [5]. A criterion for the existence of competition in a system is the reaction to the addition of a subsystem of a qualitatively new type. If a process then starts which leads to the extinction of the new subsystem or of some of the old subsystems, we have competition [5-7].

The basic structure of a dynamical competition system is the following: We have a big system of N elements, N being large ($N \gg 1$) which are grouped into s classes i = 1,2,...s, each containing N_i identical or at least similar elements. The number of elements $N_i(t)$ of certain type as well as the number s(t) of types (species, fields,...) are subject to a dynamics. The state may be described in the language of concentrations, frequencies or occupation numbers which was developed for the description of chemical, biological, ecological and social systems [17-27]. It is always useful to develop in parallel a deterministic picture on the basis of concentrations and frequencies and a stochastic model using discrete occupation numbers [19,27].

A fundamental concept in the theory of competition processes is the value of a species which means the fitness in the sense of Darwin. Competition is always based on some kind of valuation. Evidently the concept of values was first introduced by Adam Smith in the 18th century in an economical context. The fundamantal ideas of Adam Smith were worked out later by Ricardo, Marx, Schumpeter and many other economists. In another social context the idea of valuation was used at the turn of the 18th century by Malthus. Parallel to this developement in the socio-economical sciences a value concept was developed in the biological sciences by Darwin and Wallace. Wright developed the idea of fitness landscape (value landscape) which was worked out by many authors [5-7,11-16]. All these concepts are very abstract and qualitative. This led to great difficulties in the field of mathematical modelling [5-7,24-25]. Our point of view is that values are an abstract non-physical property of subsystems (species) in certain dynamical context. Values express the essence of biological, ecological, economical or social properties and relations with respect to the dynamics of the system. From the point of view of modelling and simulations values are given 'a priori', they must be considered as elements of the axioms of the dynamic models.

The whole evolution process is accompanied by valuation which occurs in each self-reproduction (self-organization) cycle. the dynamics of each cycle is essentially nonlinear [15-18] and the transition stability-instability has the character of a bifurcation. Looking at this transition from the point of view of physics a close relation to phase transitions and to critical phenomena is observed [1-4]. In principle there exist two types of self-organization corresponding to regular or chaotic attractors respectively. By an external change of the parameters the system may be brought from the regular to the chaotic regime or opposite. The critical value of the parameter will be called here a_{cr}. For this critical value of the parameter and in its neighborhood

$$| a - a_{cr} | / a_{cr} << 1 \qquad (1)$$

the system shows several peculiarities, e.g. long relaxation times, long-range spatial correlations, anomalous scaling laws, scale invariance and 1/f-noise. In the following we will call the structures occurring on the border between regular and chaotic regimes, or between two different regular or chaotic regimes, critical structures. There exist far-reaching analogies between critical structures in nonequilibrium with the critical structures (phenomena) known from equilibrium phase transitions. In order to get critical phenomena one has in general to fix the external parameters, e.g. temperature and pressure in equilibrium or the pumping rate in non-equilibrium, at the critical values. However there exists also an interesting class of systems showing "self-organized criticality". The notion of self-organized criticality (SOC) was introduced by Bak et al.[20]. The idea of SOC is that systems are kept in a critical state, without any control parameter being set explicitly. There is some evidence that SOC plays a role in evolution and especially in life phenomena [12-14,20-23]. If this turns out to be true, criticality would be one of the essential elements of the synergetics of evolution.

Let us come back now again to the valuation process which is the essential element of the selforganization cycles of evolution. The existing theory has already anticipated several value concepts such as the value of energy (i.e. entropy), the information value, the selection value in biology and the exchange value in economy. All these value concepts have several features in common [24].

(i) Values assigned to elements (subsystems) of a system or to the modes of their dynamics incorporate a certain entireness of the system, they cannot be understood by a mere view of the subsystem without its whole environment. In other words, the whole is here more than the sum of its parts.

(ii) The values are central for the structure and dynamics of the entire system; they determine the relations between the elements and their dynamical modes as well as the dynamics of these relations. Competition or selection between elements or dynamical modes are typical elements of the dynamics.

(iii) The dynamics of systems with valuation is irreversible; it is intrinsically connected with certain extremum principles for the time evolution of the values. These extremum principles may be very complex and can in only a few cases be expressed by scalar functions and total differentials.

(iv) The necessary physical condition for any form of valuation is the pumping with high-valued energy. Isolated systems show a general tendency to devaluation, what is caused already by the devaluation of the energy in the system, due to the second law.

In the simplest case the value of a subsystem (species) with respect to the competition is a real number. In other words each element (species) 1,2,...,i,...,s is assigned a number

$$V_1, V_2, \ldots, V_i, \ldots, V_s; \quad (V_i \in \Re) . \tag{2}$$

Since real numbers form an ordered set, the species are ordered with respect to their values. In such systems competition and valuation may be induced by the process of growth of species having high values and decay of species having low values (or opposite). In many cases the growth of "good" species is subjected to certain limitations. A standard case studied in detail by Fisher (1930) and Eigen (1972) is the competition by average [8]. Here all species better than the avarage over the total system, the social average

$$\langle V \rangle = \sum_{i=1}^{s} V_i N_i / N \; ; \qquad N = \sum N_i \tag{3}$$

grow and all others that are worse decay. In other words the occupation $N_i(t)$ increases if

$$V_i > \langle V \rangle \tag{4}$$

and decreases if

$$V_i < \langle V \rangle \tag{5}$$

This leads to an increase of the averaged value in time. In some cases the values are given by the distribution itself e.g.

$$V_i = \log (N_i / N) .$$

Then valuation favors equal distribution and the averages correspond to entropies. In physics, according to Boltzmann's concept, entropy maximizing is subject to the condition of constant energy and the elements are defined by a partition of the phase space of the molecules. The lowering of entropy with respect to its maximum (at given energy) expresses after Clausius, Helmholtz and Ostwald the work value of the energy in the body. The second law of thermodynamics expresses a general tendency to disorder (equipartition) corresponding to a devaluation in isolated systems.

Many competition situations in biology, ecology and economy are connected with a struggle for common raw material or food [5-10]. Here, the result of the competition can still be predicted on the basis of a set of real numbers (scalar values). In more general situations the values are not a well defined numbers but merely a property of the dynamical system [6,15,25-27]. Feistel attributes three functions to valuation:

(i) Regulating functions,

(ii) Differentiating functions,

(iii) Stimulating functions.

Valuation is absolutely central to the origin of life. This was first pointed out in the fundamental papers by Manfred Eigen and his coworkers on the self-organization of macromolecules [8,10]. Another key point is the origin of information [30-33]. We know that the existence of all living beings is intimately connected with information processing. A living system may be characterized as a natural ordered and information-processing macroscopic system [28]. Information processing we consider as a special high form of selforganization [29].

3. Valuation, Competition and Selection

Let us consider now the dynamics of the valuation process for the more complicated forms of valuation which appear first in the evolution of biological macromolecules [5-10]. The first step we have to make is the labeling of the qualitatively different elements of evolution (molecules, species, etc.). Let us assume, that the different elements form a countable set (the set of species) and let us denote them by their index of counting $i = 1,2,3, \ldots$. The counting may itself be a nontrivial problem. In most cases however, the species participating in the evolutionary game may be characterized by strings. For example let us consider the set of strings representing the binary structure of certain biomolecules (DNA, RNA, etc.)

1\$, 2\$,..., i\$, j\$,..., s\$.

A string may always be associated with an integer, its Gödel number. Let us assume that the objects (species) which are behind the strings (or the labelling numbers) are subject to certain evolutionary dynamics.

In a general sense we consider the strings i\$ etc. as the "genotypes" of the objects which are participating in our evolutionary game. All possible strings (words) may be considered as elements of an abstract metric space, the genotype space G. We assume that each genotype is connected with a set of properties forming the "phenotype" ; in this way we introduce also a phenotype space Q. The phenotype is evaluated in the evolution. Mathematically this means that any element is associated with a set of real numbers

$$(V_i^{(1)}, V_i^{(2)}, \ldots) \subset V.$$

In other words we assume a homomorphism

$$G \longrightarrow Q \longrightarrow V.$$

Here V is a real vector space. These ideas are closely connected with the concept of an evolutionary landscape, which is one of the basic concepts of the modern theory of evolution [7,11,34-36]. Our next assumption is, that for each object i either an occupation number N_i (stochastic picture) or a concentration (or fraction) x_i is defined (N_i denotes the number of representatives of the objects of kind i). The simplest model of an evolutionary dynamics is the Fisher-Eigen model which is based on the assumption that the competing objects i = 1, 2,..., s have different reproduction rates

$$E_1, E_2, \ldots, E_i, \ldots, E_s .$$

These scalar quantities now play the role of the values.The dynamics of the fractions is given by the differential equations [8]

$$\dot{x}_i = (E_i - k_o(t))\, x_i\,(t) \tag{6}$$

The decay rates follow from the normalization condition

$$\sum x_i = 1 \tag{7}$$

$$k_o(t) = \sum E_i x_i = \langle E\rangle \tag{8}$$

The resulting dynamical equation for the competition of strings

$$\dot{x}_i = (E_i - \langle E\rangle)\ x_i \tag{9}$$

shows that the species with values greater than the "social" average $\langle E\rangle$ will succeed in the competition and the others will fail. Finally only the species with the largest rate E_m will survive

$$E_m > E_i\ ;\ i = 1, 2, \ldots, s\ ;\ i \neq m \tag{10}$$

The Fisher-Eigen model is the simplest of all models of competition. It refers to an oversimplified case and in some sense one can say even that the model reflects only pseudo-competition since there is no real interaction between the species. There exist more realistic models [5-9]. Among them is the case that the objects (species) compete for a common resource x_o, which flows with constant rate Φ_o into the system. The following dynamics is assumed

$$\dot{x}_o = \Phi_o - \sum k_i\ x_o\ x_i,$$

$$\dot{x}_i = (k_i\ x_o - k'_i)\ x_i\ , \quad i = 1, \ldots, s \tag{11}$$

One can show by stability analysis that in this case the winner of the competition, i.e. the master species m, is the species with the largest value of the ratio of growth and decay rates k_i/k'_i. In other words the master species has the property

$$\frac{k_m}{k'_m} > \frac{k_i}{k'_i}\ , \qquad i = 1,\ldots,s \quad , \quad i \neq m \tag{12}$$

More general situations of a competition for common resources were elaborated by many authors [6-9]. A generalization of the Fisher-Eigen equation that includes external sources and mutations and nonlinear effects reads [27]:

$$\begin{aligned}\dot{x}_i &= (A_i^{(0)} - D_i^{(0)})\ x_i + (A_i^{(1)} - D_i^{(1)})\ x_i^2 - D_i^{(2)}\ x_i^3 \\ &+ \sum_j (M_{ij}\ x_j + B_{ij}\ x_i\ x_j - C_{ij}\ x_i\ x_j\ ) \\ &+ \sum_j \left[(A_{ij}^{(0)} + A_{ij}^{(1)} x_j)\ x_i - (A_{ji}^{(0)} + A_{ji}^{(0)} x_i)\ x_j \right] + \Phi_i\end{aligned} \tag{13}$$

The dynamics of the evolution of the species is represented by the set of parameter vectors and matrices appearing in (13). This set of parameters reflects the physics, biology, ecology, economy or sociology of the particular problem. For the moment, we assume that the various coefficients and matrix elements are known or can be empirically determined. In the sense described above the set of all parameter vectors and matrices appearing in (13) constitute the value matrix associated with a string i. The winner of the competition defined by (13) is not uniquely defined and depends in general on the matrix of values and on the initial conditions.

A generalization of (13) takes into account that the reproduction rates as well as the death rate (and possibly also the other values) depend on the age of the individuals belonging to the species i :

$$A_i^{(0)} = A_i\ (\tau)\ ,\ D_i^{(0)} = D_i\ (\tau) \tag{14}$$

In this case, under the condition of constant overall number (7) the theory yields the following eigenvalue problem [37]:

$$\int_0^\infty d\tau\ A_i(\tau)\ \exp\left[- \int_0^\tau d\ \xi\ D_i\ (\xi) - \lambda\ \tau \right] = 1 \tag{15}$$

The winner of the competition is the species with the maximal eigenvalue. In most cases early reproduction is of advantage in the competition [7]. In contrast to our earlier case, valuation is now concerned with functions of time, the aging functions.

Aging is only part of a general strategy of evolution which we call Haeckel strategy [7,35]. The essence of this strategy is the following: In the early days of life on earth, living systems consisted of one simple cell only, which had - at least in principle - infinite lifetime. Like little machines these cells mature immediately after their birth, they were able to consume free energy, to move in space, to react on external factors and to make offspring. With increasing complexity the cell organisms developed a life-cycle consisting of several periods including youth, a period of growth and learning, a period of self-reproduction and death. Aging and development is typical especially for the multi-cell organisms. This was a great achievement in evolution which made possible the formation of complex structures by individual development, learning and teaching [7,38]. The processes which lead to the formation of a new animal or plant from cells derived from one or more parents (eggs or seeds) are now studied by developmental biology. This field of science was pioneered by Haeckel who detected close relations between ontogeny and phylogeny. Haeckel's biogenetic law postulates that ontogeny recapitulates to some extent phylogeny. Modelling Haeckel strategies leads to quite complicated integro-differential equations [38]. In this context valuation appears as a functional problem.

4. Mutations, Branching and Stochasticity

As we haw shown above, self-reproduction in connection with valuation, competition and selection is an essential part of evolutionary processes. Pure physical reasons alone (natural fluctuations, external random influences) mean that any self-reproduction process is subject to errors. The copy may not be exact but only similar to the original. Modified copies introduce new elements into the system, which are then subject to valuation, competition and selection. This opens new ways to the evolutionary process and introduces the stochastics into the processes. Since any error reproduction is a stochastic choice between many possibilities, the process has branching character. At each self-reproduction process one of many branches of the paths into the future may be realized; therefore the actual chain of evolution is like a bifurcating network forming hyper-netted chains. The number of possibilities increases with increasing time. In order to to describe the complex dynamics of those processes, a stochastic language is required. The stochastic theory of evolutionary processes was pioneered by Eigen [8] and elaborated later by several authors [18,19,39]. Let us discuss here only briefly a class of stochastic models which is based on (13) [6,27]. We assume that species of different types are present or potentially present in the system. The number of types may be very large or even infinite. Each self-replication or death process changes only the number of a single type of species

$$N_i \longrightarrow N_i + 1 \quad \text{or} \quad N_i \longrightarrow N_i - 1$$

An exchange process, on the other hand, is accompanied by the change of two occupation numbers:

$$N_i \longrightarrow N_i + 1 \quad \text{and} \quad N_j \longrightarrow N_j - 1$$

Higher-order processes in which more than two complementary occupation numbers change simultaneously will not be considered. Stochastic descriptions are very important for the modelling of evolution since the description of the initial phases of innovative instabilities is possible only on the basis of stochastic models. This is because the innovation leading to a new species n is always a zero-to-one transition

$$N_n = 0 \longrightarrow N_n = 1. \tag{16}$$

Such a birth process is strongly influenced by stochastic effects. The occupation numbers N_i are functions of time. In contrast to the smooth variation of $x_i(t)$ the dynamics of $N_i(t)$ is a discrete, stochastic hopping process. We may associate the concentrations $x_i(t)$ with the ensemble averages of $N_i(t)$

$$x_i(t) = \text{const} \langle N_i(t) \rangle \tag{17}$$

where the average is performed over a large number of identical stochastic systems. The complete set of occupation numbers $N_1, N_2, \ldots, N_s$ determines the state of the system at a given time. Because of the large number of potential types of elements in typical evolutionary systems, most of the occupation numbers are zero [5-7]. The probability that the system at time t is in a particular state may be described by the distribution function

$$P(N_1, N_2, .., N_i, \ldots N_s, t).$$

The equation of motion for our stochastic process has to include the four fundamental processes of evolutionary behaviour: self-reproduction, death (decline), transition between species (mutations), and input of individuals from external sources or spontaneous generation. In self-reproduction, the species are assumed to produce exact copies of their own type as well as error copies, i.e. both identical self-reproduction and error reproduction occur. The failure rate is assumed to remain small. For identical self-replication of an element of type i, the transition probability is assumed to be given by

$$W(N_i+1 \mid N_i) = A_i^{(0)} N_i + A_i^{(1)} N_i^2 + B_{ij} N_i N_j \tag{18}$$

where $A_i^{(0)}$ is the coefficient of linear self-reproduction, $A_i^{(1)}$ measures self-amplification (second-order self- reproduction), and B_{ij} measures sponsoring or catalytic assistance from other types of strings. Error reproduction of strings i is assumed to be described by the linear relation

$$W(N_i + 1, N_j \mid N_i, N_j) = M_{ij} N_j \tag{19}$$

where the coefficient M_{ij} measures the probability that, through mutation, an element i is produced from an element j. In general species belong in an active manner to the system only for a limited time. They may die or they may be forced to leave the system. This is expressed by the probability

$$W(N_i-1 \mid N_i) = D_i^{(0)}\ N_i + D_i^{(1)} N_i^2 + D_i^{(2)} N_i^3 + C_{ij} N_i N_j \qquad (20)$$

where $D_i^{(0)}$ measures the spontaneous death rate, $D_i^{(1)}$ and $D_i^{(2)}$ express nonlinear decay processes (self-inhibition), and C_{ij} measures the suppression of elements i by elements j. The species are assumed to exchange elements. This especially may be connected with competition and selection and is assumed to be expressed by the probability

$$W(N_i + 1\ ,\ N_j - 1 \mid N_i,\ N_j) = A_{ij}^{(0)} N_j + A_{ij}^{(1)}\ N_i N_j \qquad (21)$$

The coefficients $A_{ij}^{(0)}$ amd $A_{ij}^{(1)}$ represent noncooperative and cooperative exchange, respectively. Finally, the change in occupation of a species i by inflow from the outside or by spontaneous generation is assumed to have a constant probability

$$W(N_i + 1 \mid N_i) = \Phi_i \qquad (22)$$

All coefficients in (18-22) have nonnegative values. This guarantees that the system will not enter the empty state with all $N_i = 0$. At least one element of the inflow rate must be different from zero. Averaging the stochastic process over an ensemble of systems,

(13) follows from (18-22) in the limit of high occupation numbers. It should be stressed, however, that the deterministic equations remain true only in average. Unlike the deterministic case, it is possible in the stochastic real, for instance, to pass over the barrier separating two different stable equlibria [6]. In other words, valuation is turning into a very complex stochastic phenomenon. The 'goals' of the evolutionary game are reached only with certain probability.

5. The Evolution of Information Processing - Principles of the Early Organization

The main task of a synergetics of evolution should be the understanding of the relation between selforganization and information. This includes the solution of at least two basic problems:

(i) What is the general relation between self-organization and information and how this relation may be expressed in a quantitative way?

(ii) What is the origin of information processing in evolution? How was information processing created in the process of self-organization of biomolecules?

Both of these problems seem to be unsolved so far. There exist, however, a few basic approaches. We mention especially the key papers of Eigen, Haken and Volkenstein [30-33] which reflect the main tendencies and the directions of search for final solutions.

In any case one of the key terms should be the statistical entropy concept developed by Boltzmann and Gibbs which was extended to information processing by Szilard, Shannon and Brillouin. In information theory (as in physics) the entropy is defined as the mean uncertainty per state. Let us assume that x is a set of d order parameters on the dynamic level of the description of the objects. If p(x) denotes the probability density for this set of order parameters which describe the macroscopic state, the entropy contained in the distribution (the H - function) is defined by

$$H = - \int dx \; p(x) \; \log p(x) \tag{23}$$

In the case of discrete variables i = 1,2,...,s we get the classical Shannon expression with a sum instead of the integral. In the special case that the state space is the phase space of the molecules forming the system, the Shannon entropy is (to within the Boltzmann constant) identical with the statistical Boltzmann-Gibbs entropy

$$S = k_B \; H_{ps} \tag{24}$$

Here H_{ps} is the phase space entropy

$$H_{ps} = - \int (dq\ dp\ /\ h^{f})\ \rho(q,p)\ \log \rho(q,p) \qquad (25)$$

(f-number of degrees of freedom). This shows us that the Boltzmann-Gibbs entropy is nothing other than the mean uncertainty of the location of the molecules in the phase space. Let us introduce now Haken's concept of order paramaters and conditional probabilities [30] to the physical phase space of our evolutionary system. We assume that the probability density may be represented as the product of the probability in the order parameter space and the conditional probability (Bayes formula)

$$\rho(p,q) = p(x) * \rho(p,q \mid x) \qquad (26)$$

Then a brief calculation yields

$$S = k_B\ H + S \qquad (27)$$

$$S = \int dx\ p(x)\ S(x)\ .$$

Here S(x) is the conditional statistical entropy for a given value of the order parameter

$$S(x) = - k_B \int (dp\ dq\ /\ h^{f})\ \rho(p,q \mid x)\ \ln \rho(p,q \mid x).$$

In this way we have shown that, to within a constant factor, the information entropy is that part of the statistical entropy which is connected with the order parameter distribution. In general this is a very small part of the total statistical entropy, the overwhelming part comes from the second term in (27) which reflects the entropy contained in the microscopic state; this part is not available as information. Let us give an example: The Gibbs entropy of a polynucleotide molecule of length n with 4^n possible primary structures is the sum

$$S = k\ n\ \log 4 + S \qquad (28)$$

where the second term is the standard value of the polynucleotide with fixed primary structure. The contributions to the total entropy are interchangeable in the sense discussed already by Szilard, Brillouin and many other workers [23]. Information (i.e. macroscopic order parameter entropy) may be changed into statistical-thermodynamical entropy (i.e. entropy bound in microscopic motions). The second law is valid only for the sum of both parts, the order parameter entropy and the microscopic entropy.

A similar relation as we have described for the entropy itself should hold for the entropy transfer and for the creation of information. From the point of view developed above, information transfer appears to be a special form of entropy transfer [28,29]. There are other forms of entropy transfer, such as heat conduction, which have nothing to do with information transfer, but are connected only with the microscopic motion. Creation or destruction of information is any interchange between the two contributions to the total entropy given by (27) or (28), respectiely. In the context of macromolecules the creation of information is connected with the fixing of certain primary structure in a dynamical context, i.e. by memorizing the information. Evidently some forms of entropy have a potential informational value and others have not. A necessary condition that entropy which is transferred or interchanged has an informational value is that it can be memorized. In terms of the nonlinear dynamics theory information transfer means a change of the attractor region of the order parameters in the receiving system. Correspondingly the creation of information means the freezing of the state in one region of a multi-attractor order parameter state space, which means memorizing a state. If the entropy of a liquid is transferred by heat conduction it cannot be memorized, it is not information. However when tossing a coin, one bit of informational entropy may be transferred or created. When certain primary structure of macromolecules may be fixed in a dynamical context, e.g. by self-reproduction, creation and transfer of information becomes possible.

The discussion given above may also shed some light on the open question of how the Kolmogorov-Sinai entropy is related to Gibbs' and Shannon's entropy. This entropy concept was introduced by Kolmogorov in 1958 and developed further by Sinai, Ruelle, Grassberger, Procaccia and many other researchers [40]. It is not a proper entropy but a rate of creation of information in a dynamic process. If the dynamics is preserving a measure ρ, then the Kolmogorov-Sinai expression $h(\rho)$ measures the asymptotic rate of creation of information by the dynamical map.

Positivity of the Kolmogorov-Sinai entropy in general implies that at least one of the Lyapunov exponents of the motion is positive, i.e. chaoticity is observed. In spite of the fact that only a few rigorous results about $h(\rho)$ are available, e.g. the Sinai results for rigid convex body systems, it is generally believed that the many body systems of statistical mechanics have positive Kolmogorov-Sinai entropies. In order to elucidate the role of information creation in a statistical-mechanical system let us consider a system at $t < t_o$ and with energy E in thermodynamic equilibrium and corresponding to the entropy S(E). Having so far no information about the position on the energy shell, i.e. H = 0, all the entropy is bound in microscopic motion.

$$S(E) = S_b \tag{29}$$

Let us assume now, that a measurement gives us at time $t = t_o$ a amount of information

$$\delta H = (H_{max} - H) >> \log 2 \tag{30}$$

about the microscopic state, i.e. the position of the phase point on the energy shell. Then at $t > t_o$ the chaotic character of the microscopic motion with positive Kolmogorov-Sinai entropy $h(\rho) > 0$ will lead to a decrease of the information we have and to an increase of the entropy. This process between the two parts will lead finally to the equilibrium situation, where no information is available and all entropy is bound in the microscopic chaotic motion on the energy shell.

This leads to the conjecture that devaluation takes place under the condition that the underlying microscopic motion has a positive Kolmogorov-Sinai entropy and further that the system is kept isolated. The situation is completely different if the system is pumped (i.e. is exporting thermodynamical entropy). Then the nonlinear dynamics of the order parameters may create information on the macroscopic scale which counterbalances the microscopic tendencies. These questions however still need further investigations.

The considerations given so far were very general and still cannot explain anything concrete. We have shown, however, that the creation of information is nothing unphysical but in full agreement with physical laws. There is no violation of any physical rule, including the second law of thermodynamics. A satisfying explanation of the origin of information has to be more specific, of course. In the context of this discussion there remains room only for a brief survey of the state of art.

Eigen and Schuster formulated in 1982 five principles of early organization which were worked out in their earlier papers:

(i) Formation of heteropolymers introduces an otherwise unknown richness into the structures and properties of molecules.

(ii) Autocatalysis or self-replication under conditions far from equilibrium introduces selection into molecular ensembles.

(iii) Replication errors lead to mutant distributions called quasispecies. The accuracy of the replication process sets a limit to the amount of genetic information that can be transmitted.

(iv) Cooperation between otherwise competitive self- replicating elements is introduced by higher order catalytic action in the form of positive feedback loops called hypercycles.

(v) Formation of compartments and individuation allows efficient evaluation of the relevant functional properties of translation products.

Feistel et al. developed between 1974 and 1982 a similar concept, published in less known journals and in a book [5,41-45]. This concept will be briefly surveyed, since it differs in a few details from the Eigen-Schuster concept. Beside some mathematical points (referring e.g. to the mathematical formulation of the quasispecies concept and to stochastic effects [42]) the main difference is the assumption of a very early compartmentalization. Further, instead of hypercycles the RNA-replicase cycles play in our scenario the dominant role. This argument was based on an estimate of the probabilities of the spontaneous generation of catalytic structures [43]. It was shown that very simple structures such as paths, branching systems, semicycles and small cycles have a much greater probability to come into existence than larger cyclic structures.

In our model [42-45] six stages were assumed and characterized in detail [42-45]:

(i) Physico-chemical self-organization: Spontaneous formation of polypeptides and polynucleotides, local increase of concentration in compartments (coacervates, microspheres, pores etc.), catalytic assistance of the synthesis in networks and cascades, first self-reproduction of polynucleotides (RNA) assisted by catalytic proteins (replicases), competition and selection between compartments and between replicative cycles inside compartments.

(ii) Formation of protocells and of a molecular language: Genesis of self-reproductive units consisting of RNA and proteins, division of labor between RNA and protein develops more and more, generation of elements of a molecular language, DNA takes over the role of the memory and RNA specializing in transcription functions, the building block principle for the synthesis of proteins is worked out.

(iii) Genetic code, ritualization and disvision of work: The full coding is successively replaced by a kind of stenography, the first triplett code arises and, complete, the direct chemical meaning of a sequence is replaced by a symbolic notation (ritualization), the developement of the genetic code, ritualization and the division of work lead to a minimal organism in the sense of Kaplan.

(iv) Cellular organization: The triplet code generates more and more complex structures such as membranes stabilizing the compartments, spatial separation and nonlinear advantage lead a freezing of the code, division of cells occurs, decreasing concentration of the raw material leads to metabolic chains, mobility and predator-prey relations between the cells.

(v) Genesis of autarc systems: Since the raw materials are exhausted a sharp selection pressure generates systems able to use primary sources of energy, photosynthesis is created, heterotrophy and food-webs appear, oxygen-metabolism leads higher mobility, division of labor inside the cells leads to cell compartments and especially to the cell nuclei, sexual reproduction is invented.

(vi) Morphogenesis: The metabolic products of cells in cell associations take over certain regulative functions, division of work between cells create multi-cellular organisms, certain cell groups, the neurons, specialize in information processing, first based on direct chemical or stereo-chemical relations, more and more symbolic information transfer is introduced, this leads to a new phase of ritualization [45] and finally to complex nervous systems.

In conclusion let us emphasize that most of the problems discussed here are still far from final solutions. One may hope that the elaboration of a "synergetics of evolution" will contribute to a coordinated attack on the most difficult problems such as the spontaneous generation of information processing by self-organization of matter.

References

1. H. Haken (ed.): Synergetics (Proc. Int. Symp., Elmau 1972),B.G. Teubner, Stuttgart 1973

2. H. Haken (ed.): Synergetics. A Workshop, Springer, Berlin, Heidelberg, New York 1977

3. H. Haken: Synergetics. An Introduction, 3rd. Edn. Springer Berlin, Heidelberg, New York 1983

4. H. Haken: Advanced Synergetics, Springer, Berlin, Heidelberg, New York 1983

5. W. Ebeling und R. Feistel: Physik der Selbstorganisation und Evolution, Akademie-Verlag, Berlin 1982 u. 1986

6. R. Feistel and W. Ebeling: Evolution of Complex Systems, Verlag der Wissenschaften, Berlin 1989, Kluwer Academic Publ., Dordrecht 1989

7. W. Ebeling, A. Engel and R. Feistel: Physik der Evolutionsprozesse, Akademie-Verlag, Berlin 1990

8. M. Eigen: Naturwissenschaften 58, 465(1971); M.Eigen and P. Schuster: Naturwissenschaften 64, 541(1977); 65, 341(1978)

9. J.Hofbauer and K. Sigmund: Evolutionstheorie und dynamische Systeme, Parey, Hamburg, 1984.

10. M. Eigen, J. Mc Caskill and P. Schuster: Adv. Chem. Phys. 75, 149(1989)

11. M. Conrad: Adaptability. The Significance of Variability from Molecules to Ecosystems, Plenum Press, New York 1983

12. J.D. Farmer, S.A. Kaufman and N.H. Packard: Physica 22D, 50(1986)

13. S. A. Kauffman: Physica 42D, 135 (1990)

14. S.A. Kauffman: Dynamics of Evolution. Lecture at the Workshop Complex Dynamics and Biological Evolution, Hindsgavl 1990

15. I. Prigogine, G. Nicolis and A. Babloyantz: Physics Today 25, 23,38 (1972)

16. E. Weinberger: J. Stat. Phys. 49, 1011 (1987); Correlated and Uncorrelated Fitness Landscapes and How to Tell the Difference, Biol. Cybernetics, in press

17. A.S. Mikhailov: Foundations of Synergetics I. Springer, Berlin, Heidelberg, New York 1990

18. G. Nicolis and I. Prigogine: Self-Organization in Non-Equilibrium Systems, Wiley-Interscience Publ, New York 977; Die Erforschung des Komplexen, Piper-Verlag, München/Zürich 1987

19. W. Weidlich and G. Haag: Concepts and Models of a Quantitative Sociology, Springer, Berlin, Heidelberg, New York 1983; W. Weidlich and G. Haag (eds.): Interregional Migration, Springer 1988

20. P. Bak et al.: Phys. Rev. Lett. 59, 381 (1987); Rev. A38, 364 (1988); Nature 342, 780 (1989)

21. C.G. Langton: Physica 42D, 12 (1990)

22. W. Ebeling and G. Nicolis: Europhys. Lett. 14, 191 (1991)

23. W. Ebeling: On the Relation Between Entropy Concepts. In: Festschrift on the Occasion Peter Szepfaluzys 60th Birthday (I. Kondor, ed.), World Scientific 1991; Physica A (1992) to appear

24. R. Feistel: On the Value Concept in Economy, In: Proc. Conf. MOSES (Gosen, Nov 1990), Akademie-Verlag, Berlin 1991

25. W. Ebeling: Syst. Anal. Model. Simul. 8, 3 (1991)

26. E. Mosekilde, J. Aracil, J. and P.M. Allen: System Dynamics Review 4, 14 (1988)

27. E. Bruckner, W. Ebeling and A. Scharnhorst: System Dynamics Review 5, 176 (1989)

28. W. Ebeling and M.V. Volkenstein: Physica A 163, 398 (1990)

29. W. Ebeling: Chaos, Ordnung und Information. Urania Verlag Leipzig 1989; H. Deutsch, Thun-Frankfurt a.M. 1989, 1991

30. H. Haken: Information and Self-organization. Springer, Berlin, Heidelberg, New York 1988

31. M. Eigen: Ber. Bunsenges. Phys. Chem. 80, 1059 (1976)

32. M.V. Volkenstein: Usp. Fiz. Nauk (Moscow) 143, 429 (1984)

33. M.V. Volkenstein: Entropie und Information, H. Deutsch, Thun-Frankfurt a.M. 1990

34. S.A. Kauffman and S. Levin: J. theor. Biol. 128, 11 (1987)

35. B. Derrida and L. Peliti: Evolution in a Flat Fitness Landscape, Preprint C.E.N. Saclay SPhT-026 (1990)

36. W. Fontana and P. Schuster: Biophys. Chem. 26, 123 (1987)

37. W. Ebeling: Boltzmann-, Darwin- and Haeckel-Strategies in Optimization Problems. In: H.-P. Schwefel (ed.), Proc. PPSN, Dortmund, 1990

38. W. Ebeling, A. Engel and V.Mazenko: BioSystems 19, 213(1986)

39. W. Ebeling and R. Feistel: studia biophysica 46,183(1974); Ann. Physik 34, 91 (1977); studia biophysica 71, 139 + MF 1/44-52 (1978)

40. J.S. Nicolis: Chaos and Information Processing, World Scientific, Singapore 1991

41. M. Eigen and P. Schuster: J. Mol. Evol. 19, 47 (1982)

42. W. Ebeling and R. Feistel: studia biophysica 75, 131(1979)

43. I. Sonntag et al.: Biometrical Journal 23, 501 (1981)

44. R. Feistel: studia biophysica 93, 113,121 (1983); 95, 107,133 (1983)

45. R. Feistel: Ritualisation und die Selbstorganisation der Information. In: Selbstorganisation, Jahrbuch für Komplexität (U. Niedersen, Hrsg.), Duncker & Humblot, Berlin 1990

Nonequilibrium Potentials

R. Graham and A. Hamm

Fachbereich Physik, Universität GH Essen,
Universitätsstr. 5, W-4300 Essen 1, Fed. Rep. of Germany

Abstract. We review how the familiar notion of a thermodynamic potential can be generalized for a wide class of dynamical systems continuous in time and perturbed by weak noise; how, at least in principle, the description by means of nonequilibrium potentials can be reduced to discrete maps; and we present examples of nonequilibrium potentials for the one-dimensional logistic map. The latter result is used to calculate the critical exponent for the scaling of localized noise at the period doubling bifurcation sequence observed in numerical experiments by Mayer-Kress and Haken.

We dedicate this paper to Hermann Haken on the occasion of his 65th birthday.

1. Thermodynamic Potentials and Nonequilibrium Potentials

In macroscopic classical physics one is interested in dynamical systems described by a set of evolution equations of first order in time

$$\dot{q}^{\nu} = K^{\nu}(q) \tag{1.1}$$

where q^{ν}, $\nu = 1, \ldots, n$, is a set of macroscopic variables. For simplicity we shall here restrict our attention to lumped systems without any spatial extension, and we assume that the "drift vector" $K^{\nu}(q)$ is not explicitly time-dependent. If the dynamical system (1.1) arises from thermodynamics it has a special structure. It must be consistent with the second law of thermodynamics which states that the total entropy of any closed system increases due to irreversible processes, and is not changed by reversible processes:

$$\frac{dS(q(t))}{dt} \geq 0. \tag{1.2}$$

Equation (1.1) must also be consistent with the Boltzmann-Einstein formula

$$W(q) = \text{const} \cdot \exp(S(q)/k_B) \tag{1.3}$$

which determines the probability density $W(q)$ of observing a fluctuation of the q^{ν} away from their values q_0^{ν} in thermodynamic equilibrium. In (1.2) and (1.3) the thermodynamic potential $S(q)$ is the entropy (where Boltzmann's constant is k_B) of the total system, comprising not only the subsystem of interest, whose

Springer Proceedings in Physics, Vol. 69
Evolution of Dynamical Structures in Complex Systems
Editors: R. Friedrich · A. Wunderlin

instantaneous state is described by the $q^\nu(t)$, but also all reservoirs of energy, matter, etc. to which this subsystem may be coupled. In thermodynamic equilibrium $S(q_0) = \max$. $S(q)$ is continuous, in general, but if first order phase transitions occur its first order derivatives may be discontinuous. If the reservoirs have fixed temperature T_0 and pressure P_0 then $\Delta S(q) = S(q) - S(q_0)$ may be expressed as [1]

$$\Delta S(q) = -\frac{1}{T_0} R_{\min}(q) \tag{1.4}$$

with

$$R_{\min}(q) = \left[E(q) - E(q_0) - T_0(\tilde{S}(q) - \tilde{S}(q_0)) + P_0(V(q) - V(q_0))\right] \tag{1.5}$$

where $E(q), \tilde{S}(q), V(q)$ are, respectively, the energy, entropy, and volume of the subsystem alone. $R_{\min}$ is the minimal work some hypothetical external thermally isolated system would have to perform on the subsystem in order to bring about the change from q_0 to q in a controlled way. In the absence of such a system the change from q_0 to q only happens due to a fluctuation whose probability density is given by (1.3).

The relation of (1.1) to (1.2) and (1.3) in thermodynamics is called nonequilibrium thermodynamics [2]. Let us first recall how (1.1) and (1.2) are related: Generalized thermodynamic forces $p_\nu(q)$ are introduced via

$$p_\nu(q) = \frac{\partial S(q)}{\partial q^\nu} \tag{1.6}$$

and the drift vector (1.1) is cast in the general form

$$K^\nu(q) = \frac{1}{2} Q^{\nu\mu}(q) \frac{\partial S(q)}{\partial q^\nu} + r^\nu(q) \tag{1.7}$$

with some non-negative symmetric matrix of "transport coefficients" $Q^{\nu\mu} = Q^{\mu\nu}$ and the velocity vector $r^\nu(q)$ chosen to satisfy

$$r^\nu(q) \frac{\partial S(q)}{\partial q^\nu} = 0. \tag{1.8}$$

Here and in the following the summation convention is implied. By construction the second law (1.2) is then satisfied

$$\frac{dS(q(t))}{dt} = \frac{1}{2} Q^{\nu\mu}(q) p_\nu(q) p_\mu(q) \geq 0. \tag{1.9}$$

In thermodynamics (1.8) is usually satisfied by taking $r^\nu(q)$ in the special form

$$r^\nu(q) = \frac{1}{2} A^{\nu\mu}(q) \frac{\partial S(q)}{\partial q^\mu} \tag{1.10}$$

with an antisymmetric matrix $A^{\nu\mu}(q) = -A^{\mu\nu}(q)$. Then (1.7) is Onsager's well known linear relation

$$K^\nu(q) = \frac{1}{2}L^{\nu\mu}(q)p_\mu(q)\ ; \quad L^{\nu\mu} \equiv Q^{\nu\mu} + A^{\nu\mu} \tag{1.11}$$

between thermodynamic forces p_μ and thermodynamic fluxes $\dot{q}^\nu$. As the first term of (1.7) increases the total entropy it is irreversible, i.e. it must change sign compared to $\dot{q}^\nu$ under time reversal; as the second term of (1.7) does not change the total entropy it is reversible and must transform like $\dot{q}^\nu$ under time reversal. Indeed, this behavior is guaranteed by the Onsager-Casimir symmetry relations for $L^{\nu\mu}$ [3].

The relation between (1.1) and (1.3) in thermodynamics is less direct, because (1.1) is a deterministic set of evolution equations neglecting fluctuations, $k_B = 0$, while (1.3) refers to just such fluctuations. Still, a relation exists — the well known fluctuation-dissipation relation. It fixes the strength of the small random perturbations of the deterministic dynamical system (1.1) due to thermodynamic fluctuations. (The validity of a Markovian description of these fluctuations is assumed here.) Expressed in terms of the Fokker-Planck equation satisfied by the conditional probability density $P(q|q',t)$ to observe q at time t, if q' was observed at $t = 0$ the fluctuation-dissipation relation implies [4]

$$\frac{\partial P}{\partial t} = -\frac{1}{2}\frac{\partial}{\partial q^\nu}\left[L^{\nu\mu}(q)\left(\frac{\partial S(q)}{\partial q^\mu} - k_B\frac{\partial}{\partial q^\mu}\right)P\right]. \tag{1.12}$$

For $k_B = 0$, eq. (1.1) determines the characteristics of (1.12). For $k_B \neq 0$ eq. (1.3) gives a time-independent solution of (1.12).

These facts are, of course, well known. We recall them here because an important part of this structure is preserved outside thermodynamics in a large class of dynamical systems perturbed by weak noise. For a review see [5]; for mathematical work see [6].

Let us consider such systems described by the stochastic differential equations (in the Ito calculus)

$$\dot{q}^\nu = K^\nu(q) + \sqrt{\eta}\, g^\nu{}_i(q)\,\xi^i(t) \tag{1.13}$$

with the Gaussian white noise ($i = 1, \dots m$)

$$\langle \xi^i(t)\rangle = 0 \quad \langle \xi^i(t)\,\xi^j(t')\rangle = \delta^{ij}\delta(t-t') \tag{1.14}$$

and define $Q^{\nu\mu} = g^\nu{}_i(q)g^\mu{}_j(q)\delta^{ij}$. We adopt the summation convention also for latin indices. In (1.13) η is a formal parameter, analogous to k_B, measuring the noise strength. For $\eta = 0$ eq. (1.13) reduces to (1.1). The time-independent (stationary) probability density asymptotically for small noise may be represented as

$$W(q) \simeq Z(q)\exp\left(-\frac{\Phi(q)}{\eta}\right). \tag{1.15}$$

Here $\Phi(q)$ is a continuous function which may have discontinuities in its first (and higher order) derivatives, which we call nonequilibrium potential, and

$Z(q)$ is a prefactor. What is now the relation between (1.13) and (1.15)? Equation (1.15) must be a time-independent solution $\partial W/\partial t = 0$ of the Fokker-Planck equation corresponding to (1.13)

$$\frac{\partial W}{\partial t} = \frac{\partial}{\partial q^\nu}\left[-K^\nu(q) + \frac{1}{2}\frac{\partial}{\partial q^\mu}Q^{\nu\mu}(q))W\right] \tag{1.16}$$

from which

$$K^\nu(q)\frac{\partial \Phi}{\partial q^\nu} + \frac{1}{2}Q^{\nu\mu}(q)\frac{\partial \Phi}{\partial q^\nu}\frac{\partial \Phi}{\partial q^\mu} = 0 \tag{1.17}$$

and a homogeneous linear first order partial differential equation for $Z(q)$ follow. Therefore, one can still write

$$\begin{aligned} K^\nu(q) &= -\frac{1}{2}Q^{\nu\mu}(q)\frac{\partial \Phi}{\partial q^\mu} + r^\nu(q) \\ r^\nu(q)\frac{\partial \Phi}{\partial q^\nu} &= 0 \end{aligned} \tag{1.18}$$

generalizing (1.7), (1.8) in terms of a nonequilibrium potential. As a consequence (1.9) also has a generalization: For $\eta = 0$

$$\frac{d\Phi(q(t))}{dt} = -\frac{1}{2}Q^{\nu\mu}(q)\frac{\partial \Phi}{\partial q^\nu}\frac{\partial \Phi}{\partial q^\mu} \leq 0. \tag{1.19}$$

Therefore, $\Phi(q_0) = \min$ determines the steady states of the system for $\eta = 0$, and $\Phi(q)$, like the total entropy in thermodynamics, can serve as a Lyapunov function. What are the differences to thermodynamics? In general $\Phi(q)$, unlike the total entropy, cannot be related to the minimal work like in (1.4). In order for such a relation to exist, the system under study should at least have a well-defined temperature T_0 in its stationary state. In many applications this is, in fact, the case (for an example see [7]), even though the stationary state in all other respects may differ strongly from a state of thermodynamic equilibrium. However, the most important difference to thermodynamics is the fact that, in general, the Φ-decreasing processes and the processes leaving Φ invariant cannot be identified with irreversible and reversible processes, respectively. (However, for a very early special case where the difference disappears [8] see [9]). Therefore, unlike in thermodynamics, the form (1.18) with (1.17) hardly places any restrictions on the admissible drift vectors $K^\nu(q)$. Rather, these equations represent a local condition, (1.17), which must be satisfied by the nonequilibrium potential for given $K^\nu(q)$. In some sense, the situation is therefore reversed, compared to thermodynamics: There the thermodynamic potential is known or assumed to be of a given form, and K^ν, $Q^{\nu\mu}$ are determined consistent with that knowledge while in the present case K^ν and also $Q^{\nu\mu}$ are considered as known, and Φ must be determined. Then the singular part (for $\eta \to 0$) of (1.15) is known. One may use this knowledge e.g. to obtain asymptotic estimates of mean first-exit times $\tau(\partial G)$ from an attractor A at a local minimum $\Phi = \Phi(A)$ of Φ across a boundary ∂G surrounding A within its domain of attraction,

$$\tau(\partial G) \sim \inf_{q \in \partial G} \exp\left[\frac{\Phi(q) - \Phi(A)}{\eta}\right] \tag{1.20}$$

Equation (1.20) generalizes the Arrhenius law of thermodynamics.

2. Functional Integral and Minimum Principle

The conditional probability density $P(q|q', t)$ is the Green's function of (1.16). It can be written as a functional integral [10]

$$P(q|q',t) = \int D\mu[q] \exp\left(-\frac{1}{\eta}\int_{q(0)=q'}^{q(t)=q} L(q,\dot{q},\eta)\,d\tau\right) \tag{2.1}$$

with a suitable formal measure $D\mu$ on the space of continuous paths and a Lagrangian $L = L_0(q, \dot{q}) + O(\eta)$ with

$$L_0(q,\dot{q}) = \frac{1}{2}Q^{-1}_{\nu\mu}(q)(\dot{q}^\nu - K^\nu(q))(\dot{q}^\mu - K^\mu(q)). \tag{2.2}$$

We have not written out the terms of first and higher order in η which will not be needed in the following (and whose explicit form depends on the choice adopted for $D\mu$). In (2.2) it is assumed that $Q^{\nu\mu}(q)$ is non-singular, and $Q^{-1}_{\nu\mu}$ is the matrix-inverse of $Q^{\nu\mu}$; we shall here restrict ourselves to this case, for simplicity. The integral in (2.1) is taken over all paths $q(\tau)$ with $q(0) = q'$ and $q(t) = q$.

The time-independent probability density satisfies, for arbitrary $t > 0$

$$W(q) = \int dq'\, P(q|q',t)W(q'). \tag{2.3}$$

In the limit $\eta \to 0$, using (1.15), (2.1), (2.2), the integral in (2.3) and the functional integral in (2.1) may be carried out in saddle point approximation and we obtain

$$\Phi(q) = \inf_{q'}\left(\inf_{\substack{q(\tau)\\ q(0)=q',q(t)=q}} \int_0^t L(q,\dot{q})\,d\tau + \Phi(q')\right). \tag{2.4}$$

In the limit $t \to \infty$ the infimum over q' is attained in one of the attractors A_i of the unperturbed dynamical system $\dot{q}^\nu = K^\nu$,

$$\Phi(q) = \min_i\left(\inf_{\substack{q(\tau)\\ q(0)\in A_i,q(\infty)=q}} \int_0^\infty L(q,\dot{q})\,d\tau + \Phi(A_i)\right) \tag{2.5}$$

since for any point q' in the domain of attraction of A_i we must have $\Phi(q') \geq \Phi(A_i)$ due to (1.19) and

$$\inf_{\substack{q(\tau)\\ q(0)\in A_i, q(\infty)=q}} (\ldots) \le \inf_{t'\ge 0} \inf_{\substack{q(\tau)\\ q(0)\in A_i, q(t')=q', q(\infty)=q}} (\ldots). \tag{2.6}$$

The action integral in (2.5) is independent of the amount of time the path $q(\tau)$ initially spends in the attractor A_i. Therefore (2.5) can be replaced by

$$\Phi(q) = \min_i \left(\inf_t \inf_{\substack{q(\tau)\\ q(0)\in A_i, q(t)=q}} \int_0^t L(q,\dot q)\, d\tau + \Phi(A_i) \right). \tag{2.7}$$

The constants $\Phi(A_i)$ must be determined in such a way, that $\Phi(q)$ is continuous in the repellers whose stable manifolds form the separatrices which separate the domains of attraction of the A_i [11]. A general graph-rule for the determination of the $\Phi(A_i)$ is given in [6].

The variational problem of finding

$$\Phi_i(q) = \inf_t \inf_{\substack{q(\tau)\\ q(0)\in A_i, q(t)=q}} \int_0^t L(q,\dot q)\, d\tau \tag{2.8}$$

can be solved by standard methods: After introducing the generalized momentum $p_\nu = \partial L_0/\partial \dot q^\nu$ and the Hamiltonian

$$H_0(q,p) = p_\nu \dot q^\nu - L_0(q,\dot q) = \frac{1}{2} Q^{\nu\mu}(q) p_\nu p_\mu + K^\nu(q) p_\nu, \tag{2.9}$$

the minimizing $q(\tau)$ in (2.8) satisfies the canonical equations

$$\dot q^\nu = \frac{\partial H_0}{\partial p_\nu} \quad , \quad \dot p^\nu = -\frac{\partial H_0}{\partial q^\nu}. \tag{2.10}$$

The infimum over t implies that we are only interested in solutions with $H_0(q,p) = 0$. Equation (2.8) can therefore be expressed as

$$\Phi_i(q) = \int_{A_i}^q p_\nu\, dq^\nu. \tag{2.11}$$

The equation (1.17) is the Hamilton-Jacobi equation for the present approach.

3. Discrete Time Systems

Many conceptual problems concerning dynamical systems of the form (1.1) are greatly simplified if a suitable Poincaré surface of section can be found [12]: Assume that there is a hypersurface Σ in q-space such that the flow generated by (1.1) intersects Σ transversely and brings back any point $x \in \Sigma$ to some point $y = F_p(x) \in \Sigma$ after a first return time $\Theta(x)$. The $(n-1)$-dimensional discrete-time dynamical system given by the Poincaré map $F_p(x)$ contains much of the features of the original deterministic continuous time system.

Now we want to apply a similar technique to simplify the study of a system with random perturbations as given by (1.13). There is of course no longer a deterministic relation between two successive passages through Σ, but a probabilistic one:

The one-step transition probability density $p(y|x)$ to find the next transversal through Σ at $y \in \Sigma$ having started at $x \in \Sigma$ completely characterizes the noisy Poincaré map. For weak noise, applying a saddle point approximation to the path integral formulation of transition probability densitites (2.1) one obtains the following asymptotic form:

$$p(y|x) \sim \exp\left(-\frac{\varrho_p(x,y)}{\eta}\right) \tag{3.1}$$

with

$$\varrho_p(x,y) = \inf_{\theta} \inf_{\substack{q(\tau) \\ q(0)=x,q(\theta)=y \\ q(t)\notin\Sigma \text{ for } 0<t<\theta}} \int_0^\theta L_0(q,\dot{q})\,d\tau. \tag{3.2}$$

The variational problem of determining $\varrho_p(x,y)$ can, in principle, again be solved by the canonical formalism (2.10).

In the following, we do not confine ourselves to the special noisy maps derived from continuous time systems but turn to the general case of Markovian random sequences with the following asymptotics of one-step transition probabilities [16]

$$p(y|x) \sim \exp\left(-\frac{\varrho(x,y)}{\eta}\right). \tag{3.3}$$

Here, $\varrho(x,y)$ is an arbitrary non-negative function only subject to one condition relating the random sequence to a deterministic dynamical system given by a map $F(x)$. This condition reads

$$\varrho(x,y) = 0 \iff y = F(x). \tag{3.4}$$

It is easy to check with (3.2) that $\varrho_p(x,y)$ satisfies (3.4) with $F(x) = F_p(x)$.

For computational reasons it is sometimes useful to make the connection between the deterministic and the random system more explicit by writing down a stochastic difference equation for the random system:

$$x_{n+1} = F(x_n) + \xi_n(x_n). \tag{3.5}$$

Comparing with (3.3) one derives that the generally state-dependent random variable $\xi_n(x)$ is distributed – independently for different values of n – according to the probability density

$$\psi(\xi,x) \sim \exp\left(-\frac{1}{\eta}\varrho(x,F(x)+\xi)\right). \tag{3.6}$$

Early studies of such systems were reported in [13-15]. For recent studies by means of non-equilibrium potentials see [17-23].

In analogy to (1.15) the asymptotic form of the invariant density $\hat{w}(q)$ of the noisy discrete time system can be expressed in terms of a function $\hat{\Phi}(q)$:

$$\hat{w}(q) \sim \exp\left(-\frac{\hat{\Phi}(q)}{\eta}\right), \tag{3.7}$$

and we call the function nonequilibrium potential again. A minimum principle for $\hat{\Phi}$ can be derived by imitation of the procedure in Sect. 2.

The new point of departure, superseding equation (2.1), is the asymptotic form of the N-step transition probability density, obtained by $(N-1)$-times applying the Chapman-Kolmogorov equation to (3.3):

$$p(q|q', N) \sim \int \prod_{i=1}^{N-1} dq_i \exp\left(-\frac{1}{\eta}\sum_{i=0}^{N-1} \varrho(q_i, q_{i+1})\right) \tag{3.8}$$

where $q_0 = q'$ and $q_N = q$.

The result (2.7) is now replaced by

$$\hat{\Phi}(q) = \min_j \left(\inf_N \inf_{\substack{(q_i) \\ q_0 \in A_j,\, q_N = q}} \sum_{i=0}^{N-1} \varrho(q_i, q_{i+1}) + \hat{\Phi}(A_j) \right), \tag{3.9}$$

where the A_j are the attractors of the deterministic system. At this point we should mention that the nonequilibrium potential obtained from (2.7) or (3.9) is a continuous, but in general not continuously differentiable function. This, in retrospect, compromises the saddle point approximation leading to (2.4). Nevertheless the results (2.7) and (3.9) (with a slightly changed characterization of the sets A_j), can be proved rigorously under quite weak conditions – we refer to [6,16,22].

If for every fixed q^μ there exists a function $v_q^\nu(p)$ such that

$$v_q^\nu\left(\frac{\partial}{\partial u}\varrho(q, q+u)\right) = u^\nu, \tag{3.10}$$

then we can cast the minimization over sequences $q_0, \ldots, q_N$ in (3.9) into a canonical formalism again:

With the Hamiltonian

$$H(q,p) = p_\nu v_q^\nu(p) - \varrho(q, q + v_q(p)) \tag{3.11}$$

the canonical equations

$$\begin{aligned} q_{n+1} - q_n &= \frac{\partial H}{\partial p}(p_{n+1}, q_n), \\ p_{n+1} - p_n &= -\frac{\partial H}{\partial q}(p_{n+1}, q_n) \end{aligned} \tag{3.12}$$

supply minimizing sequences.

There is even an analogue to the Hamilton-Jacobi equation (1.17):

$$\hat{\Phi}_i\left(q + \frac{\partial H}{\partial p}(q,p)\right) - \hat{\Phi}_i(q) = p\frac{\partial H}{\partial p}(q,p) - H(q,p) \tag{3.13}$$

when q and p are related by

$$p + \frac{\partial H}{\partial q}(q,p) = \frac{\partial \hat{\Phi}_i}{\partial q}(q). \tag{3.14}$$

However, here we are confronted with a disadvantageous feature of the construction of nonequilibrium potentials for discrete time systems, compared with continuous time systems: While for continuous time the minimizing paths obtained from (2.10) are continuous curves, the minimizing paths obtained from (3.12) are sequences of points which may freely jump around in phase space. This is the reason why we cannot obtain a partial differential equation by the Hamilton-Jacobi arguments but have to put up with the clumsy functional equation (3.13), where the values of $\hat{\Phi}_i$ at two distinct points are compared. And worse: if $\hat{\Phi}_i$ has a discontinuity in its first derivative in between those two points, eq. (3.13) is even incorrect. But since such discontinuities are quite common, the Hamilton-Jacobi approach is very ineffective for discrete time systems — unlike for continuous time.

A second remark concerning differences between continuous and discrete time is again related to the lacking continuity of paths in the latter case: For continuous time systems, nonequilibrium potential wells indicate how large the influence of the random perturbations has to be in order to allow the system to escape from inside the well. This does not necessarily hold for discrete time systems since it may be possible to jump out of the well just by the deterministic dynamics, namely if there are disconnected attractors. Thus, potential maxima which separate components of the same attractor cannot be interpreted as potential barriers. This is especially evident for the nonequilibrium potential $\hat{\Phi}$ of a Poincaré map F_p: By the construction at the beginning of this section it is easy to check that $\hat{\Phi} = \Phi|_{\Sigma}$. Thus, what seems to be a well in $\hat{\Phi}$ may be just a cross section through a groove in Φ that winds round to intersect Σ the next time at a remote place.

Notwithstanding the lacking continuity of paths, nonequilibrium potentials for discrete time systems have their own virtues compared with the continuous time counterparts: They are in general easier to compute, and they display the typical effects of nonlinear dynamics in lower dimension.

4. Nonequilibrium Potentials for Some Quadratic Maps

We now turn to an application of the method described in the preceding section: we exploit nonequilibrium potentials to explain some numerical results [14,15] on the influence of noise on the logistic model, given by the family of maps of the interval

$$F_\mu(x) = 1 - \mu x^2. \tag{4.1}$$

We concentrate on two special values of the parameter μ: $\mu_T = 1.7548$, where a periodic attractor of period three coexists with a strange repeller, and $\mu_\infty = 1.4011\ldots$, the limit of the period doubling sequence.

The numerical observations in [14,15] were depicted as follows: given a small threshold value χ and noise whose strength is characterized by its standard deviation σ, the sets $B_\chi(\sigma)$ were defined where the invariant density $\hat{w}(x)$ of the noisy system exceeds the threshold,

$$\hat{w}(x) > \chi \iff x \in B_\chi(\sigma). \tag{4.2}$$

The figures 1–3 in Ref. [14] and 11, 21 in Ref. [15] were generated by plotting $B_\chi(\sigma)$ against σ.

For bounded noise one can choose $\chi = 0$, but for noise which allows large deviations, $\chi = 0$ is inadequate because of $\hat{w}(x) > 0$ on the whole interval. Thus, we take a small $\chi > 0$.

How are the plots of $B_\chi(\sigma)$ related to nonequilibrium potentials? Let us assume that (3.3) is satisfied by the noisy system. Then (3.7) tells us that the sets $B_\chi(\sigma)$ are, for sufficiently weak noise and small threshold χ, level sets of the nonequilibrium potential:

$$B_\chi(\sigma) \approx \{x : \hat{\Phi}(x) < C_\chi \cdot \eta(\sigma)\} \tag{4.3}$$

where C_χ is a positive constant depending on the threshold value, and $\eta(\sigma)$ denotes the conversion between the standard deviation of the noise, σ, and the strength parameter, η, involved in (3.3).

Therefore, a plot of the borders of the set $B_\chi(\sigma)$ versus σ for small σ is expected to be identical to the graph of $\hat{\Phi}$ versus x after a reparametrization $\sigma \to C_\chi \cdot \eta(\sigma)$.

In order to test this statement, the results of a computation of the nonequilibrium potentials for μ_T and μ_∞ are shown in Figs. 1 and 2 with the following choice of $\varrho(x,y)$ in (3.3):

$$\varrho(x,y) = \varrho_2(x,y) = \frac{1}{2}\,|y - F(x)|^2. \tag{4.4}$$

This choice means that the random perturbations are roughly Gaussian. It is not possible, however, to work consistently with a strictly Gaussian distribution of the noise because the noise must not throw the system out of the interval. A convenient way to avoid this problem was proposed in [13]: Set $p(y|x) = 0$ if y does not lie in the interval. Only if y is in the interval take $p(y|x)$ Gaussian, modified by a normalization factor, which is x-dependent. The normalization factor, however, does not harm the asymptotics (3.3) with (4.4).

With (4.4) we obtain from (3.10) and (3.11)

$$H(q,p) = \frac{1}{2}p^\mu p_\mu + (F^\mu(q) - q^\mu)p_\mu. \tag{4.5}$$

Our Figs. 1 and 2 were obtained in the following way: The Hamilton equations (3.12) corresponding to (4.5) were iterated for suitable initial conditions near the attractor. For the resulting sequences (q_i) the sums $\sum_{i=0}^{N} \varrho_2(q_i, q_{i+1})$ were calculated and plotted versus q_{n+1} which led to the dots in the figures. According to (3.9) the nonequilibrium potential is the lower envelope of the dots.

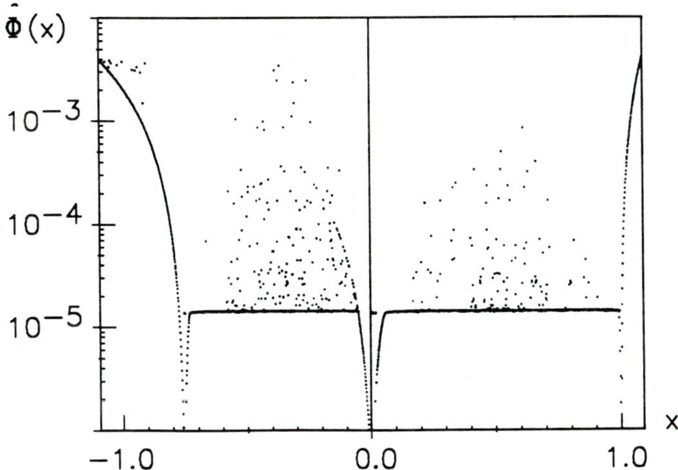

Fig. 1: The nonequilibrium potential (logarithmic scale) of the logistic map with $\mu = 1.7548$, which has a stable period three orbit

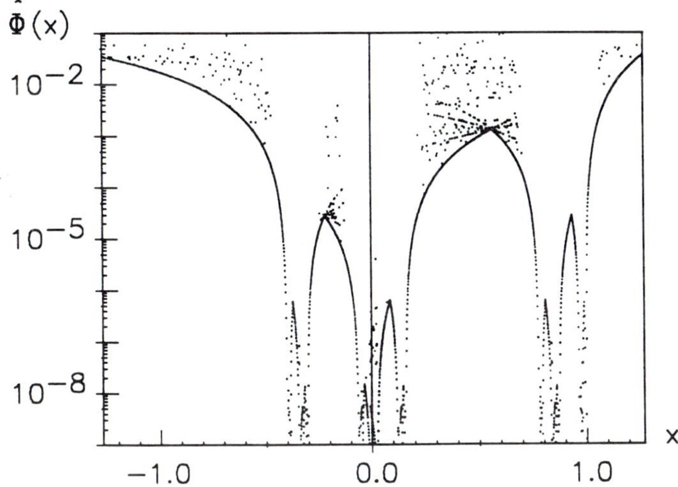

Fig. 2: The nonequilibrium potential (logarithmic scale) of the logistic map at the limit of the Feigenbaum period doubling sequence ($\mu \approx 1.401$). The Feigenbaum attractor is indicated below the abcissa

There is a clear similarity in shape between the nonequilibrium potential in Fig. 1 and the observations in Fig. 3 of Ref. [14] and Fig. 11 of Ref. [15] for the period three window, as well as between Fig. 2 and the observations in [14,15] (their Figs.2 and 21, respectively) for the period doubling accumulation point.

For $\mu = \mu_T$ (Fig. 1) the most remarkable features are the two plateaus that range over the two intervals embedding the fractal repeller between the three stable periodic points, which form the potential minima. That $\hat{\Phi}$ = const on those intervals is not only a numerical result but can also be shown by rigorous arguments [22]. The idea of a proof is roughly as follows: The potential at every point y of the two intervals has at most the value of the potential at the interval edges, since there is a deterministic orbit of F_{μ_T} which starts arbitrarily close to an interval edge and comes arbitrarily close to y. On the other hand, the potential at y cannot be less than the value at the interval edge, because every minimizing sequence starting from the stable periodic orbit has to come arbitrarily close to an interval edge before it goes to y. The fact that $\hat{\Phi}$ = const on the two intervals explains the observation of [14,15] that for maps in the period three window the ability to resolve the period 3 dissapears very abruptly as the noise strength is increased.

For $\mu = \mu_\infty$ (Fig. 2) the structure of the nonequilibrium potential is quite different. The Feigenbaum "attractor", where the potential is minimal, is a Cantor set. In each gap of the Cantor set there is a rather sharp potential maximum at the unstable periodic point which lies in the gap. The maxima at all points of one and the same unstable 2^n-periodic orbit have equal height, say $\hat{\Phi}^{(n+1)}$. For increasing n, i.e. for increasing period, the values of $\hat{\Phi}^{(n+1)}$ decrease. The maxima of the nonequilibrium potential correspond to the "band mergences" as described in the context of Fig. 3 of Ref. [14] and Fig. 21 of Ref. [15]. Thus the shape of the nonequilibrium potential shows how the structure generated by periodic orbits of smaller and smaller periods are blurred by increasing noise strength. Comparison of the plateau height in Fig. 1 with the peak heights in Fig. 2 indicate that the low-period structures of period doubling survive much longer than the structures of the period 3 window, in accordance with the observations from [14,15].

The most interesting detail in Fig. 2 is the scaling behavior of the potential maxima: The ratio $\hat{\Phi}^{(n)}/\hat{\Phi}^{(n+1)} = \lambda$ is constant, or strictly speaking, converges to a constant value for large n. In the logarithmic plot (Fig. 2) this means a constant difference between the peak heights. The value of λ obtained from the figure is $\lambda = 44$.

We sketch how to derive the exact value for λ: Suppose that we know a minimizing sequence $(q_0, \ldots, q_N)$ leading from the critical point $x_0 = 0$ of F_∞ to the unstable $2^{(n_0-1)}$-periodic point which is nearest to x_0, i.e.

$$\hat{\Phi}^{(n_0)} = \sum_{j=1}^{N} \varrho_2(q_{j-1}, q_j) = \sum_{j=1}^{N} \frac{1}{2} |q_j - F_\infty(q_{j-1})|^2. \tag{4.6}$$

From Feigenbaum's scaling theory we know that the sequence $((-\alpha)^{-n_1}q_0, (-\alpha)^{-n_1}q_1, \ldots, (-\alpha)^{-n_1}q_N)$ with $\alpha = 2.5\ldots$ leads to the gap with $\hat{\Phi}^{(n_0+n_1)}$ nearest to x_0 and that

$$\hat{\Psi}^{(n_0,n_1)} := \sum_{j=1}^{N} \frac{1}{2} |(-\alpha)^{-n_1}q_j - F_\infty^{2^{n_1}}((-\alpha)^{-n_1}q_{j-1})|^2 \sim \alpha^{-2n_1}\hat{\Phi}^{(n_0)}. \tag{4.7}$$

This sum, however, is not $\sum_{j=1}^{N} \varrho_2((-\alpha)^{-n_1}q_{j-1},(-\alpha)^{-n_1}q_j)$, but it can be used to estimate the corresponding ϱ_2-sum for a new sequence of length $(1 + N \cdot 2^{n_1})$ derived form the above one by inserting between $(-\alpha)^{-n_1}q_{j-1}$ and $(-\alpha)^{n_1}q_j$ a solution of the Hamilton equations (3.12) for (4.5) of length $2^{n_1} - 1$. The result of the estimate is

$$\hat{\Phi}^{(n_0+n_1)} \approx \hat{\Psi}^{(n_0,n_1)}/D^{(n_1)} \tag{4.8}$$

with

$$D^{(n_1)} := \sum_{j=1}^{2^{n_1}} \prod_{k=j}^{2^{n_1}-1} |F'_\infty(F^k_\infty(x_0))|^2. \tag{4.9}$$

Following an idea of [24], the scaling behavior of (4.9) can be investigated with a thermodynamic formalism:

$$D^{(n_1)} \sim \alpha^{-2n_1} \cdot \exp(n_1 2\mathcal{F}(-2)), \tag{4.10}$$

where $\beta\mathcal{F}(\beta)$ is the so called free energy of the Feigenbaum attractor, which is a numerically well studied function [25]. Inserting (4.7) and (4.10) we obtain

$$\lambda = \frac{\hat{\Phi}^{(n)}}{\hat{\Phi}^{(n+1)}} = \exp(2\mathcal{F}(-2)) \approx (6.619)^2 \tag{4.11}$$

in agreement with our numerical measurement.

The role of the above constant for the noise scaling behavior is well-known [26], [27] and the "band merging" points in Fig. 21 of Ref. [15] show the ratio 6.619 as expected. However, the numerical data of [14] showed a different ratio of the "band merging"points, namely approximately 8.5, which has not been explained by the former methods.

An explanation can be given by our present methods: First we note that Mayer-Kress and Haken [14] did not study Gaussian perturbations, but localized noise, equidistributed on an interval of width σ. This does not correspond to (3.3) for any choice of $\varrho(x,y)$. Nevertheless, we can also treat this type of noise by representing the localized noise as the $r \to \infty$ limit of noise satisfying (3.3) with

$$\varrho(x,y) = \varrho_r(x,y) = \frac{1}{r}|y - F(x)|^r, \quad (r > 1) \tag{4.12}$$

which generalizes (4.4). For arbitrary r we have the Hamiltonian

$$H_r(q,p) = \frac{r-1}{r}|p|^{\frac{r}{r-1}} + (F(q) - q) \cdot p. \tag{4.13}$$

Our above strategy for determining $\hat{\Phi}^{(n)}/\hat{\Phi}^{(n+1)}$ still works and results in

$$\lambda_r = \frac{\hat{\Phi}^{(n)}}{\hat{\Phi}^{(n+1)}} = \exp\left(r\mathcal{F}\left(-\frac{r}{r-1}\right)\right). \tag{4.14}$$

To compare with the results of [14] we have to take into account the different parametrization of the noise strength as explained in (4.3). Since $\eta(\sigma) \sim \sigma^r$ for (4.12), the appropriate ratio to look at is

$$\left[\frac{\hat{\Phi}^{(n)}}{\hat{\Phi}^{(n+1)}}\right]^{1/r} = \exp\left(\mathcal{F}\left(-\frac{r}{r-1}\right)\right). \tag{4.15}$$

The limit $r \to \infty$ has the following numerical value:

$$\exp(\mathcal{F}(-1)) \simeq 8.490 \tag{4.16}$$

in full agreement with the observations of [14].

Acknowledgements

We would like to thank Tamás Tél for useful discussions. This work was supported by the Deutsche Forschungsgemeinschaft through its Sonderforschungsbereich 237 "Unordnung und große Fluktuationen".

References

[1] L. D. Landau, E. M. Lifshitz: *Statistical Physics* (Pergamon, Oxford, 1958)
[2] S. R. De Groot, P. Mazur: *Non-equilibrium Thermodynamics* (North-Holland, Amsterdam, 1962)
[3] L. Onsager: Phys. Rev. **37** (1931), 405; **38** (1931), 2265; H. B. G. Casimir: Rev. Mod. Phys. **17** (1945), 343
[4] M. S. Green: J. Chem. Phys. **20** (1952), 1281
[5] R. Graham: In *Noise in nonlinear dynamical systems, Vol. 1*, ed. by F. Moss, P. V. E. McClintock, (Cambridge University Press, Cambridge, 1989)
[6] M. I. Freidlin, A. D. Wentzell: *Random Perturbations of Dynamical Systems* (Springer Verlag, Berlin, 1984)
[7] R. L. Kautz: Phys. Rev. **A38** (1988), 2066
[8] R. Graham, H. Haken: Z. Physik **243** (1971), 289; **245** (1971), 141
[9] H. Haken: Phys. Rev. Lett. **13** (1964), 326
[10] R. Graham, *Springer Tracts in Mod. Phys., Vol. 66* (Springer Verlag, Berlin, 1973)
[11] R. Graham, T. Tél: Phys. Rev. **A33** (1985), 1322
[12] A. J. Lichtenberg, M. A. Lieberman: *Regular and Stochastic Motion* (Springer Verlag, Berlin, 1983)
[13] H. Haken, G. Mayer-Kress: Z. Phys. **B43** (1981), 185
[14] G. Mayer-Kress, H. Haken: J. Stat. Phys. **26** (1981), 149
[15] J. P. Crutchfield, J. D. Farmer, B. A. Huberman: Phys. Rep. **92** (1982), 46
[16] Yu. Kifer: *Random Perturbations of Dynamical Systems* (Birkhäuser, Boston, 1988)
[17] P. Talkner, P. Hänggi: *loc. cit. [5], Vol. 2*
[18] P. Grassberger: J. Phys. **A22** (1989), 3283
[19] P. D. Beale: Phys. Rev. **A40** (1989), 3998
[20] P. Reimann, P. Talkner: Helv. Phys. Acta **63** (1990), 845; **64** (1991), 947; and to be published

[21] R. Graham, A. Hamm, T. Tél: Phys. Rev. Lett. **66** (1991), 3089

[22] A. Hamm, R. Graham: J. Stat. Phys. **66** (1992)

[23] R. Graham, A. Hamm: In *From Phase Transitions to Chaos, Topics in Modern Statistical Physics* ed. by G. Györgyi, I. Kondor, L. Sasvári, T. Tél (World Scientific, Singapore, 1992)

[24] E. B. Vul, Ya. G. Sinai, K. M. Khanin: Usp. Math. Nauk **39** (1984), 3,3 (Engl. transl.: Russ. Math. Surv. **39** (1984), 3,1)

[25] Z. Kovács: J. Phys. **A22** (1989), 5161

[26] J. Crutchfield, M. Nauenberg, J. Rudnick: Phys. Rev. Lett. **46** (1981), 933

[27] B. Shraiman, C. E. Wayne, P. C. Martin: Phys. Rev. Lett. **46** (1981), 935

Part II

Lasers and Nonlinear Optics

Spatio-Temporal Instabilities in Nonlinear Optical Systems

M. Brambilla[1,2], *M. Cattaneo*[1], *L.A. Lugiato*[1], *R. Pirovano*[1], *C. Pitzen*[1], and *F. Prati*[1,2]

[1]Dipartimento di Fisica dell' Università di Milano, Via Celoria 16, I-20133 Milano, Italy
[2]Physik Institut, Universität Zürich, CH-Zürich, Switzerland

1. Introduction

Pioneered by Hermann Haken, the field of optical instabilities has shown a flourishing development in the last fifteen years. One of the culminating points in this trend was the experimental observation [1] of Lorenz-like chaos as predicted by Haken's analogy between the single-mode laser model and the Lorenz model [2].

The great majority of theoretical investigations in this field were carried out in the plane-wave approximation, i.e. assuming that the electric field is uniform in all planes orthogonal to the direction of propagation. In this approximation the transverse Laplacian which describes diffraction drops out of the field equation, and one focuses on the behaviour of the system as a function of time and, possibly, of the longitudinal spatial variable. However, the last decade witnessed an increasing tendency to drop the plane-wave approximation and to study the effects which arise in the structure of the electric field transverse with respect to the direction of propagation. As a matter of fact, the transverse Laplacian (i.e. diffraction) plays, in the case of nonlinear optical systems, the same role as diffusion in nonlinear chemical reactions or in biology. Therefore one has the possibility of studying in optics phenomena of spontaneous pattern formation and transformation that are familiar in these fields, with the additional advantage of the speed in time evolution and data acquisition, which is typical of optical systems. Thus, Transverse Nonlinear Optics becomes a new chapter of Synergetics [3,4].

Transverse phenomena were studied both in passive systems without population inversion [5-14] and in active systems as lasers [15-29]. Their analysis has revealed, for example, the existence of solitonic patterns [5], of vortex structures [22-24,26,30,31], of hexagonal arrays [7,9], of space-time chaos [13,14]. All these results extended the analogy of optical systems with hydrodynamics, started by Haken in ref. 2 , to cover a vast spectrum of phenomena. In addition, this analogy can be substantiated from a formal viewpoint by constructing a close connection between the laser equations and those which govern the motion of a compressible fluid [24]. Other formal links between optics and the other nonlinear systems are provided by the laser Ginzburg-Landau equation [26] and by the laser Kuramoto-Sivashinsky equation [33].

Springer Proceedings in Physics, Vol. 69
Evolution of Dynamical Structures in Complex Systems
Editors: R. Friedrich · A. Wunderlin

This paper continues, in a sense, a report presented at the sixtieth birthday of Hermann Haken [34]. We will discuss some phenomena in lasers, as for example the formation of crystals of vortices or the onset of dynamical patterns. In the case of passive systems, we will demonstrate the existence of a spatio-temporal instability [35], which leads to a chaotic behaviour closely similar to the Lorenz-Haken chaos [1,2]. We will show that the presence of vortices in the electromagnetic field can give rise to special mechanical effects in the motion of appropriately detuned neutral atoms which interact with the field [36]. In addition, we will mention some research [37] intended to realize optical associative memories based on transverse effects in lasers. In this system the laser is used as the discriminator element, as suggested by Haken [38,39].

2. Description of the Model

We consider a ring laser with spherical mirrors, assuming that the length of the active region is much smaller than the Rayleigh length of the cavity. For a cylindrically symmetrical cavity the transverse profile of the cavity modes is described by the functions

$$A_{pl}(\rho,\varphi) = \sqrt{\frac{2}{\pi}}(2\rho^2)^{|l|/2}\left[\frac{p!}{(p+|l|)!}\right]^{1/2} L_p^{|l|}(2\rho^2)e^{-\rho^2}\, e^{il\varphi} \quad , \tag{2.1}$$

where $p = 0, 1, \ldots$ is the radial index and $l = 0, \pm 1, \ldots$ is the angular index, ρ denotes the radial coordinate normalized to the beam waist w, and $L_p^{|l|}$ are Laguerre polynomials of the indicated argument.

We assume that the active medium is a homogeneously broadened system of two-level atoms with linewidth $\gamma_\perp$, and that the excited region has a gaussian transverse shape of radius r_p, i.e. the transverse configuration of the equilibrium population inversion is described by the function

$$\chi(\rho) = \exp(-2\rho^2/\psi^2) \quad , \quad \psi = 2r_p/w \quad . \tag{2.2}$$

An important property of the Gauss-Laguerre modes is that their frequency depends on the transverse mode indices p and l via the combination $2p+|l|$, a situation that produces mode degeneracy.

We suppose that the free spectral range is much larger than the transverse mode separation [19], so that we can select modes corresponding to only one value of the longitudinal index.

We introduce the normalized slowly varying envelope of the electric field which can be expanded as follows

$$F(\rho,\varphi,t) = \sum_{pl} f_{pl}(t)A_{pl}(\rho,\varphi) \quad . \tag{2.3}$$

The modal amplitudes f_{pl} obey the time evolution equations [19]

$$\frac{df_{pl}}{dt} = -k\left[(1+ia_{pl})f_{pl} - 2C\int_0^{2\pi} d\varphi \int_0^{\infty} d\rho\, \rho A^*_{pl}(\rho,\varphi)P(\rho,\varphi,t)\right] \quad , \tag{2.4a}$$

where C is the pump parameter, k is the cavity linewidth; a_{pl} denotes the difference between the frequency of the mode of indices p, l and that of the fundamental TEM_{00} mode, normalized to k. P is the normalized slowly-varying envelope of the atomic polarization. Equation (2.4a) must be coupled with the atomic Bloch equations, which read

$$\frac{\partial P}{\partial t} = \gamma_\perp \left[F(\rho, \varphi, t) D(\rho, \varphi, t) - (1 + i\delta_{AC}) P(\rho, \varphi, t) \right] \quad , \tag{2.4b}$$

$$\frac{\partial D}{\partial t} = -\gamma_{\|} \left[\mathcal{R}e \left(F^*(\rho, \varphi, t) P(\rho, \varphi, t) \right) + D(\rho, \varphi, t) - \chi(\rho) \right] \quad , \tag{2.4c}$$

where D is the normalized population inversion, $\gamma_{\|}$ is its relaxation rate and δ_{AC} is the detuning of the atomic transition from the frequency of the TEM_{00} mode, normalized to $\gamma_\perp$. In (2.4b,c) it is understood that F is given by the expansion of (2.3).

3. The Case of a Degenerate Family

In this section we assume that the atomic line is on resonance with a frequency-degenerate family of transverse modes such that $2p + |l| = q$, with q fixed. We suppose, in addition, that all the other cavity modes either suffer from large losses, or their frequency separation from the atomic line is much larger than the atomic linewidth; therefore only the modes belonging to the frequency-degenerate family take part in the laser emission. In this case the sum in (2.3) is restricted to the modes of the active family. By choosing as reference frequency the frequency of the degenerate family $2p + |l| = q$, we can set $a_{pl} = 0$ and $\delta_{AC} = 0$ in (2.4a,b).

3.1 Stationary Patterns, Multistability and Variational Principle

Figs. 1a-d show the different stable stationary emission intensity patterns that are found in the case $q = 2$ by varying the parameters C and ψ; the dark parts correspond to large intensity. Tamm and Weiss [23] realized a ring Na_2 laser which excites selectively the frequency degenerate family $q = 2$, and observed stationary patterns which are in very good agreement with those shown in Fig. 1. In particular, if one starts with the laser operating with the single-mode state of Fig. 1a, and increases gradually the pump parameter C, the system exhibits a continuous transition to the multimode configuration of Fig. 1b, which amounts to a process of spontaneous breaking of the cylindrical symmetry as predicted in [18].

In general there exist regions in the parameter space $(2C, \psi)$ where two or more transverse patterns coexist, thus giving rise to the phenomenon of 'spatial multistability'; for $q = 2$ one finds regions where the pattern of Fig. 1c coexists with that of Fig. 1b or 1d. This multistability is different from the usual optical

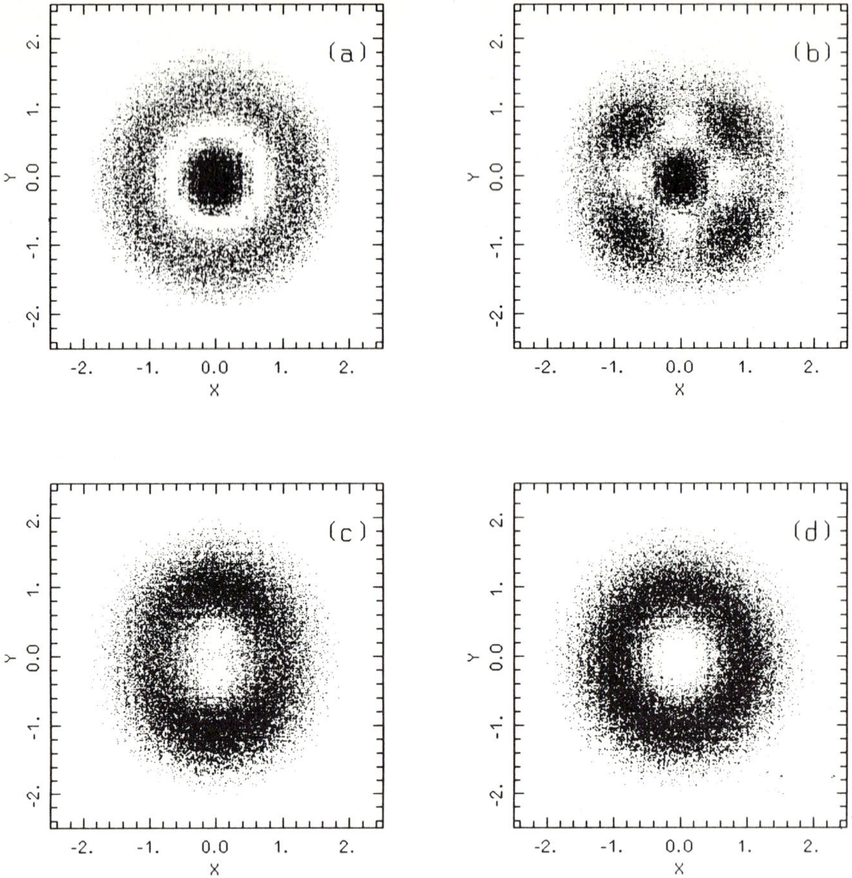

Figure 1. Intensity distribution in the transverse plane for the stable stationary solutions of the family $2p + |l| = 2$. a and d correspond to the single modes $p = 1, l = 0$ and $p = 0, l = \pm 2$, respectively; b and c are multimode configurations. Four, two and one vortices are visible in Figs. 1b, 1c and 1d, respectively

multi- or bi-stability because here the total intensity of the transmitted beam varies little among different coexistent patterns, while what varies radically is the spatial distribution of the radiation.

Let us now introduce the functional [24]

$$V = \int_0^{2\pi} d\varphi \int_0^{\infty} d\rho\, \rho \Big[|F(\rho,\varphi)|^2 - 2C\chi(\rho) \ln\big(1 + |F(\rho,\varphi)|^2\big) \Big] \quad , \tag{3.1}$$

where it is understood that the field F is given by the sum of (2.3) restricted to the modes of the degenerate family. Thus, V is in fact a function of the mode amplitudes f_{pl} and of their complex conjugates f_{pl}^* with $2p + |l| = q$.

By using the orthonormality of the modes A_{pl}, one verifies easily that the stationary equations that are obtained by setting equal to zero the time derivatives in (2.4) can be written in the form

$$0 = \frac{\partial V}{\partial f_{pl}^*} \tag{3.2}$$

and complex conjugates.

Hence the stationary solutions correspond to the stationary points of the functional V, which therefore plays the role of a generalized free energy in this system which lies far from thermal equilibrium [3]. A necessary condition for the stability of a stationary solution is that it corresponds to a minimum of V; this condition becomes sufficient in the good cavity limit $k \ll \gamma_{\|}, \gamma_{\perp}$ in which the atomic variables can be adiabatically eliminated so that one obtains the set of equations for the modal amplitudes

$$\frac{df_{pl}}{d\tau} = -\frac{\partial V}{\partial f_{pl}^*} \tag{3.3}$$

together with their complex conjugates; the normalized time τ is equal to κt.

This principle determines the values of the moduli $|f_{pl}|$ of the modal amplitudes in the steady states, as well as their relative phases. In order to find the stationary states, it is essential to allow for *complex* values of the amplitudes f_{pl}, because in all multimode stationary solutions at least one of the relative phases is different from zero.

We note that the potential V is directly a functional of the field F, so that its expression (3.1) does not depend on the particular choice of the basis for the modal expansion.

3.2 Associative Memory Operation

A relevant feature that this class of systems has proved to be capable of exploiting is the possibility of acting as a device in which one can store information and which is able to recognize an external 'object' on the basis of its memory. Such a device should be able to perform an associative memory function; strictly this means that when the system is given a partial or ambiguous representation of a memory stored pattern, it is capable to 'remember' (reconstruct) the whole object.

The process of reconstruction implies the capability of the system to 'decide' which among the objects stored in its memory has the maximum similarity to the proposed pattern, hence the system must enhance the dominant features of the ambiguous input pattern and delete all other elements. We do not intend to introduce a general concept of similarity, including for example scale invariance, but we simply state that two patterns are most similar when they are superimposable.

In our case we are interested in comparing a given pattern A with the set of patterns B_1, B_2, ..., B_N, that can be achieved by the laser in a regime of multistability of order N; in this way we want to decide which of them is the most similar to A. Therefore the similarity parameter is conveniently defined as the normalized inner product between the slowly varying envelope of the input field $A(\rho, \varphi)$ and that of each stable state $B_n(\rho, \varphi)$ which together constitute the memory of our device

$$S(A, B_n) = \frac{\left| \int_0^{2\pi} d\phi \int_0^{\infty} d\rho\, \rho\, A(\rho, \phi)\, B_n^*(\rho, \phi) \right|}{\left(\int_0^{2\pi} d\phi \int_0^{\infty} d\rho\, \rho\, |B_n(\rho, \phi)|^2 \right)^{1/2} \left(\int_0^{2\pi} d\phi \int_0^{\infty} d\rho\, \rho\, |A(\rho, \phi)|^2 \right)^{1/2}} \tag{3.4}$$

$n = 1 ... N$. The normalizing factor is introduced to make S independent of the total intensity of the patterns B_n and A.

Each stationary solution B_n has a certain basin of attraction in the phase space of the system. If the proposed pattern A is thought of as the initial condition for the electric field, the laser will approach the stationary state $B_{\overline{n}}$ such that the considered initial condition belongs to the basin of attraction of $B_{\overline{n}}$. The characterization of the basin of attraction is in general an extremely difficult task, but in our case we could prove [37] that the similarity parameter introduced in (3.4) also furnishes the criterion to decide which basin of attraction the proposed pattern A belongs to. Precisely, $\overline{n}$ is the value of n which corresponds to the maximum value of the similarity parameter $S(A, B_n)$.

This amounts to saying that the laser has recognition capability.

In order to create the initial configuration A we use an external signal, which is injected into the laser while this is below threshold and we extinguish it as soon as the laser is taken above threshold, so that the stationary states of the system coincide with those of the free-running laser.

These results show that the laser can operate as an associative memory, in which the 'objects' in memory are the coexisting stationary solutions of the laser itself. This system is, however, not very interesting because one wants the objects stored in memory to be real images, and not patterns as e.g. those shown in Fig. 1. This problem can be overcome by combining the laser with a linear system, formed by an appropriate arrangement of lenses, holograms and pinholes [37]. The images in memory I_n are stored in one of the holograms, in such a way that there is a one-to-one correspondence among images I_n and stationary states of the laser B_n. The task of the linear part is to transform an arbitrary image I which is offered to the system into a pattern A which is injected into the laser. The system is devised in such a way that the values of the similarity parameter S_n of A with the stationary configurations B_n are respectively proportional to the values of S for the original image I when it is compared with the corresponding images in memory I_n.

3.3 Optical Vortices

With the exception of the pattern of Fig. 1a, all other stable stationary solutions of the case $2p + |l| = 2$ display isolated points in the transverse plane, where both the real and the imaginary part of the electric field envelope vanish simultaneously, so that the radiation intensity is zero. In addition, if one considers a closed counterclockwise loop l which surrounds one and only one of these points S, one has that the total variation of the phase of F over the loop is an integer multiple of 2π; precisely if we set $F = |F|e^{i\Phi}$, we have

$$\Delta\Phi = \int_l \nabla\,\Phi\,\cdot\,dl = \pm m\,2\pi \quad , \tag{3.5}$$

where m is a positive integer. We call these structures 'optical vortices' [31] or 'phase singularities', because they correspond to singularities of the vector field $\nabla\,\Phi$. The number $\pm m$ is called the 'charge' of the singularity. In most stationary solutions the optical vortices are arranged in the form of crystal-like arrays (Fig. 1). Fig. 2 shows the field lines for the gradient of the phase of the electric field in the case of the pattern of Fig. 1b. A vortex-like structure for these field lines is evident.

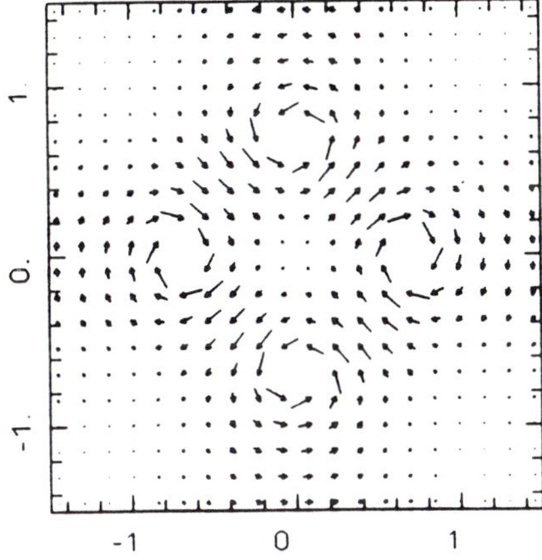

Figure 2. Plot of the electric field phase gradient for the same pattern of Fig. 1b. The analogy with the velocity field of a fluid in presence of vortices is evident

4. Laser Hydrodynamics

In this section we go beyond the case of a single degenerate family of modes and we let the electric field be a linear combination of all the transverse modes of the cavity having the same longitudinal index. In the limit $k \ll \gamma_\perp, \gamma_\|$, in which one can eliminate adiabatically [3] the atomic variables, the equation that governs the dynamical evolution of the electric field is [19]

$$\frac{\partial F}{\partial \tau} = -\left[1 - ia_{01}\left(\frac{1}{4}\nabla_\perp^2 - \rho^2 + 1\right)\right]F + 2C\frac{(1 - i\delta_{AC})F}{1 + \delta_{AC}{}^2 + |F|^2}\chi(\rho) \quad , \qquad (4.1)$$

where we set $\tau = kt$.

In order to reformulate (4.1) in the form of 'hydrodynamical equations' we put $F = |F|e^{i\Phi}$ and we introduce a 'mass density' $\sigma = |F|^2$ and a 'velocity' $\mathbf{v} = \frac{a_{01}}{2}\nabla\,\Phi$. With some algebraic manipulations, Eq. (4.1) and its complex conjugate can be cast in the form [24]

$$\frac{\partial \sigma}{\partial \tau} + \nabla \cdot (\sigma \mathbf{v}) = 2\sigma\left[\frac{2C}{1 + \delta_{AC}{}^2 + \sigma}\chi(\rho) - 1\right] \qquad (4.2)$$

$$\frac{\partial \Phi}{\partial \tau} = -\frac{a_{01}}{4}|\nabla\,\Phi|^2 + a_{01}\left[\frac{\nabla^2\sqrt{\sigma}}{4\sqrt{\sigma}} + 1 - \rho^2\right] - \frac{2C\delta_{AC}}{1 + \delta_{AC}{}^2 + \sigma}\chi(\rho)\,. \qquad (4.3)$$

We can compare (4.2) and (4.3) with the fundamental equations of hydrodynamics, namely the equation of continuity for the mass density σ

$$\frac{\partial \sigma}{\partial \tau} + \nabla \cdot (\sigma \mathbf{v}) = 0 \qquad (4.4)$$

and the Bernoulli equation for the velocity potential Φ (such that $\mathbf{v} = Q\nabla\,\Phi$, where the constant Q is introduced because Φ is assumed dimensionless)

$$\frac{\partial \Phi}{\partial \tau} = -\frac{Q}{4\pi}|\nabla\,\Phi|^2 + \mathcal{P}\frac{2\pi}{Q}\,, \qquad (4.5)$$

which involves the pressure term $\mathcal{P}$. From the comparison between these two pairs of equations we can conclude that in the optical case the total 'mass' is not conserved because of the presence of the two dissipative terms that appear in the r.h.s. of (4.2), while the analogy with the Bernoulli equation is complete if we put $Q = \pi a_{01}$ and define a 'pressure'

$$\mathcal{P} = \frac{a_{01}^2}{2}\left[\frac{\nabla^2\sqrt{\sigma}}{4\sqrt{\sigma}} + 1 - \rho^2\right] - \frac{a_{01}}{2}\frac{2C\delta_{AC}}{1 + \delta_{AC}{}^2 + \sigma}\chi(\rho) \quad . \qquad (4.6)$$

In this way we have established a formal analogy between lasers and hydrodynamics, in the sense that all the relevant physical quantities that describe a fluid, i.e. mass density, velocity and pressure, can be derived from the modulus and the phase of the electric field emitted by a laser.

5. Mechanical Effects on Neutral Atoms

It is well known that neutral atoms interacting with a resonant coherent field $E = \sqrt{I}\exp(\mathrm{i}\Phi)$ experience mechanical forces, generically called dipole and scattering forces. The force for a two-level atom at rest is given by the sum of two terms, one depending on the gradient of the field intensity ($\vec{\alpha} = \nabla\, I/I$)

and the other depending on the gradient of the field phase ($\vec{\beta} = \nabla\ \Phi$) [40]

$$\vec{f} = -\hbar\Omega \frac{s}{1+s}\vec{\alpha} + \hbar\frac{\Gamma}{2}\frac{s}{1+s}\vec{\beta}\ , \tag{5.1}$$

where Ω is the detuning between the laser field and the atomic transition, Γ is the natural linewidth and s is the saturation parameter.

In our case, the adoption of a beam having a particular transverse structure showing vortices leads to new and interesting effects for confinement and cooling in the transverse plane. The vector fields $\vec{\alpha}$ and $\vec{\beta}$ in the case of a doughnut laser beam (as shown in Fig. 1d) are drawn in Fig. 3. As one can see from the plot of $\vec{\alpha}$ and from (5.1), the choice of positive values of the atomic detuning Ω makes it possible to confine atoms in low field regions. The contribution of β to the forces is related to the existence of vortical forces around the minima of intensity.

We have studied in particular the interaction of atoms with beams having a transverse configuration corresponding to Figs. 1b and 1d, both for a travelling and a standing wave longitudinal configuration. In order to take into account the motion of the atoms we have derived the correction of (5.1) up to first order in the transverse velocities.

In the case of the travelling wave we have proved that it is possible to achieve transverse confinement of atoms having transverse velocities of the order of some hundred cm/sec. Anyway the travelling wave configuration is unable to provide confinement of the atoms in the direction of propagation of the laser beam; this is the reason which led us to consider a standing wave constituted by two counterpropagating beams having the same transverse configuration.

Our numerical simulations showed that it is possible to achieve cooling and 3-dimensional trapping, without using three couples of counterpropagating beams as is usually done in optical molasses [41].

Figs. 4a-b refer to the case of a single Na atom in a standing wave with the transverse configuration of Fig. 1d, injected with a longitudinal velocity

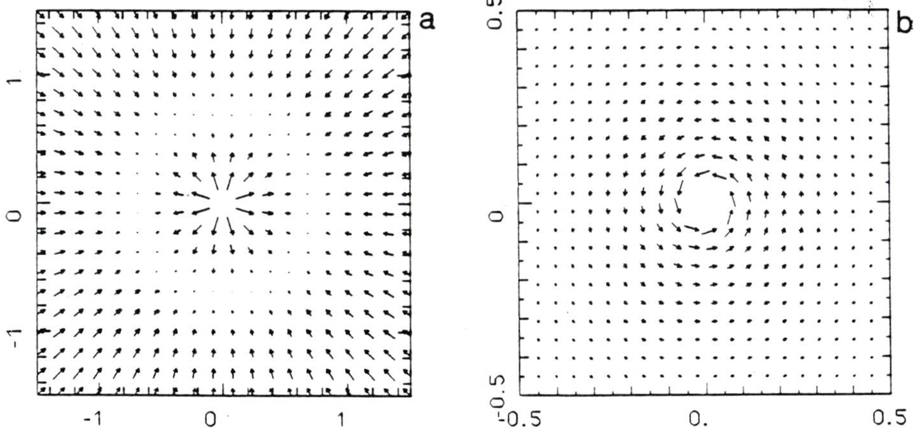

Figure 3. Plot of the gradient of the intensity $\vec{\alpha}$ (a) and of the gradient of the phase $\vec{\beta}$ (b) for the same pattern of Fig. 1d

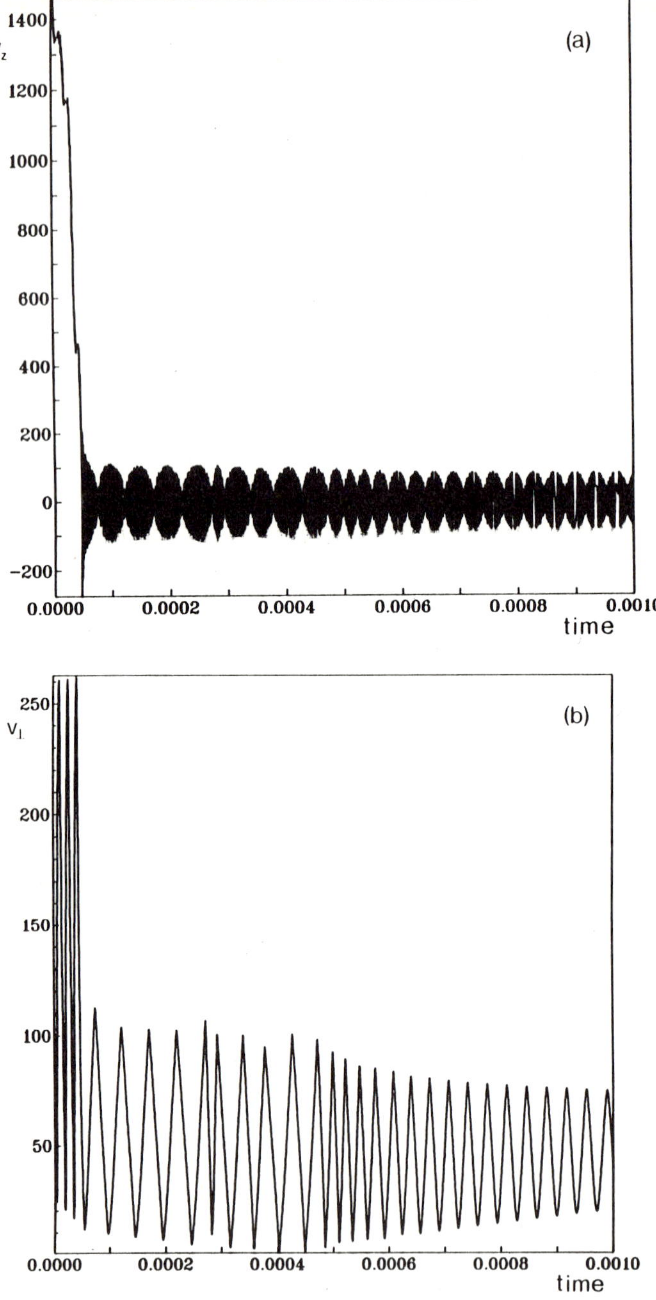

Figure 4. Time evolution of the longitudinal (a) and transverse (b) velocity of a Na atom moving in the electric field formed by two counterpropagating beams having the same transverse configuration of Fig. 1d

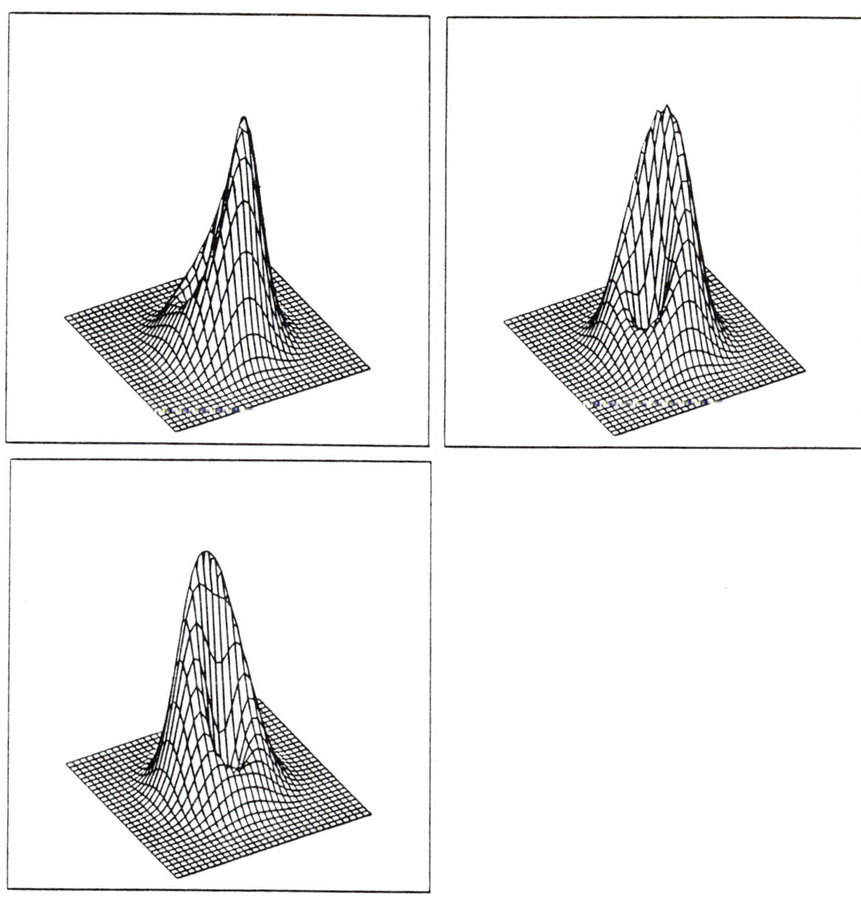

Figure 5. Intensity profiles of the rotating pattern described by Eq. (6.2). The sequence clearly shows the rotation and the existence of a vortex

of 1400 cm/sec and a transverse velocity of 100 cm/sec. The total power of the field is 800 mW and the detuning Ω is 50 times the natural linewidth Γ. The longitudinal velocity is drastically reduced until the atom is trapped around a node of the standing wave (Fig. 4a), while the reduction factor for the transverse velocity is about 1/2 (Fig. 4b). Simulations of ensembles of atoms confirmed that these effects are not due to particular realizations. One limit of these results is the absence of quantum fluctuations, so that no predictions on the stability of the traps can be made at this stage; this essential point will be investigated in the future.

6. Rotating Patterns, Creation and Annihilation of Vortices

When the cavity modes in play are frequency-degenerate, in the long term the system usually approaches a stationary state. This is no longer true in general

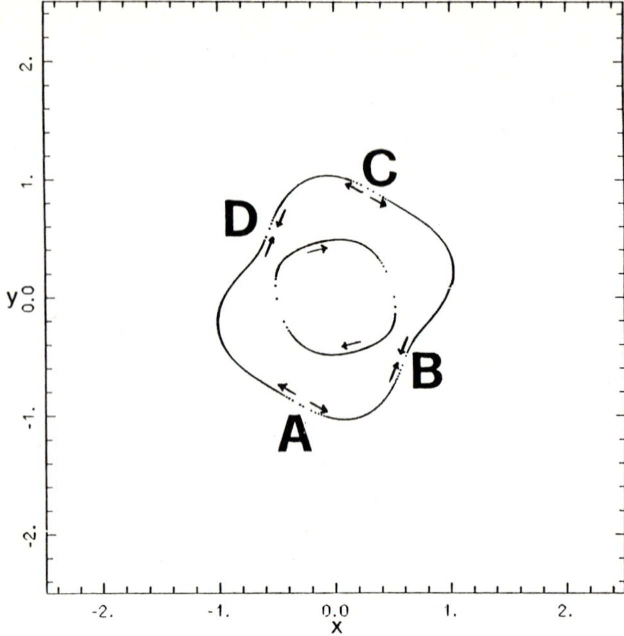

Figure 6. Motion of the vortices in the transverse plane in a case of creation and annihilation. Pairs of vortices of opposite charges are periodically created at points A and C and annihilated at points B and D

when the atomic gain line simultaneously excites modes of different frequencies. In this case, the laser approaches a stationary state only when the modes activate the phenomenon of cooperative frequency locking [16,19], otherwise the system realizes a dynamical state. In practice there is an unlimited variety of such states, of increasingly higher complexity when the transverse dimension of the laser is increased. Here we illustrate only two cases that arise from the interaction of a small number of modes; a more detailed discussion can be found in [42]. The simplest interesting dynamical state is that of a regularly rotating pattern. A case of this sort can be realized for example when the atomic line activates simultaneously the frequency-degenerate families $2p + |l| = 0$ and 1. Assume that the laser realizes a combination of the modes $p = 0, l = 0$ and $p = 0, l = 1$, so that the electric field has the form [28]

$$F(\rho, \varphi, t) = e^{-\rho^2} \left[f_{00} + \rho f_{01} e^{i(\varphi - \delta\omega\, t)} \right] \quad , \tag{6.1}$$

where $\delta\omega = k a_{01}$. Hence the intensity pattern is

$$|F(\rho, \varphi, t)|^2 = e^{-2\rho^2} \left[|f_{00}|^2 + \rho^2 |f_{01}|^2 + 2\rho |f_{00}||f_{01}| cos(\varphi - \delta\omega\, t - \varphi_0) \right] \tag{6.2}$$

and displays a regular rotation with angular velocity $\delta\omega$. Fig. 5 shows the rotating pattern for a selected value of the parameters. The structure displays a single vortex which covers a circle during the rotation. Rotating patterns of this kind have been observed experimentally both by Tredicce and collaborators in a CO_2 laser and by Weiss and collaborators in a Na_2 laser [42].

More complex patterns arise if the atomic line activates the modes of the three frequency-degenerate families with $2p + |l| = 0, 1$ and 2. In the case of Fig. 6 the structure displays creation and annihilation of vortices; one of them covers a small loop around the origin, while on a larger loop outside it, there are two points (A,C) where a pair of vortices of opposite charge is created, and two points (B,D) where a pair is annihilated. The motion of the vortices undergoes significant accelerations (decelerations) when the two vortices of a pair are near to an annihilation (creation) point; this suggests the existence of attractive forces between optical vortices with charges of opposite sign, similar to the forces described in Ref. 32 in the case of the Ginzburg-Landau equation.

A detailed analysis of the motion of vortex pairs after creation or before annihilation shows that the distance r between them varies as the square root of time and correspondingly a force proportional to $1/r^3$ acts on the vortices.

7. Passive Systems

The generalization of the model equations for the laser to the case of an atomic medium without population inversion (absorber) is readily obtained if we consider injecting into the resonator an external driving field with frequency ω_{in} and with the shape of the gaussian mode $p = 0, l = 0$. The slowly varying envelope of the input field is

$$F_{in}(\rho, \varphi) = Y\, A_{00}(\rho, \varphi) \quad , \tag{7.1}$$

and we suppose that the reference phase is chosen in such a way that Y is real. We include this expression in the Maxwell equation for the electric field interacting with the atomic medium, and we make use of the same approximations introduced in Sect. 2; one has

$$\frac{df_{pl}}{dt} = -k\Bigg[(1 + i\theta + ia_{pl})f_{pl} - Y\,\delta_{p.0}\,\delta_{l.0} + 2C\int_0^{2\pi} d\varphi \int_0^{\infty} d\rho\, \rho A^*_{pl}(\rho,\varphi)P(\rho,\varphi,t)\Bigg] \quad , \tag{7.2}$$

where $\theta = (\omega_{00} - \omega_{in})/k$ is the cavity detuning. Equation (7.2) must again be coupled with the atomic Bloch equations (2.4b,c), where the atomic detuning is now $\Delta = (\omega_a - \omega_{in})/\gamma_\perp$ (since the reference frequency is ω_{in}) instead of δ_{AC} and $\chi(\rho) = 1$ because here we are not describing an optical pumping but a homogeneous absorber.

If one introduces the limitation of considering the dynamical system in which only three modes contribute to the dynamics, namely the $p = 0, l = 0$ and $p = 0, l = \pm 1$ modes (which means that we consider only the families $q = 0$ and $q = 1$), one can guarantee the existence of the single-mode $p = 0, l = 0$

solution and to cast its analytic formula

$$y_1 = x_1 \left\{ \left[1 + \frac{2C}{x_1^2} \ln \frac{1 + \Delta^2 + x_1^2}{1 + \Delta^2} \right]^2 + \left[\theta - \frac{2C\Delta}{x_1^2} \ln \frac{1 + \Delta^2 + x_1^2}{1 + \Delta^2} \right]^2 \right\}^{1/2} , \tag{7.3}$$

where $x_1 = \sqrt{\frac{2}{\pi}} |f_{00}^{st}|$ and $y_1 = \sqrt{\frac{2}{\pi}} Y$. Multimode stationary solutions cannot in general be described in analytic form and we obtained them as a result of the dynamical evolution of the systems.

Keeping our analysis restricted to the three-mode dynamics, we can describe how the onset of the modes of family $q = 1$ can destabilize the single-mode TEM_{00} stationary solution (7.3), giving rise to more complex spatio-temporal structures; actually an approximated linear stability analysis can be performed and various regimes are found [35].

A first class of them is made up by stationary regimes where all the modes are locked to the frequency ω_{in} of the input beam; it is possible to discriminate two kinds of these regimes: one where only cylindrically symmetrical modes ($l = 0$) are present, and the other where asymmetrical modes also contribute. We shall refer to the first kind of structures as $S1$, while the second kind will be designated by $S2$ and they can be seen as the result of an instability of the $S1$ structures.

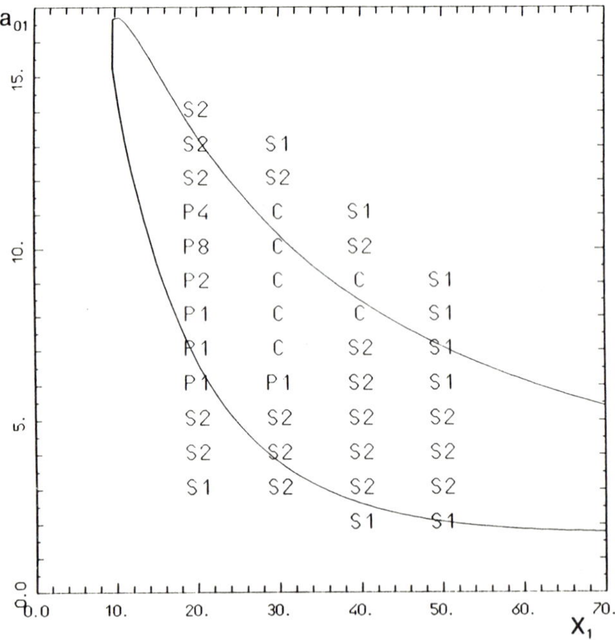

Figure 7. Instability domain of the single-mode solution (7.3) in the (x_1, a_{01}) plane. The values of the other parameters are $C = 100$, $\Delta = 10$, $\theta = -2$, $\gamma_{\parallel} = \gamma_{\perp}$. The meaning of the lettering is explained in the text

7.1 Rotating Patterns

These patterns arise in the good cavity limit $k \ll \gamma_\perp, \gamma_\|$, and they are qualitatively identical to those reported in Sect. 3.3, though here the TEM_{00} component is much more relevant, due to the injection of the gaussian external field.

7.2 Dynamical Regimes

A rich variety of dynamical patterns has been found in the bad cavity case $k \approx \gamma_\perp, \gamma_\|$. In Fig. 7 the solid line shows the domain of instability predicted by the analysis of the three-mode model, while the letters indicate the type of regime reached with the numerical integration of the dynamical equations, in which we have considered the model extended to the case of three families $q = 0, 1, 2$. The meaning of $S1$ and $S2$ has already been explained, the symbol Pn, with $n = 1, 2, 4, 8$ refers to periodic regimes where the total intensity is periodic with period n , while C denotes a chaotic regime. As one can see there are extended areas of chaotic behaviour (which can also be found with only three modes); a period doubling cascade can be identified and one notes that when the mode separation is increased the system tends to reach the $S1$ configuration, because all other higher order families are disfavoured.

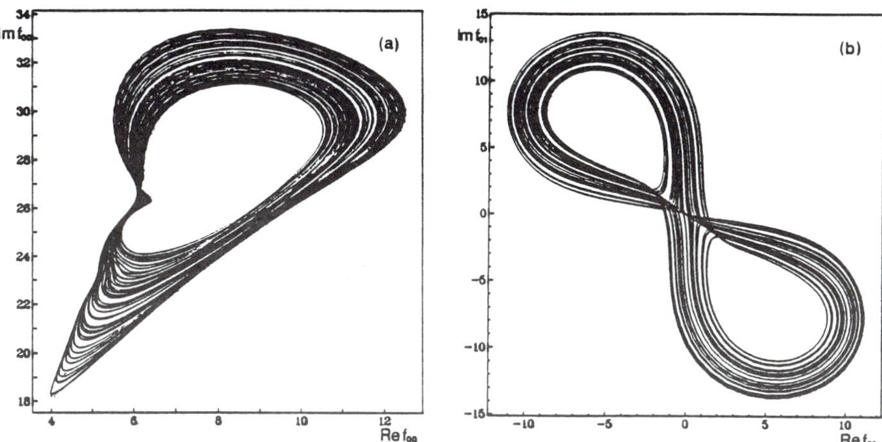

Figure 8. Example of Lorenz-like chaotic behaviour: (a) projection of the trajectory on the $(\mathcal{R}e\ f_{00}, \mathcal{I}m\ f_{00})$ plane, (b) projection of the trajectory on the $(\mathcal{R}e\ f_{01}, \mathcal{I}m\ f_{01})$ plane

that when the mode separation is increased the system tends to reach the $S1$ configuration, because all other higher order families are disfavoured.

The presence of unstable behaviour outside the instability domain and of stable regimes inside it should not be surprising, as we are comparing the numerical results relative to the six-mode model with the stability analysis of the three-mode model, which is the only one that can be treated analytically.

Figs. 8a,b show the case of chaotic behaviour; in Fig. 8a the projection of the trajectory in the phase space on the hyperplane ($\mathcal{R}e\ f_{00}, \mathcal{I}m\ f_{00}$) is depicted, while Fig. 8b shows the projection on the hyperplane ($\mathcal{R}e\ f_{01}, \mathcal{I}m\ f_{01}$). Here a behaviour similar to that of the Lorenz-Haken model [2] is evident in the temporal evolution of the complex modal amplitude f_{01}. As a matter of fact, in the case of Fig. 8 the asymmetrical modes experience gain and build up from zero, behaving as laser modes. This kind of chaotic behaviour is well established over a wide range of parameters in the model. We note that this behaviour is robust in the sense that it persists even when modes belonging to higher order families are included in the dynamical equations.

Acknowledgments

This work has been carried out in the framework of the ESPRIT Basic Research Action on 'Transverse Optical Patterns' (TOPP).

References

1. C.O.Weiss and J.Brock , Phys. Rev. Lett. **57** , 2804 (1986)
2. H.Haken, Phys. Lett. **53** A, 77 (1975)
3. H.Haken, *Synergetics: An Introduction* 3rd Edn., Springer-Verlag Berlin 1983
4. H.Haken, *Advanced Synergetics*, Springer-Verlag Berlin 1983
5. J.H. Moloney, in *Chaos, Noise and Fractals*, ed. by R. Pike and L.A. Lugiato, Hilger, Bristol 1988
6. W.J. Firth, in *Instabilities and Chaos in Quantum Optics*, ed. by N. B. Abraham, F.T.Arecchi and L.A.Lugiato, Plenum, New York 1988
7. W.J.Firth, "ECOOSA '90 - Quantum Optics", ed. by M.Bertolotti and E.R.Pike, Insitute of Physics, Bristol, 1991, p. 173
8. L.A.Lugiato and R.Lefever, Phys. Rev. Lett. **58**, 2209 (1987)
9. G.Grynberg, E.Le Bihan, P.Verkerk, P.Simoneau, J.R.R.Leite, D.Bloch, S.Le Boiteux and M.Duclay , Opt. Comm. **67** , 363 (1988)
10. G.Giusfredi, J.F.Valley, R.Pon, G.Khitrova and H.M.Gibbs , J. Opt. Soc. Am. B **5** , 1181 (1988)
11. S.Akhmanov, M.A.Vorontsov and V.Y.Ivanov, JETP Lett. **47**, 707 (1988)
12. M.LeBerre, E.Ressayre and A.Tallet, Phys. Rev. A **25**, 1604 (1982)
13. F.T.Arecchi, In Ref. 7, p. 257
14. F.T.Arecchi, G.Giacomelli, P.L.Ramazza and S.Residori, Phys. Rev. Lett. **65**, 2531 (1991)
15. P.Hollinger and C.Jung , J. Opt. Soc. Am. B **2**, 218 (1985)
16. L.A.Lugiato, C.Oldano and L.M.Narducci, Journ. Opt. Soc. Am. B **5** 879 (1988)

17. L.A.Lugiato, F.Prati, L.M.Narducci, P.Ru, J.R.Tredicce and D.K.Bandy Phys. Rev. A **37** , 3847 (1988)
18. L.A.Lugiato, F.Prati, L.M.Narducci and G.L.Oppo, Opt. Comm. **69**, 387 (1989)
19. L.A.Lugiato, G.L.Oppo, J.R.Tredicce, L.M.Narducci and M.A.Pernigo, Journ. Opt. Soc. Am. B **7**, 1019 (1990)
20. J. R. Tredicce, E. J. Quel, A.M. Ghazzawi, C. Green, M.A. Pernigo, L.M. Narducci and L.A. Lugiato, Phys. Rev. Lett. **62** , 1274 (1989)
21. C.Tamm, Phys. Rev. A **38**, 5960 (1988)
22. V.Klische, C.O.Weiss and B.Wellegehausen, Phys. Rev. A **39**, 919 (1989)
23. M.Brambilla, F.Battipede, L.A.Lugiato, V.Penna, F.Prati, C.Tamm and C.O.Weiss, Phys. Rev. A **41**, 5090 (1991)
24. M.Brambilla, L.A.Lugiato, V.Penna, F.Prati, C.Tamm and C.O.Weiss, Phys. Rev. A **41**, 5114 (1991)
25. F.T.Arecchi, R.Meucci and L.Pezzati, Phys. Rev. A **42**, **65**, 2531 (1990)
26. G.L.Oppo, L.Gil, W.J.Firth, In Ref. 7, p. 191
27. C.Green, G.B.Mindlin, E.J.D'Angelo, H.G.Solari and J.R.Tredicce, Phys. Rev. Lett. **65**, 3124 (1990)
28. M.Brambilla, M.Cattaneo, L.A.Lugiato and F.Prati, In Ref. 7, p. 133
29. L.A.Melnikov, S.Tatarkova, C.N.Tatarkov, Journ. Opt. Soc. Am. B **7**, 1286 (1990)
30. M.Berry, *Singularities in Waves and Rays*, in *Physics of Defects*, R.Balian et al. eds., Les Houches session **XXXV**,
31. P.Coullet, L.Gil and F.Rocca, Opt. Comm. **73**, 403 (1989)
32. S. Rica and E. Tirapegui, Phys. Rev. Lett. **64**, 878 (1990)
33. R. Lefever, L.A. Lugiato, K. Wang, N.B. Abraham and P. Mandel, Phys. Lett. **135**, 254 (1989)
34. L.A.Lugiato, L.M.Narducci and R.Lefever, in *Lasers and Synergetics*, ed. by R.Graham and A.Wunderlin, Springer-Verlag Berlin 1987, p. 53
35. M.Brambilla, G.Broggi and F.Prati, submitted to Physica D
36. M.Brambilla, C.Pitzen, F.Prati and L.A.Lugiato, in preparation
37. M.Brambilla, L.A.Lugiato, M.Pinna, F.Prati, C.Pagani, P.Vanotti, M.Y.Li and C.O.Weiss, in preparation
38. H. Haken, in *Computational Systems - Natural and Artificial*, ed. by H. Haken, Springer-Verlag Berlin 1987, p. 2
39. A. Fuchs and H. Haken, in *Neural and Synergetic Computers*, ed. by H. Haken, Springer-Verlag Berlin 1988, p. 16
40. J.P.Gordon and A.Ashkin, Phys. Rev. A **21**, 1606 (1980)
41. S.Chu, J.E.Bjorkholm, A.Ashkin and A. Cable, Phys. Rev. Lett. **57**, 314 (1986)
42. M.Brambilla, M.Cattaneo, L.A.Lugiato, R.Pirovano, F Prati, A.J.Kent, G.-L.Oppo, A.B.Coates, C.O.Weiss, C.Green, E.J.D'Angelo and J.R.Tredicce, in preparation

The Laser with a Saturable Absorber: A Paradigm for the Study of Laser Instabilities

D. Dangoisse, D. Hennequin, M. Lefranc, and P. Glorieux

Laboratoire de Spectroscopie Hertzienne, associé au CNRS, Université de Lille I, F-59655 Villeneuve d'Ascq Cedex, France

Abstract. From single-mode to multimode operation, the CO_2 laser with a saturable absorber displays a variety of phenomena which provide a large basis of experimental facts illustrating the power of the concepts of nonlinear dynamics. Recent experiments on spatio-temporal dynamics confirm once more the efficiency of the various methods introduced by this new approach.

1. Introduction

In 1975, Haken established the parallel between laser equations and the Lorenz model [1]. Together with the development of nonlinear dynamics, this triggered the interest of laser physicists in Western countries for instabilities in lasers, rejoining the Russian community which had considered similar problems in masers [2]. Among the many systems which display instabilities, the CO_2 laser containing a saturable absorber (LSA) appears as one of the richest in terms of the variety of phenomena it exhibits. This occurs because of the many control parameters available which act on different nonlinear terms and also because the absorber parameters may be altered in a wide range by changing the nature of the absorber. Moreover this system is sufficiently well defined so that a stimulating interaction between theoreticians and experimentalists can lead to rapid progress in various directions [3] and also a large amount of experimental data is available since the LSAs have been investigated from the very beginning of laser studies.

As a matter of fact, it was noticed very quickly that when a suitable absorber is inserted inside the cavity, the laser no longer emits cw radiation but it tends to oscillate in a time dependent way. Tentative explanations for these oscillations were derived on the basis of sophisticated models [4]. The advent of the methods of nonlinear dynamics in the field of lasers brought a considerable simplification and provided global pictures of the mechanisms from which they originate. Reviews of the work on LSA instabilities were

Springer Proceedings in Physics, Vol. 69
Evolution of Dynamical Structures in Complex Systems
Editors: R. Friedrich · A. Wunderlin

published earlier [5]. In this paper, we present the progress of the experimental investigations of the CO_2 LSA with emphasis on the experimental works subsequent to the above mentioned reviews. For developments of the LSA theory, the reader is referred to [5, 6]. The experiments about the influence of noise on the LSA bistability and dynamics [7] will not be described here, as their connection to the mechanisms of the instabilities is rather intricate.

In a first step, the purely temporal dynamics of the single mode LSA was reinvestigated and the many signal shapes found in these experiments could be interpreted as resulting from the interaction of a Hopf and a homoclinic bifurcation [8]. In parallel to the explanation in terms of bifurcations, progress has recently been made on the characterization of the chaotic regimes, in particular through the use of symbolic dynamics [9]. This could be performed either by looking for grammatical rules for symbols or by establishing the template describing the links between unstable periodic orbits embedded inside the strange attractor associated with the chaotic regime.

All the above-mentioned phenomena were obtained in single mode, single line LSAs. A natural extension of this work is to consider, in a second step, how they are altered by multimode operation of the laser. The simplest case corresponds to bimode operation on two different transverse modes [10]. Starting from a situation of simple self-pulsing regime and varying suitable control parameters, alternation of periodic and quasiperiodic regimes following a Farey hierarchy in a truncated Devil's staircase could be observed when the LSA oscillates on two modes. The three-mode regimes can be interpreted in the same picture[11].

Interest has recently been attracted by pattern formation in lasers as reviewed by C.O. Weiss [12] and the laser operating on two transverse modes can be considered as one of the simplest cases of transverse dynamical effects in lasers. In that respect, the LSA also provided interesting contributions, such as the first observation of bistability between different families of modes or its possible use for pattern recognition.

The following sections of this paper will follow this more or less historical approach. Single mode instabilities are discussed in the next section while the third section summarizes the results obtained in bi- and tri-mode operations. Spatial structures are considered in the last section.

2. Single mode instabilities and dynamics of the LSA

For a LSA emitting on a single line and a single mode, the output intensity may be constant, periodic or chaotic depending on the parameter values. Various kinds of time dependence have been observed but until the advent of nonlinear dynamics, little attention was paid to the particular shape of the signals [13]. Apart from the standard ON and OFF states where the laser emits cw and no radiation respectively, practically two kinds of signal shapes have been experimentally observed : either the cw signal is modulated with various amplitudes or the laser emits bursts of radiation [14]. In the latter regime, the laser typically emits pulses composed of a large initial pulse followed by several peaks of lower intensity. These pulses may be followed by a time during which the laser output is in the OFF state. The various regimes may be characterized by the number of small peaks and their periodicity. In most cases, these pulses are periodic but we have found between the periodic regions, small domains of chaotic behavior [15].

Examples of such peaks are given on Figure 1 which illustrates how the various regimes are related in the (cavity detuning δ, absorber pressure p) parameter plane. Note that the detunings are given with respect to the central tuning of the cavity of the laser without absorber. As CH_3I is known to have several absorption lines overlapping with the relevant CO_2 laser line, the absorption coefficient may not be simply related to the cavity detuning.

At low absorber pressures, i.e. low absorption and high saturability, the laser is always in the ON state, i.e. it delivers a cw power, whereas it is in the OFF (no power) state for absorber pressures above 150 mTorr in the conditions of Figure 1. In between these two limits, the cw state destabilizes through different scenarios. For central tuning, small amplitude periodic oscillations superimpose onto the cw intensity and the modulation depth increases with the absorber pressure until the laser eventually stops at high pressures. This is a clear case of a supercritical Hopf bifurcation which for instance occurs at p = 78 mTorr for $\delta = 0$.

On the other hand for large detunings ($\delta = -25$ MHz), the time dependent regimes appear composed as described above, i.e. a large initial pulse followed by a variable number n of oscillations of smaller intensity. For large detunings, these regimes are periodic and will be denoted as $P^{(n)}$. In the conditions of Figure 1, n is smaller than 4, but regimes with very large n (15 - 40) have been observed in other LSAs such as the $CO_2 + SF_6$ LSA [7].

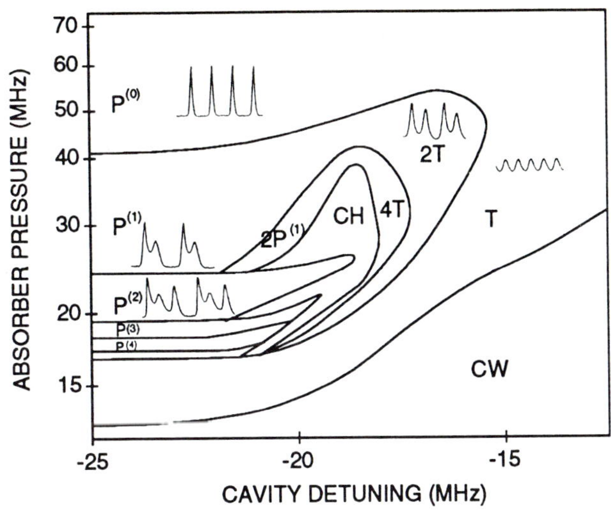

Figure 1 : Phase diagram of the CO_2 + CH_3I LSA in the (cavity detuning - absorber pressure) plane. The notation are those introduced in ref. 16. The shapes in the diagram are those of the signals.

In the region between these two well-defined transition schemes, there is a wide variety of other scenarios involving periodic and chaotic regimes. For instance at δ = -10 MHz and increasing the absorber pressure, the laser undergoes a cascade of period-doubling bifurcations followed by chaotic regimes ($65 < p < 80$ mTorr), then a reversed (4T-2T-T) cascade is observed. The situation is even more complicated at δ = -15 MHz where alternation of periodic $P^{(n)}$, period-doubled $2^m P^{(n)}$ and chaotic regimes occurs.

The general organization of the various periodic and chaotic regimes depends on the particular absorber and on the laser parameters. To understand the origin of such a transition scenario, it is interesting to plot phase portraits of the laser evolution. This is easily achieved using e.g. the time delay technique. The reconstructed trajectories evidence the existence of a saddle-focus corresponding to the (now unstable) steady-state intensity. Here the saddle (focus) corresponds to the attracting (repulsive) direction. As the control parameters are changed, the corresponding stability coefficients evolve. Shilnikov demonstrated that a homoclinic bifurcation could lead to chaos under some conditions on the sign and relative values of these coefficients [17]. Details of the corresponding scenarios were given in [18]. According to their results, at the homoclinic bifurcation there is an accumulation of nT periodic regimes which evolve individually towards chaos through period-doubling cascades. At the end of each cascade, chaos appears as a sequence composed of signals with various periods ranging from T to nT. Our observations follow these predictions in all points excepted three: (i) their theory predicts a behavior symmetrical with respect to the accumulation point while in experiments the $P^{(n)}$ type regimes are found on one side only of the bifurcation, (ii) the steady-state is not unstable but there can be bistability between the time dependent $P^{(n)}$ and the cw regimes and (iii) the amplitude of the oscillations around the focus never shrinks to zero. The reason for this puzzle has recently been found in the work of Gaspard and Wang [19]. What is observed is in fact a homoclinic tangency to an unstable periodic orbit. This cycle is created in a subcritical Hopf bifurcation which transforms into a supercritical bifurcation observed in other regions of the parameter space. This obviously explains points (ii) and (iii) but also accounts for the observation that the $P^{(n)}$ type regimes exist on one side only of the bifurcation. Note that bifurcation diagrams depend on the sense of variation of the control parameters since different dynamical and/or static regimes can coexist in this laser [13].

When other parameter ranges are explored, the main ingredients of the LSA dynamics remain the same: the global picture results from the coexistence and possible interaction through e.g. heteroclinic connections of (un)stable fixed points, (un)stable limit cycles and a saddle-focus point corresponding to the ON state [20]. In particular the question of whether the long period of zero intensity signal corresponds to an heteroclinic connection between the saddle-focus and the zero intensity (OFF) point which is a saddle above threshold, was extensively discussed. It appears now that this part of the signal corresponds to a motion in the $I = 0$ plane but not necessarily in the vicinity of the $I = 0$ solution [20].

Progress has been made not only on the structure of the bifurcations leading to chaos and their evolution with several parameters but the analysis of chaos itself has recently been the subject of recent advances mostly related to the introduction of symbolic dynamics in the analysis of experiments. Coding data was extensively used in theoretical works but surprisingly that technique was applied to chemical reactions only [21] and it has only recently been introduced in the field of laser chaos. The standard method is based on the partition of e.g. the first return map of points in the Poincaré section. Codes are associated with each part, typically two, of the map. Different techniques are then used to find and eventually quantify the amount of order in the sequence of codes. Then, two kinds of information, metric or topological, can be extracted from these results. In the following, we describe how methods proposed by Badii, Destexhe and by Mindlin et al. could provide more insight into the chaotic dynamics of the LSA [22-24].

Destexhe and Badii proposed two complementary approaches: (i) the signal is considered as "stochastic", i.e. as a Markov chain, and its degree of correlation is measured through its Markov order, (ii) the "order" of the signal is evaluated through the hierarchy of symbols needed to represent the data. An obvious coding of the signals delivered by the LSA appears because of their particular shapes. As the signal is composed by large initial spikes followed by a

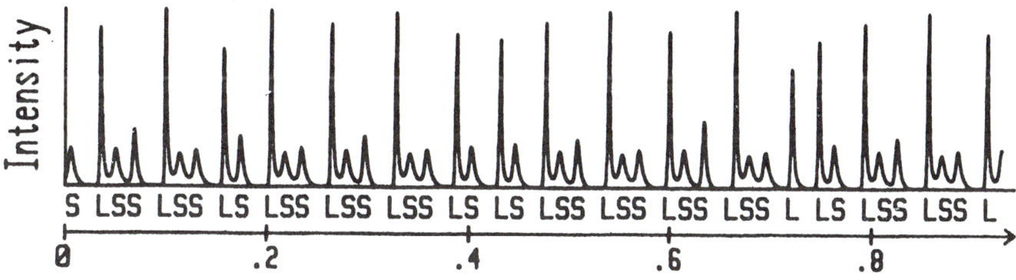

Figure 2 : Example of a chaotic signal delivered by the LSA and its coding by symbols (L and S).

variable number of small oscillations, the first stage could be to code the signal as a sequence of letters L (large) and S (small). It is then easily found that a more efficient coding is provided by hypersymbols which are simply the number of S preceding each L, i.e. the number n of turns that the trajectory makes around the saddle-focus following the large initial spike [9]. This coding appears to be very robust and can be supported by more technical arguments. Figure 2 gives an example of a chaotic signal and the corresponding coding.

Large sequences of signals have been coded and analyzed. Statistical tests indicate that in the LS coding, the chaotic sequence of Figure 2 may be considered as a Markov process of order 2 while the n-coding leads to Markov order 0 and therefore it contains all the memory of the signal and is particularly appropriate for the construction of the logic tree suggested by Badii.

The construction of this tree indicates how the sequences of different symbols are organized. A search for sequences of symbols is made based on two criteria : their existence and their periodicity. The periodicity of these hypersymbols reflects the occurrence of unstable periodic orbits inside the chaotic attractor. The optimum situation corresponds to infinite chains with a large periodicity, for instance Badii worked on chains of 10^8 symbols and a periodicity of about 20. Unfortunately it is difficult to sample experimental chains of more than 10^4 symbols with stable parameters. Therefore we have shown that the periodicity has to be reduced to 6 for 10^6 points and 5 for 10^4. The code with

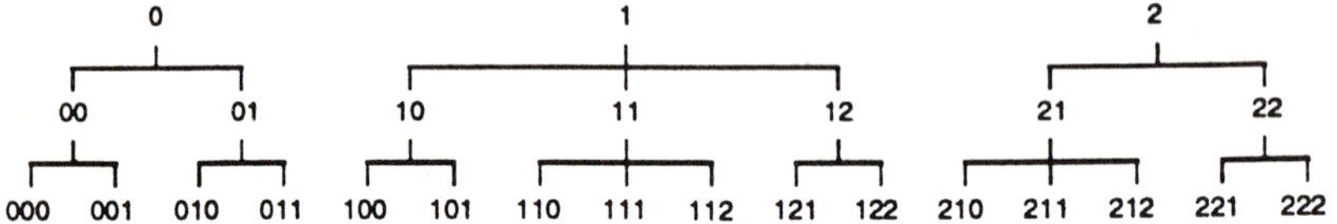

Figure 3 : First three levels of the logic tree linking the periodic sequences of symbols.

the n-symbols also reduces the maximum allowed periodicity. So we were able to build the tree up to level 3. Such a tree as deduced from numerical data on the LSA is given in Figure 3. At the first levels, symbols 0, 1 and 2 are obtained. Starting at level 2, some grammatical rules limit the possible combinations. For example, the combinations 02 and 20 do not appear in the tree. At level 3, a larger number of words disappear. In fact some of the words are not only nonperiodic, but are even forbidden. For example the absence of 120 denotes the rule "the word 12 is never followed by 0". The complexity of these rules can be measured using the "metric complexity" C_1 [22]: if the knowledge of level m permits to predict exactly the combinations present at level m + 1, that means that no grammatical rule exists and the system may be considered as "simple", corresponding to a null metric complexity.

Figure 4 represents the evolution of this metric complexity as a function of the level in the logic tree in the same conditions as those of Figure 2 and in the corresponding experimental situation, i.e. in a chaotic regime close to the $P^{(2)}$ region. Note that the final value is in excellent agreement with the theoretical one. Measurements of C_1 allow one to quantify evolution inside the chaotic regions [24].

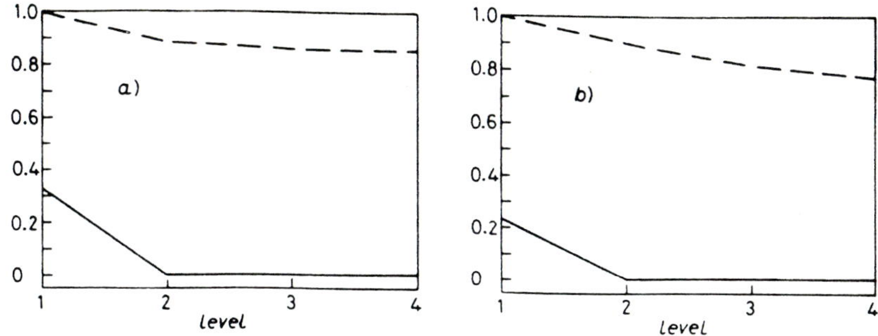

Figure 4 : Evolution of the metric complexity as the level of computation in the logic tree is increased. The dashed lines follow the total probability of the level, i.e. the filling factor a) numerical simulations and b) experimental data in the corresponding situation.

Recently Papoff et al. managed to apply the technique proposed by Mindlin et al. to obtain qualitative information on the chaotic attractor[25]. By extracting from chaotic time series data, the low-period unstable orbits, they could obtain topological information about the "template" which supports the strange attractor. This very promising technique allows one to determine whether previously proposed models are compatible with the data.

3. Two transverse mode dynamics.

The possibility of multimode operation further increases the richness of the LSA. Considering the many different dynamical regimes mentioned in the preceding section, the multimode instabilities should first be considered in a situation as simple as possible. Two-transverse mode dynamics was also recently studied both experimentally and theoretically by Tanii et al. who also found antiphase motion between the two modes [26]. In addition, they considered the situation in which the passive Q-switching on each mode is chaotic and as expected they found larger correlation dimensions and more complex return maps for the two-mode case with respect to the single mode.

For sake of simplicity, we first investigated the case of operation on two different transverse modes and set the laser such as when single mode, it operates in the $P^{(0)}$ regime, i.e. produces single spike pulses. Then by changing one control parameter it can go from oscillation on mode 1 to oscillation on mode 2 with a wide range of bimode operation between the regions where the laser is single mode. As the emission pattern of the laser depends on the emitting mode, it is possible to set the detector at places where it preferentially detects one mode. Then this mode appears to deliver large signals while those of the other mode appear as much smaller. We have used this technique to assign the signals to each mode and checked it with e.g. heterodyne

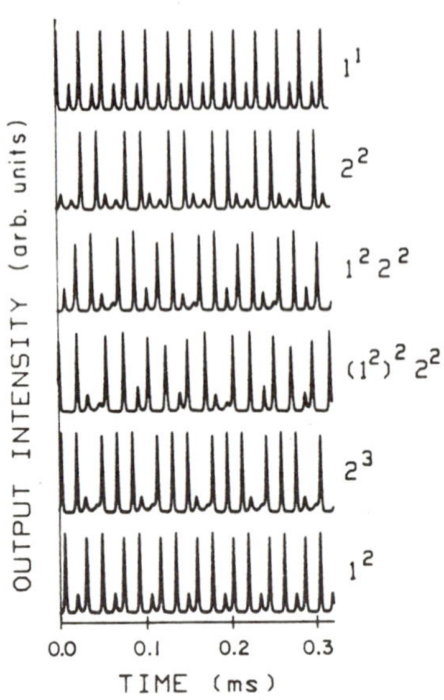

Figure 5 : Examples of periodic regimes and antiphase observed in the bimode LSA.

detection. With such settings of the LSA, the transition between the two modes occurs through a succession of periodic regimes separated by nonperiodic ones.

In the periodic regimes, the LSA dynamics exhibit antiphase motion, they are governed by the Farey arithmetic and lie on a staircase as shown on Figures 5 and 6. First they display antiphase dynamics, i.e. modes 1 and 2 do not coexist except for very small times. The laser switches from mode 1 to mode 2 and vice versa: oscillation on one mode kills the other one. Note that here the total intensity never gets close to zero as this occurs for the single mode LSA instabilities.

In order to find the organization of these various periodic regimes, we will call r^p a periodic regime formed by p large pulses similar to those emitted by mode 1 followed by r small pulses similar to those of mode 2. A periodic regime can be composed of several r^p symbols as shown on Figure 5. Then we define a firing number as $F = (\Sigma p_i) / (\Sigma (p_i + r_i))$. Figure 6 shows the succession of firing numbers observed in our experiment when the cavity length is varied so that the laser changes from mode 2 to mode 1 from the left to the right

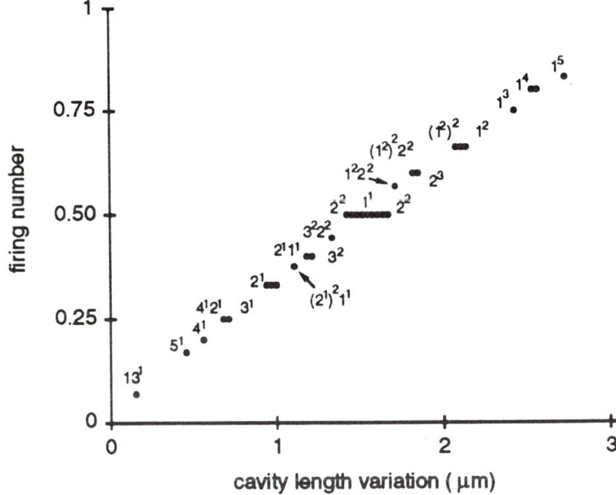

Figure 6 : Staircase of the firing number of the periodic states observed in the bimode LSA.

in the Figure. It appears that they are organized following a staircase, where the changes in period occur through a monotonic variation of the firing number.

It must be noticed that the different basic periodic regimes observed in our experiment may be deduced using the Farey arithmetic from the two limit regimes 1^0 and 0^1. The Farey sum of the firing numbers associated with the latter gives 0/1 + 1/1 = 1/2, the simplest state associated with this number is 1^1. The next steps of the tree may be determined in the same way. All the basic regimes of the first four levels of the tree have been observed in our experiments, as reported on Figure 6, together with some states of higher levels.

The plot of Figure 6 has a staircase structure that is strikingly similar to that encountered in the evolution towards chaos of quasiperiodic systems. Here, the number of stairs is finite and the variable is the firing number instead of the winding number for quasiperiodicity. In this situation, the dynamics of the LSA mostly appears as related to the coupling of two oscillators, each associated with one laser mode.

Nonperiodic regimes were also observed between the periodic regions. They do not correspond to

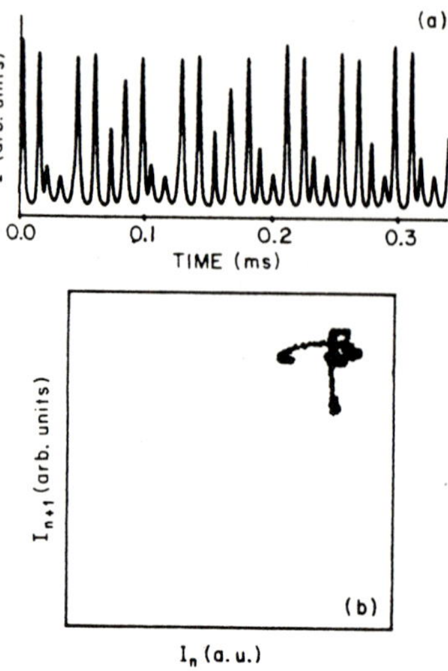

Figure 7 : Chaotic itinerancy observed in the two-mode LSA. In (a), total intensity of the laser as a function of the time; in (b), first return map (I_n, I_{n-1}) for intensity of one of the mode.

quasiperiodicity since the spectrum of the signal exhibits a continuous background.

Moreover, as shown on Figure 7, the first return map of successive points of the Poincaré section of the reconstructed attractor in the $(I,\dot{I},\ddot{I})$ phase space does not indicate the existence of a closed cycle but has an accumulation point in the upper right corner and this piecewise first return map suggests the existence of chaos. In that regime and because the antiphase motion is preserved, one can consider that the laser chaotically switches between the two patterns associated with each mode, thereby providing the simplest case of chaotic itinerancy [27].

4. Spatial structures in the LSA.

In the above section the bimode operation of the LSA has been considered from a purely temporal point of view. As the two modes correspond to different transverse modes, spatial dynamics are here associated with this bimode

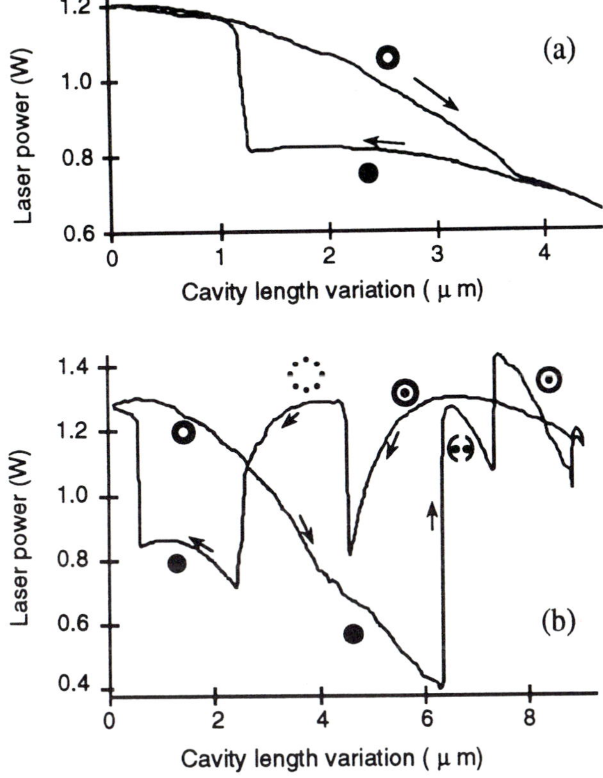

Figure 8 : Bistability between families of modes for the CO_2 + SF_6 LSA on the 10P24 line. The transverse mode structures are schematically indicated. (a) and (b) correspond to two different sets of parameters.

operation. We have therefore reinvestigated the LSA with spatial resolution. In the first experiments, the steady-state structures were investigated. The design of the laser cavity has been modified to allow for many-transverse-mode operation.

Figure 8 shows diagrams on which the evolution of the mode structure versus the cavity length is shown in two limit cases. The situation in figure 8a has been selected to shed light on the simplest situation, namely that in which bistability between two modes is obtained. Note that contrary to what was observed in standard lasers [12], bistablity occurs here between modes of different families, according to the terminology of Lugiato [28]. In most cases, the situation is much more complicated and multistability is routinely observed as shown in figure 8b.

Because each mode has a different resonance frequency and many SF_6 lines are overlapping with the CO_2 laser gain line, a detailed quantitative theory is difficult but for lasers such as the LSA considered here where only a few modes are involved, the approach of Lugiato et al. in which the field is expanded on the basis of the empty cavity modes appears to be adequate.

Acknowledgements
We benefited from discussions with P. Mandel, T. Erneux, C. Baesens, C. Tresser, A. Arneodo. E. Louvergneaux, A. Bekkali, F. Papoff, E.Arimondo, B. Zambon are gratefully thanked for collaborations at different stages of this work.

References

[1] H. Haken, Phys. Lett. 53A,77 (1975).

[2] see e.g. A. N. Oraevsky, Molekulyarnye Generatory (molecular-beam masers) (Nauka, Moscow, 1964); Ya. I. Khanin, Dinamika Kvantovykh Generatorov (Dynamics of Quantum Generators) (Sov. Radio, Moscow, 1975)

[3] see e.g. the special issues of J. Opt. Soc. Am. B2 (1985) and B5 (1988).

[4] I. Burak, P.L. Houston, D.G. Sutton and J.I. Steinfeld, IEEE J. Quant. Electron. QE-7, 73 (1971) ; J. Dupré, F. Meyer and C. Meyer, Rev. Phys. Appl. 10, 285 (1975).

[5] N.B. Abraham, P. Mandel and L.M. Narducci. in Progress in Optics XXV, ed. E. Wolf, (North Holland, Amsterdam 1988) p.1; A. Bekkali, F. Papoff, D. Dangoisse and P. Glorieux, J. Phys. Coll. 49C2, 349 (1988).

[6] Proc. Conf. Nonlinear Dyn. Opt. Syst., E. Garmire, P. Mandel eds (Opt. Soc. Am., Washington, 1990).

[7] D. Hennequin, F. de Tomasi, L. Fronzoni, B. Zambon and E. Arimondo, Opt. Commun. 70, 253 (1989)

[8] M. Lefranc, D. Hennequin and D. Dangoisse, J. Opt. Soc. Am. B8, 239 (1991)

[9] D. Hennequin and P. Glorieux, Europhys. Lett. 14, 237 (1991).

[10] D. Hennequin, D. Dangoisse and P. Glorieux, Opt. Commun. 79, 200 (1990); D. Hennequin, D. Dangoisse and P. Glorieux, Phys. Rev. A42, 6966 (1990); K. Tanii, M. Tachikawa, F-L. Hong, T. Tohei and T. Shimizu, in ref. [6] p.389.

[11] D. Dangoisse, D.Hennequin and P.Glorieux, unpublished

[12] C.O. Weiss, this volume.

[13] E. Arimondo, F. Casagrande, L. Lugiato and P. Glorieux, Appl. Phys. 30B, 57 (1983); M.L. Asquini and F. Casagrande, Nuovo Cimento, 2D, 917 (1983); A.Jacques, P.Glorieux, Opt. Commun. 40, 455 (1982); E. Arimondo and E. Menchi, Appl. Phys. B37, 55 (1985).

[14] M. Tachikawa, K. Tanii, M. Kajita and T. Shimizu, Appl. Phys. B39, 83 (1986); K. Tanii, M. Tachikawa, M. Kajita and T. Shimizu, J. Opt. Soc. Am. B5, 244 (1988); E. Arimondo, P. Bootz, P. Glorieux and E. Menchi, J. Opt. Soc. Am. B2, 193 (1985); M. Tachikawa, K. Tanii and T. Shimizu, J. Opt. Soc. Am. B 4, 387 (1987).

[15] A. Bekkali, F. Papoff, D. Dangoisse and P. Glorieux, J. Phys. Coll. C2, 349 (1988); B. Zambon, F. de Tomasi, D. Hennequin and E. Arimondo, Phys. Rev. A40, 782 (1989); M. Tachikawa, F.L. Hong, K. Tanii and T. Shimizu, Phys. Rev. Lett. 60, 2266 (1988).D. Hennequin, F. de Tomasi, B. Zambon and E. Arimondo, Phys. Rev. A37, 2243 (1988).

[16] F. de Tomasi, D. Hennequin, B. Zambon and E. Arimondo, J. Opt. Soc. Am. B 6, 45 (1989)

[17] L.P. Shil'nikov,, Sov. Math. Dokl. 6, 163 (1965); Math. USSR Sbornik 10, 91 (1970).

[18] P. Glendinning and C. Sparrow, J. Stat. Phys. 35, 645 (1984). P. Gaspard, R. Kapral and G. Nicolis, J. Stat. Phys. 35, 697 (1984).

[19] P.Gaspard, X.J.Wang, J.Stat.Phys.48, 151 (1987)

[20] B. Zambon in ref. [6], p. 512.

[21] e.g. J.S. Turner, J.C. Roux, W.D. McCormick and H.L. Swinney, Phys. Lett. 85A, 9 (1981).A. Destexhe, G. Nicolis and C. Nicolis,in "*Measures of Complexity and Chaos*", p.63, NATO ASI Series, N.B.Abraham, A.M.Albano, A.Passamente, P.E.Rapp editors, Plenum Press (1989). A. Destexhe, Phys. Lett.143A, 373 (1990)

[22] R. Badii in "*Measures of Complexity and Chaos*", p.63, NATO ASI Series, N.B.Abraham, A.M.Albano, A.Passamente, P.E.Rapp editors, Plenum Press (1989).

[23] A. Dextexhe, Phys. Lett. A 143, 373 (1990)

[24] G.B.Mindlin, X.J. Hou, H.G. Solari, R. Gilmore and N.B. Tufillaro, Phys. Rev. Lett.64, 2350, (1990)

[25] F. Papoff, A. Fioretti, E. Arimondo, G.B. Mindlin, H. Solari, R. Gilmore, Phys. Rev. Lett., to be published

[26] K. Tanii, M. Tachikawa, T. Tohei, F.-L. Hong and T Shimizu, Phys. Rev. A, to be published

[27] K. Otsuka, Phys. Rev. Lett. 67, 1090 (1991)

[28] L.Lugiato, this volume

Dislocations in Laser Fields

*C.O. Weiss, Chr. Tamm, and K. Staliunas**

Physikalisch-Technische Bundesanstalt, Bundesallee 100,
W-3300 Braunschweig, Fed. Rep. of Germany

Abstract. Screw dislocations or vortices have recently been observed in laser fields and calculated from the laser equations. These defects are particle-like solutions. In the experiments the dislocations are found to form "molecules" or even "crystals" due to the mutual interaction and the forces of gradients in the background field. Most interesting structure formation in aggregation of defects can therefore be expected on the one hand, and interesting statistical ("thermodynamic") properties of chaotically moving defects - a two-dimensional defect "gas" - on the other hand. At present the most interesting results come from comparisons of two complementary approaches: the direct solution of the complex Ginzburg-Landau equation to which the laser equations can be reduced in the simplest case and mode expansions.

1. Introduction

In recent years general insights of nonlinear physics have been successfully tested experimentally in optics. The unique properties of optical systems, in particular the fast time scales, the possibility of precisely controlling the system's degrees of freedom might also in the future prove useful to advance nonlinear physics in general.

Ties between optics, and in particular lasers, and fluids e.g. and nonlinear physics in general have been known for some time.

In 1975 it was shown by Haken [1] that the equations for a homogeneously broadened laser in the plane-wave, mean field approximation, are isomorphic to the Lorenz equations describing conceptually Bénard convection in fluids. The experimental proof that lasers can show the chaotic dynamics of the Lorenz model was given in the 1980s [3]. The 1980s were a particularly fertile period for investigations of dynamical nonlinear optical systems. After the first finding of intrinsic chaotic dynamics for a laser [2], a large number of experiments verified concepts of nonlinear physics in optical systems, mostly various types of lasers [3].

Springer Proceedings in Physics, Vol. 69
Evolution of Dynamical Structures in Complex Systems
Editors: R. Friedrich · A. Wunderlin

Good agreement between the plane-wave models and experiments was often found, somewhat surprisingly, since the plane-wave approximation is obviously not valid in experiments where the simplest field structures are Gaussian beams.

2. Optical Defects

Lugiato finally approached this problem directly by considering cases which involve higher order transverse modes [4]. One simple clear prediction is phase-locking of transverse modes close in frequency [4]. Since this case has a relation to the Turing instability which has been considered to explain among other things the patterns of animal skin (zebra), an experimental verification was considered worthwhile. The simplest case of nearly degenerate modes in a laser occurs for the pair of TEM_{01} and TEM_{10} modes which are close but not equal in frequency in laser resonators whose rotational symmetry is broken (e.g. by the astigmatism of Brewster windows).

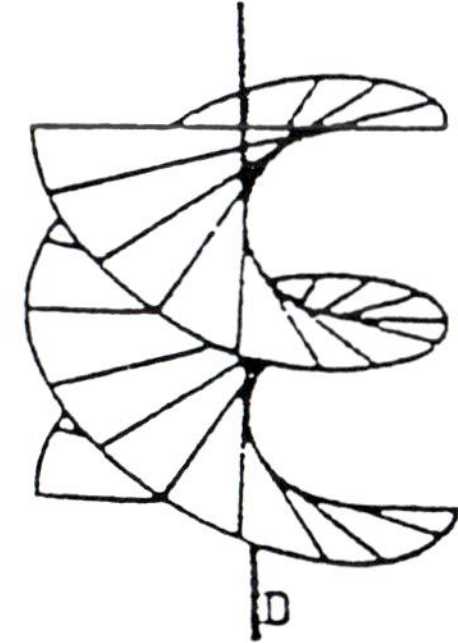

Fig. 1. The equiphase surface of the $TEM_{01}*$ mode

The phase-locking between TEM_{01} and TEM_{10} Hermite modes was experimentally demonstrated [5] and it was realized that the phase-locked wave has a helical structure of the wave front [6], Fig. 1. It followed that for the emission of such a wave the laser must break the inversion symmetry. This was experimentally proven by showing the left-right helicity bistability of the laser [6]. Fig. 2 shows the experimental set-up and the hysteresis between the two states. Coullet then showed numerically - considering the phase-symmetry breaking at the laser threshold - the spontaneous creation of defects (vortices) at the laser threshold [7]. The calculation was performed for a laser with very large Fresnel number (Fig. 3).

One can give an intuitive explanation of the appearance of vortices in lasers: if the laser cross

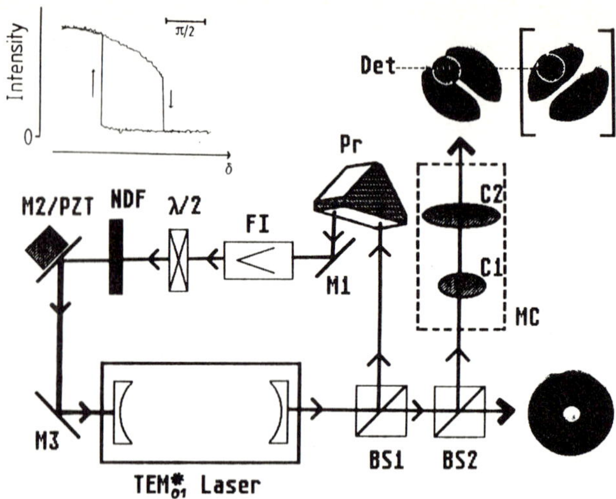

Fig. 2. Experimental set-ups: BS1, BS2, beam splitters; C1, C2, cylindrical lenses, forming the transverse-mode converter MC; Det, detector monitoring the local intensity in the bistable transverse intensity pattern - the detector position coincides with the nodal line of one of the patterns; Pr, rectangular reflecting prism - the plane of reflection of Pr is rotated out of the drawing plane by 45°; FI, Faraday isolator "optical diode"); $\lambda/2$, $\lambda/2$-retardation plates; NDF, neutral density filter; PZT; piezoceramic transducers; M1, M2, M3, mirrors. For simplilcity, mode matching and beam expanding lenses are not shown. Inset shows the local intensity at the detector position Det as a function of the optical feedback path length δ etc.

section is large, the optical field will build up independently, from noise, in different regions of the cross section, when the laser is switched on. The phase of the field will then be broken to different values in the different regions. However, since diffraction couples adjacent points in the cross section, there must be short range correlation in the presence of long range decorrelation, a situation usually requiring the existence of defects. (This is completely analoguous to the appearance of cosmic strings in the early universe or the generation of superfluid vortices when the critical temperature is crossed.)

Then, it follows that defects or vortices can spontaneously appear when a laser is switched on if

$$t_{diff} > t_{field}$$

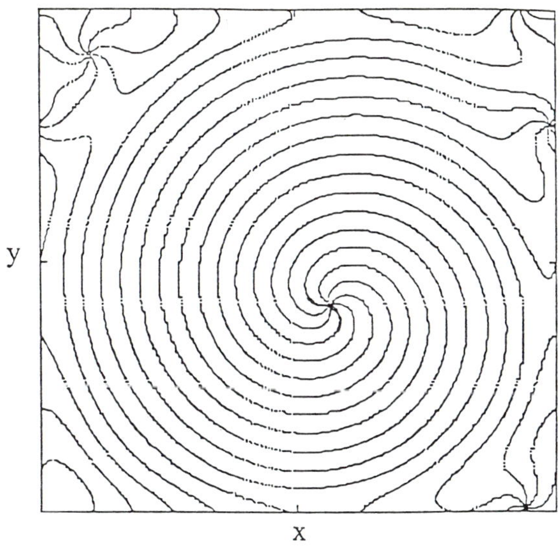

Fig. 3. Equiphase lines of the field of a large Fresnel number laser (distances between two equiphase lines is $\pi/4$)

where t_{diff} is the time needed for diffraction to communicate phase information across the laser cross section. t_{field} is the build-up time of the field inside the laser resonator. Evidently, t_{diff} is proportional to the Fresnel number of the laser, while t_{field} decreases with increasing small signal gain.

Consequently for the spontaneous creation of defects or vortices a large Fresnel number laser with high gain of the medium is necessary, which shows immediately why vortex formation does not usually occur in lasers: For typical continuous lasers the gain is not very high and the Fresnel number is usually small. Excimer lasers, however, operate in the regime typical for vortex creation to occur, as probably a number of other pulsed high-power lasers do.

It was then realized that waves of the $TEM_{01}{}^*$ type (Fig. 1) and the spontaneously created defects have the same structure, with helical wavefront, zero intensity and a pole of the phase gradient at the centre [8]. A photograph of a laser beam, Fig. 4, taken before [9], was thus speculated to show an array of defects [10].

It was then possible to show:
1) that simple superposition of Hermite modes generates arrays of defects, which are arranged in a manner perfectly reminiscent of a two-dimensional crystal (Fig. 5), the equiphase lines resembling field lines between charged particles [11];

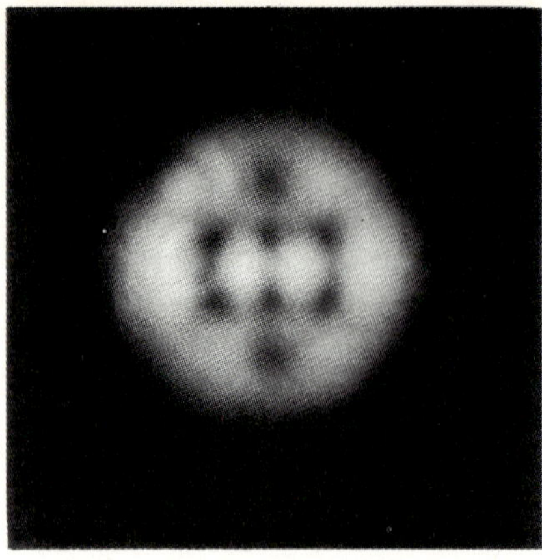

Fig. 4. Laser beam cross-section of the emission of an optically excited Na_2-laser. Assuming that half of the dark spots correspond to right-hand helical points and the other half to left-hand helical points permits one to visualize the structure as two "unit cells" of an "ionic crystal"

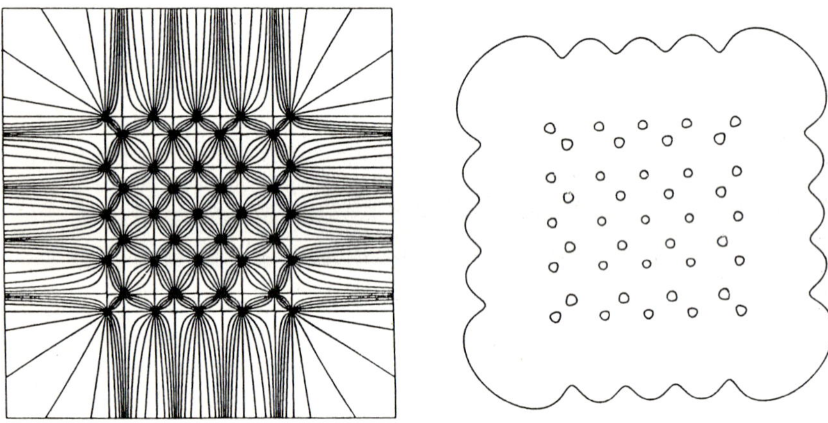

Fig. 5. 10 % intensity contour and equiphase lines of a defect crystal constructed by superpositon of (5,4) Hermite modes

2) that simple superposition of certain Laguerre modes generates field defects of multiple topological charge, i.e. the circulation of the phase gradient around the point of zero intensity is:

$$\oint_C \vec{\nabla}\Phi(\vec{r})\cdot d\vec{s} = 2\pi\cdot N \qquad (1)$$

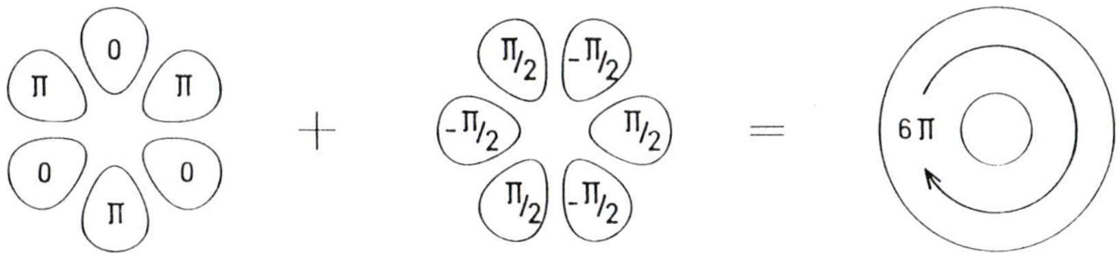

Fig. 6. The superposition of two Laguerre (0,3) modes results in a triply charged defect

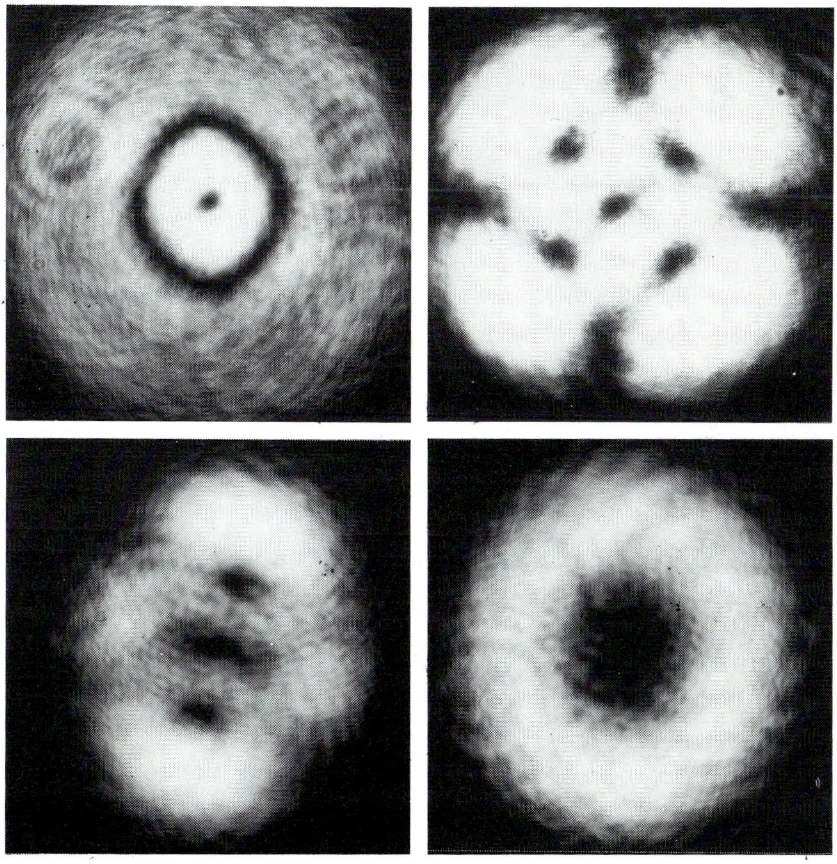

Fig. 7. Steady state emission patterns of a Na_2-laser in the transverse mode family of index 3. The dark spots correspond to defects, the lower pattern is a triply charged defect

(Fig. 6) [11]. N is called the topological charge. These multiply charged defects are the analogue of the quantitized superfluid vortices. Such "defect crystals" and "multiply charged defects" were then found as steady-state solutions of the laser equations and were also observed experimentally [8]; Fig. 7.

In the experiments indication of some repulsion between two defects of equal charge was found so that it was speculated upon the existence of lattice vibrations ("phonons") of the defect crystals and on the possible "melting" of these crystals or the dissociation, when the lattice vibrations are strongly excited. In the meantime, numerical calculations [12] have shown indeed the existence of the lattice vibrations or the occurrence of resonances (even undamped ones) in the lattice.

As the defects in the crystals appear as zero-dimensional interference fringes (interference nodes) one might doubt the possibility of destroying the crystals without destroying the "particles" themselves. The prediction of a vortex circling around the optical axis of a laser as the simplest case of a free, moving "particle" was therefore tested experimentally [13]. The prediction gives the circling frequency of the vortex as the transverse laser mode spacing. In the laser used (a homogeneously broadened optically pumped, travelling wave Na_2-laser with controllable pump profile) this mode spacing is 100 MHz so that it is not possible to take two-dimensional snapshot pictures. Instead two detectors in different positions of the laser beam cross-section were used. Fig 8 shows the arrangement of the detectors and the signals observed at the different positions. These signals provide clear enough evidence that an intensity minimum moves around the optical axis. The rotation frequency was 6 % below the empty resonator transverse mode spacing and a change of 20 percent from this value was observed to occur on changing experimental conditions such as the Na_2-density, which we speculate is due to the change of the background phase due to the presence of a vortex (see Sect. 3). This evidence of a freely moving vortex makes the possibility of "dissociating the vortex "crystals"" more realistic.

For stationary or very slowly moving defects there is a possibility to detect the position and the charge of a vortex by an interference technique [14] since the vital structure of the vortex is a phase structure. Fig. 9 shows interference fringes with a laser field containing two stationary vortices. At the point where an interference fringe terminates is a vortex located. In fact, a gas of moving vortices has been detected with this interference technique in the field emitted by a pumped photorefractive resonator of large Fresnel number [15].

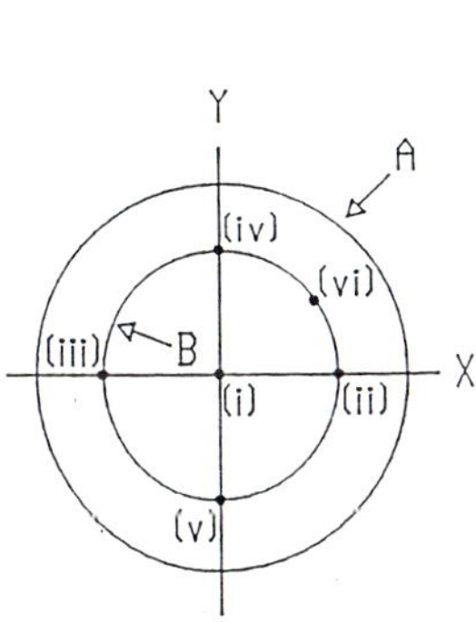

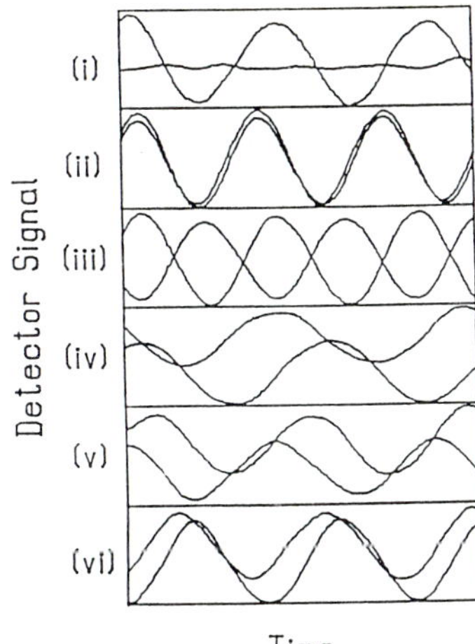

Fig. 8. Detection of a vortex circling about the optical axis in a Na_2-laser (circling frequency is $\approx$ 100 MHz).
a) Positions of detectors in the laser beam cross-section. A reference detector is fixed at position II; a second detector measures at the positions I-VI.
b) Signal of the detector at the positions I-VI. For comparison the signal of the reference detector at position II is given in each figure.

Fig. 9. Interference pattern of a wave containing two defects of charge +1 and -1 with a plane wave inclined with respect to the dislocated wave. The number of fringes merging into one indicates the magnitude, and the merging direction the sign of the topological charge of the defects

3. Lasers and Hydrodynamics

The similarity of the appearance and behaviour of defects in lasers with phenomena in fluids, superconductors and superfluids, finds its explanation in the possibility of reducing the laser equations in a limiting case to the complex Ginzburg-Landau equation (CGLE), as well as to hydrodynamical equations. The Maxwell-Bloch equations (MBE) provide a semiclassical description of the fields in a laser [16]:

$$\frac{\partial E}{\partial t} = -(i\omega_C + \varkappa)E + \varkappa P + id\varkappa\nabla^2 E; \qquad (2a)$$

$$\frac{\partial P}{\partial t} = -(i\omega_A + \gamma_\perp)P + \gamma_\perp ED; \qquad (2b)$$

$$\frac{\partial D}{\partial t} = -\gamma_\parallel((D - D_0) + (E^*P + P^*E)/2); \qquad (2c)$$

here $E(\vec{r},t)$, $P(\vec{r},t)$ and $D(\vec{r},t)$ are two-dimensional envelopes of electric field, polarisation and population inversion correspondingly, $\kappa, \gamma_\perp$ and $\gamma_\parallel$ are decay rates of these variables, $D(r,t)$ is the population inversion in absence of radiation, and the frequencies ω_A and ω_C are atomic gain line and cavity mode central frequencies. The vector $\vec{r}$ describes the position in the plane (r,ϕ) which is transverse to the laser beam propagation direction z.

The term $id\kappa\nabla^2 E$ in the equation (2a) is the diffraction term, ($d=r_0^2/(4FT)$ is the diffraction constant, where $F = r_0^2/\lambda L$ is the cavity Fresnel number, r_0 is the mean beam radius, L is the full cavity length, λ is the wavelength of the emitted wave, and T is the cavity mirror transmittivity). Due to this term information is transferred in the laser beam cross-section and collective effects causing long distance order in a plane transverse to the laser beam arises.

In the "good-cavity" limit, when both the induced polarisation and population are decaying much faster than the electric field ($\kappa \ll \gamma_\perp, \gamma_\parallel$), the MBE system can be reduced to the complex Ginzburg-Landau equation (CGLE) [7,17,18]:

$$\frac{\partial A}{\partial t} = \mu_1(\vec{r})\cdot A + id\cdot\nabla^2 A - (1+i\beta)|A|^2 A; \qquad (3)$$

where $A(\vec{r},t)$ is an order parameter corresponding to the electric field. The spatially dependent function $\mu_1(r)$ is the spatial gain profile, and $\beta=(\omega_C-\omega_A)/\gamma_\perp$ is the normalized cavity detuning value.

The CGLE in form of (3) displays instability in case of negative detuning β: every perturbation of the field A selffocuses and leads to blow-up. (It must be mentioned that the case $\beta < 0$ is significant because the activation of transverse modes in a laser requires a detuning of negative sign). To prevent this blow-up a diffusion term besides the diffraction term must be introduced in (3). The origin of this term is the small diffusion of polarisation and population inversion (always present in a laser), as well as finiteness of atomic gain line width. The diffusion parameter d_1 can be considered as positive, small compared with the diffraction parameter ($d_1 \approx d \cdot \kappa/\gamma_\perp$ for most class-A-lasers), and independent on cavity detuning. Then the CGLE can be rewritten in the following form:

$$\frac{\partial A}{\partial t} = (\mu_1(\vec{r})+i\mu_2(\vec{r}))A + (d_1+id)\nabla^2 A - (1+i\beta)|A|^2 A; \quad (4)$$

The spatially inhomogneneous phase gain profile $\mu_2(r)=-ar^2$ is introduced phenomenologically to reflect the presence of curved mirrors in a laser, where $a=Ff/(r_0{}^2T)$ is the phase gain and f represents the cavity mirror curvature (f=0 for a plane mirror cavity, f=1 for confocal cavity, f=2 for concentric cavity etc.)

The model (4) describes best the fields in class A lasers (after the classification in [3]). A Na_2-laser, for instance, is representative for class A. The material variables of a class B laser (e.g. a CO_2-laser), however, are not fast enough to be enslaved by the optical field, and two equations are required for a correct description (for class C lasers, all three equations, correspondingly).

It is known that the equation (4) in the homogeneous field case possesses vortex solutions [7,19-25]. A single isolated vortex is stationary (because of spatial symmetry) and has the following asymptotic form near the core ($r \to 0$):

$$A(r,\phi,t) = \lambda r \cdot \exp\left[i(-\Omega t + m\phi)\right]; \quad (5)$$

The parameter $m = \pm 1$ is the sign of the topological charge of the vortex, Ω is the vortex rotation frequency - the field frequency change due to the presence of the vortex (for class A lasers $\Omega \approx \beta$), and the parameter r_{core} defines the dimension of vortex core ($r_{core} \approx d$).

The term "vortex" comes into optics from hydrodynamics, where such structures are common [26,27]. To show the correspondence between the vortices in optics and hydrodynamics we rewrite the CGLE (4) in terms of the field intensity $\rho = |A|^2$ and phase $\Phi = \arg(A)$:

$$\frac{\partial\rho}{\partial t} = 2\rho\mu_1 + d_1\left[\nabla^2\rho - \frac{(\vec{\nabla}\rho)^2}{2\rho} - 2\rho(\vec{\nabla}\Phi)^2\right] - 2d\cdot\vec{\nabla}(\rho\cdot\vec{\nabla}\Phi) - 2\rho^2; \tag{6a}$$

$$\frac{\partial\Phi}{\partial t} = \mu_2 + d_1\left[\nabla^2\Phi + \nabla\Phi\cdot\frac{\vec{\nabla}\rho}{\rho}\right] + d\left[\frac{\nabla^2\rho}{2\rho} - \frac{(\vec{\nabla}\rho)^2}{4\rho^2} - (\vec{\nabla}\Phi)^2\right] - \beta\cdot\rho; \tag{6b}$$

Then, if the intensity of optical field ρ is considered as a fluid density, and the gradient of the field phase as velocity $\nabla\Phi = \vec{u}$, the equations (6) can be rewritten in the following form:

$$\frac{\partial\rho}{\partial t} + \vec{\nabla}(\rho\vec{u}) = 2\mu_1\rho - 2\rho^2 + \rho\cdot\frac{d_1}{d}\left[\frac{\vec{\nabla}^2\ln(\rho)}{2} + \frac{(\vec{\nabla}\ln(\rho))^2}{4} - \vec{u}^2\right]; \tag{7a}$$

$$\frac{\partial\vec{u}}{\partial t} + \vec{\nabla}\left(\frac{\vec{u}^2}{2}\right) = \vec{\nabla}\mu_2 + \vec{\nabla}\left[\frac{\vec{\nabla}^2\ln(\rho)}{4} + \frac{(\vec{\nabla}\ln(\rho))^2}{8}\right] - \beta\vec{\nabla}\rho + \frac{d_1}{2d}\cdot\vec{\nabla}\left[\vec{\nabla}(\rho\cdot\vec{u})/\rho\right]; \tag{7b}$$

The transverse coordinate r has been normalized here: $r^2/2d \rightarrow r^2$, and also the phase gain parameter: $2ad \rightarrow a$.

The equation (7a) is a fluid continuity equation: the left hand side of it is the full derivative of the density (the optical field energy density) along the flow trajectory, and the right hand side terms reflect the presence of spatially distributed sources and sinks (the pump and saturation).

The equation (7b) corresponds to an Euler equation for viscous and compressible fluids. The left hand terms correspond to a full acceleration of the fluid, the right hand terms mean correspondingly the diffusion due to viscosity, the external force density gradient, and the pressure gradient.

The subsequent simplification can be made by eliminating adiabatically the density from the system (7):

$$\rho = \mu_1 - \frac{1}{\rho}\cdot\vec{\nabla}\left(\frac{\vec{\rho}\nabla\Phi}{2}\right); \tag{8a}$$

$$\frac{\partial\Phi}{\partial t} + \frac{(\vec{\nabla}\Phi)^2}{2} = \nu\cdot\nabla^2\Phi - V(r) + \nu\cdot\vec{\nabla}\Phi\cdot\vec{\nabla}\ln(\rho) + \frac{\nabla^2\ln(\rho)}{4} + \frac{(\vec{\nabla}\ln(\rho))^2}{8}; \quad (8b)$$

The viscosity parameter now is $\nu = (d_1/d + \beta)2$, and the function $V(r) = \beta\mu_1 - \mu_2$ is the effective external force potential defined by the inhomogeneous pump and inhomogeneous phase gain (due to presence of curved mirrors) in the laser. The depth of the potential well V(r) roughly corresponds to a laser cavity Fresnel number $\Delta\Phi = a\cdot r_0^2 = Ff/T$, where r_0 is the laser beam radius. The potential well depth ratio to the system viscosity parameter $\Delta\Phi/\nu = ar_0^2/\nu = Ff/(T\nu)$ corresponds to a Reynolds number Re of our hydrodynamical system ($Re = Lv/\nu$, where L and v are characteristic length and velocity in the hydrodynamical system).

Equation (8b) describes the dynamics of a dissipative fluid (the dissipative term is present); simultaneously it possesses the quantized vortex type solutions (5) characteristic for superfluids. In the viscous fluid case vortices are also generated - so-called "dissipative vortices" [26]. These dissipative vortices are either nonstationary (decaying in time), or stationary but supported by external forces. In both cases the vortices have a core of finite dimension (proportional to viscosity parameter ν). In our case the vortices are nondissipative (like those in superfluidity), but the viscosity affects the interaction of few vortices.

The fact that the laser equations (2) can be reduced to hydrodynamical equations (7,8) allows us to treat the laser as a "laser fluid" enclosed in some potential well. It allows us to find a multitude of analogies between optics and hydrodynamics. Firstly, due to the external force the fluid tends to gather in the centre of the potential well. On the other hand, the diffusion and the internal pressure (roughly proportional to fluid density) tends to homogenize the fluid density distribution. As an equilibrium of these two forces the fluid density profile is formed, corresponding to the light beam profile in a laser.

Numerical simulations and analytical investigation show [28] that the density profile is roughly Gaussian, while the radial flow velocity profile is more complicated. (In the general case the radial flow is always present due to spatially distributed sources and sinks.) In Fig. 10 the phase profiles are depicted depending on cavity Fresnel number F, and cavity detuning β.

Every localized structure in the fluid drifts together with the fluid flow. The drift direction for a non-rotating localized structure coincides with fluid flow direction and is directed towards or away from the centre of the potential well; depending on the parameters of the system. In a particular parameter range (as depicted in Fig. 11) there exist potential minima away from the centre of the cavity, which corresponds to a location of the localized structure non-symmetrically with respect to the centre of the fluid pool.

A vortex in a fluid drifts with the flowing fluid like every other localized structure. To find the motion laws for a vortex, we assume that the phase Φ as well as the intensity ρ is perturbed in equation (8) by the presence of the vortex:

$$\Phi(\vec{r},t)=\Phi_0(\vec{r},t) + \Phi_v(\vec{r}-\vec{r}_v,t) + \Delta\omega t; \qquad (9a)$$

$$\rho(\vec{r},t)=\rho_0(\vec{r},t)\cdot\rho_v(\vec{r}-\vec{r}_v,t); \qquad (9a)$$

here Φ_0 and ρ_0 are the background phase and background intensity on which the vortex fields Φ and ρ_V are superimposed. The vortex rotation frequency changes by a value $\Delta\omega$ due to background field gradients, and the position r_V (t) of the vortex core is a function of time. Then, after inserting (9) into (8), after some mathematical manipulation, one obtains the following expressions for the vortex drift velocity:

$$\vec{v}_v = \vec{\nabla}\Phi_0 + \frac{1}{2}\cdot(\vec{m}\times\vec{\nabla}\ln(\rho_0)) + 2\nu\cdot(\vec{m}\times\vec{\nabla}\Phi_0) - \nu\cdot\vec{\nabla}\ln(\rho_0); \qquad (10)$$

where m is a unit vector which is transverse to the plane (r,φ) (m is parallel to a vortex line in a three-dimensional space). It is directed upwards (along the z-axis) for a vortex with a positive topological charge and downwards for a negative charge. The relation
$\nabla\ln(\rho) = 2(\vec{m}\times\vec{\nabla}\Phi)$ between density and phase near the vortex core has been used to obtain the cross terms in (10). Whereas the origin of the longitudinal drift term (directed along the phase gradient) is obvious, the transverse drift term (directed perpendicularly to density and phase gradients) requires interpretation. The origin of this term is the following: The fluid density near the vortex core is zero. Consequently the vortex can be treated as a "hollow", rotating object. If it is placed in the background fluid with inhomogeneous density (and pressure correspondingly), it is

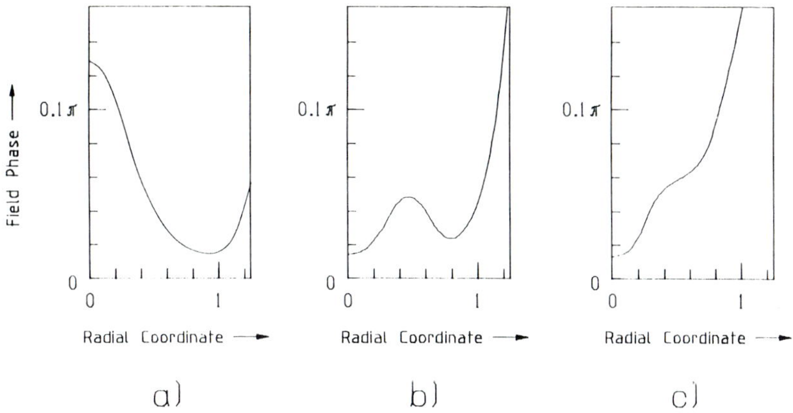

Fig. 10. The mean field phase in a laser beam for different Fresnel numbers of the cavity and different detunings ß.
(after numerical integration of (4) with f = 0.25. d_1/d = 0.1, T = 0.25).
a) F = 5, ß = -4; b) F = 3, ß = -2.5; c) F = 3, ß = 2

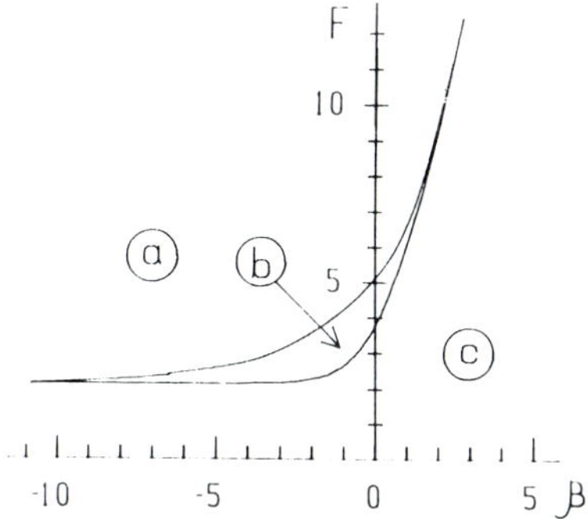

Fig. 11. The regions in the phase plane (ß,F) corresponding to the different shapes of mean field: (a) one maximum is present in centre of the cavity, (b) the maximum is at non-zero distance from cavity centre, (c) no maxima at all. (After numerical integration of (4) with f = 0.25, d_1/d = 0.1, T = 0.25)

subject to an Archimedian force (Archimedian forces push the objects of lower density out of the regions of high pressure). Additionally the rotation causes a so-called Magnus drift [27] of the vortex, directed perpendicular to the applied force direction.

As the consequence of the small viscosity in our case, the vortex drift caused by the background fluid density gradient has a a small component directed along the direction of applied force as well as a component directed perpendicularly to the phase gradient.

The vortex motion law (10) was tested in numerical simulations by generating the inhomogeneous background fields and observing the motion of the vortex due to these field gradients. The simulations confirmed that the vortex starts moving with a constant velocity (corresponding to (10)) instantaneously after the field gradients are "switched on". No acceleration justifying the idea of an inertia for a vortex was found, at least in the parameter range corresponding to a class A laser.

Thus, according to the vortex motion law (10), the vortex can spiral towards the optical axis of the cavity, circle around it at a particular distance or spiral away from the axis, eventually leaving the beam - depending on parameters of the laser.

It is interesting to find the angular frequency of the vortex circling (or spiralling) about the optical axis. If the distance between the vortex and the laser beam centre is r_v, then we obtain:

$$\omega_v = \frac{|\vec{v}_\perp|}{r_v} = \frac{|\vec{\nabla}\ln(\rho)|}{2r_v} = \frac{2}{r_0^2}\ ; \tag{11}$$

if it is supposed that the laser beam is of Gaussian intensity distribution with a beam halfwidth equal to r_0. The frequency difference between adjacent transverse modes of an optical cavity is $\Delta\omega = 2/r_0^2$ exactly the same as the vortex circling frequency, [4,17].

This result can be understood by noting that a vortex located outside the centre of the cavity can be considered as a superposition of TEM_{00} mode (a Gaussian mode) and TEM_{01}* mode (a "doughnut"). The beating of these two modes (intermode spacing frequency $\Delta\omega$) causes the circling of the vortex with the same frequency around the optical axis of the beam.

In Fig. 12 the vortex drift trajectories are given for two different Fresnel numbers and zero detuning ($\beta=0$), illustrating the different types of motion (see Sect. 2).

Such a vortex circling has been observed experimentally, and was interpreted as a simultaneous emission of the cavity modes TEM_{00} and TEM_{01}* [13]. The present interpretation considers a vortex drifting due to the inhomogeneities of the background fluid, and this leads to the same result.

In the case of either large Fresnel number ($F>1$) of a cavity, or large negative detuning ($\beta<0$), a few

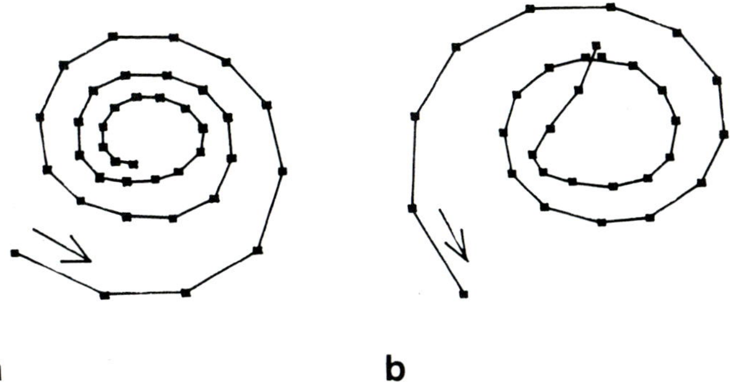

Fig. 12. The trajectories of a vortex in a laser beam "trap". (After numerical integration of system (4) with $\beta = 0$, $d_1/d = 0.1$).
a) a = 50, b) a = -20

vortices of the same charge can be placed in the laser beam cross-section. Each vortex is affected by a force directed towards the centre of the beam due to the fluid flow, and simultaneously by repulsive forces from the other vortices. As a result of these forces, a steady state pattern, a vortex crystal which can rotate around the centre of the beam, can form. Periodic or even chaotic movement of the "trapped" vortices (vortex "gas") is also possible. The corresponding behaviour has been observed in superfluids [27], as well as in an optical system [15]. The viscosity in (8) depends on the detuning of the cavity β, if diffraction and diffusion are considered constant. When β is positive and large (it corresponds to detuning into a region where no transverse modes are present - Fig. 13), then no modes are activated. In our hydrodynamical analogue the fluid is too viscous even for a single vortex to be placed inside the potential well. The vortex is pushed out and only the trivial transverse pattern (a Gaussian beam) is generated.

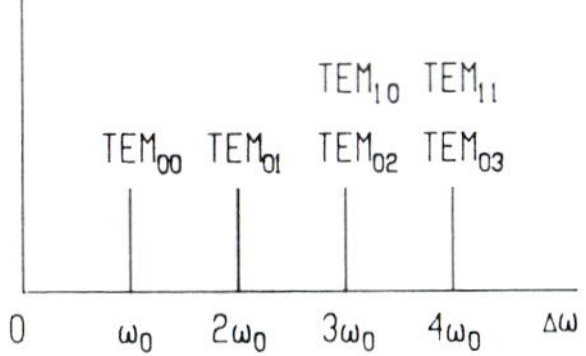

Fig. 13. Laser cavity transverse mode families. Only one longitudinal mode family is present. (Mean field description).

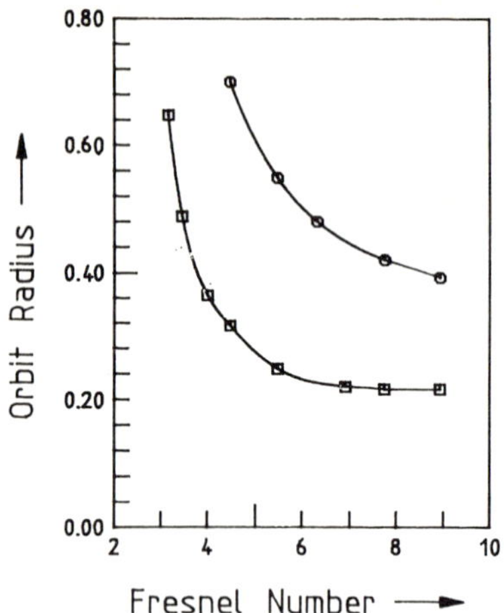

Fig. 14. Vortex circling radius in a few vortex system in a laser beam "trap". The vortices were initially situated symmetrically with respect to optical axis of the beam.
($f = 0.25$, $\beta = 0$, $d_1/d = 0.1$, $T = 0.25$)
□ - Two vortex case;
○ - Four vortex case

The negative detuning decreases the effective viscosity of the fluid and allows the number of vortices in the potential well to increase. In lasers this means tuning to a region of richer transverse mode families (see Fig. 13) and, consequently, the possibility of more complicated structures in the laser beam cross-section.

Fig. 14 gives the distance between two (and four) symmetrically arranged vortices as a function of the cavity Fresnel number. This example shows how the complexity of stable patterns depends on the cavity Fresnel number and consequently on the Reynolds number of the hydrodynamical system. The existence of bistability between the two- and the four-vortex state (which was also found experimentally [8]) can be seen from Fig. 14.

It would be interesting to calculate the forces between the vortices as a function of separation, and then, knowing the single vortex motion laws in a laser beam, to find the equilibrium distances between rotating vortices in, for example, a two vortex rotating pattern. Unfortunately, the calculation of a general expression for the interaction force between two vortices is still an unsolved problem. Some attempts have been made [20-25], but it seems that the forces (especially in the "far-from-core" limit) are functions not just of the separation between vortices, but also depend on the history of the system.

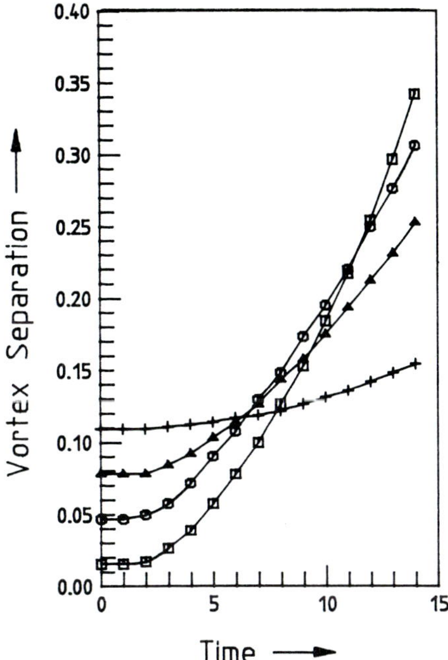

Fig. 15. The temporal evolution of the separation of two vortices for different initial separations. (After numerical integration of (4) in the case of flat external pump, $f = 0$, $\beta = 0$, $d_1/d = 0.1$, $T = 0.25$)

Results of our numerical simulations (Fig. 15, where time evolutions of vortex separations are presented), hint at such behaviour: vortices, initially separated by a distance d, accelerate in a radial direction until they obtain an asymptotic velocity. Such behaviour is characteristic for the undamped movement of a particle with inertia in a repulsive potential. Consequently vortex motion cannot be described by an equation like $\vec{v}_i = f(\vec{r}_i, \vec{r}_j)$, where $j=1,2,\ldots$ $(j = i)$ is the index of the vortex with which the i-th vortex is interacting. A differential equation of at least second order is required, and the vortices must be treated as objects with more than half a degree of freedom per spatial dimension, as in the case of an overdamped particle dynamics.

4. Multistability and Pattern Recognition

A system with a very high dissipation has only one basin of attraction; thus there is only one asymptotic (and stationary) state for the system. On the other hand, when there is no dissipation, each initial condition defines a separate trajectory so that the system has an infinite number of "states". Consequently at some intermediate dissipation the system has a finite number of coexisting attractors which are approached from different initial conditions. As a consequence it is not surprising that a laser can have coexisting steady states. In the case of transverse phenomena, the different coexisting states have different optical field distributions. A simple case was mentioned in Sect. 2, the bistability between the helical waves which break the inversion symmetry. More complicated cases of multistability involving "crystals" with two and more vortices which show a higher multiplicity of states have been shown to exist, both numerically and experimentally [8].

A multistable laser has a very direct capability of recognizing patterns. If the laser is switched on, it approaches at random one of the coexisting steady

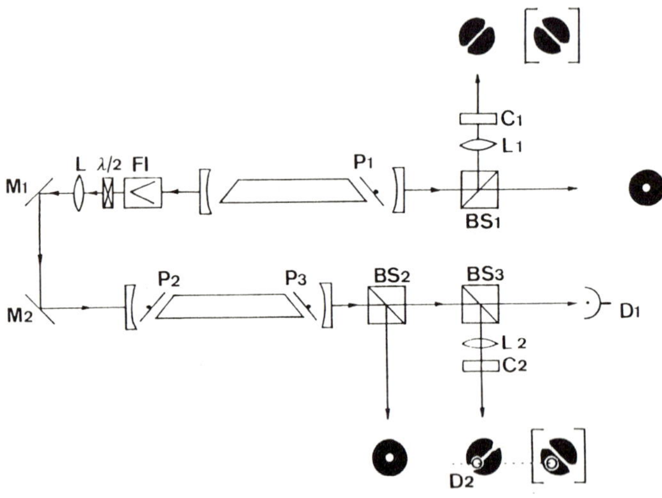

Fig. 16. Experimental arrangement for proof of a simple pattern recognition capability of lasers.
A "master" laser emitting either a right or a left handed helical wave injects a second "recognizing" laser with right or left- handed helical wave emission, as coexisting laser states. The helical wave emission is forced by absorbing dots in the lasers (P_1, P_2, P_3). Faraday isolator FI, waveplate $\lambda/2$. lens L, serve to isolate the "master" laser and mode-match the radiation from the "master" into the "recognizer". BS_1 permits to a) monitor the overall shape of the "doughnut" (helical wave) and b) with astigmatic imaging, to see the helicity of the wave.
BS_2 with BS_3 serve the same purpose for the "recognizer" laser. D_1 is used for control of the resonator frequency of the "recognizer" laser a relative to the "master" laser frequency.

states. If, however, radiation corresponding to one of the coexisting states is injected into the laser the build-up of the laser field starts from this field and consequently approaches finally the state corresponding to the injecion field.

It is plausible (and has also been proven numerically [29]) that if the injected field is a mixture of the laser states or a noisy or distorted state, the field developing in the laser will be the one whose similarity (i.e. its scalar product or correlation coefficient) with the injected field is the largest.

In order to demonstrate the pattern recognition capability of a laser, a simple demonstration was performed using the set-up shown in Fig. 16. One laser capable of emitting either a left-hand helical wave or a right-hand one injects another laser which is repetitively switched on and off. The handedness of the wave is detected by an astigmatic mode converter [6]. A high level then corresponds to left-handedness, a low level corresponds to right-handedness. Zero level obviously corresponds to the injected (recognizing) laser off.

Technically the isolator in conjunction with the waveplate defines the direction of light propagation and thus which laser is the "master" and which the "recognizer". Lenses match the "master" beam to the recognizer's resonator. Evidently, for light to be injected into the recognizer's resonator, the latter must be tuned to the master radiation frequency. This is achieved by a resonator-frequency-feedback-control circuit. For details see [29].

Fig. 17a shows the switching on and off of the recognizing laser without any injection. The laser, as expected, approaches the two states randomly. With a left-hand helical wave injected (Fig. 17b) the recognizing laser approaches reliably the same helicity equally as for right-hand helicity injection (Fig. 17c). The intensity of the wave injected in these experiments was about 5% of the intensity of the recognizing laser. Experiments in which mixtures of helical waves are injected are in progress as are experiments with a larger number of coexisting states. Evidently, each stable state may be transformed by a hologram into a desired image (e.g. a letter), however, for large numbers of coexisting states it may be possible to inject coherent images directly.

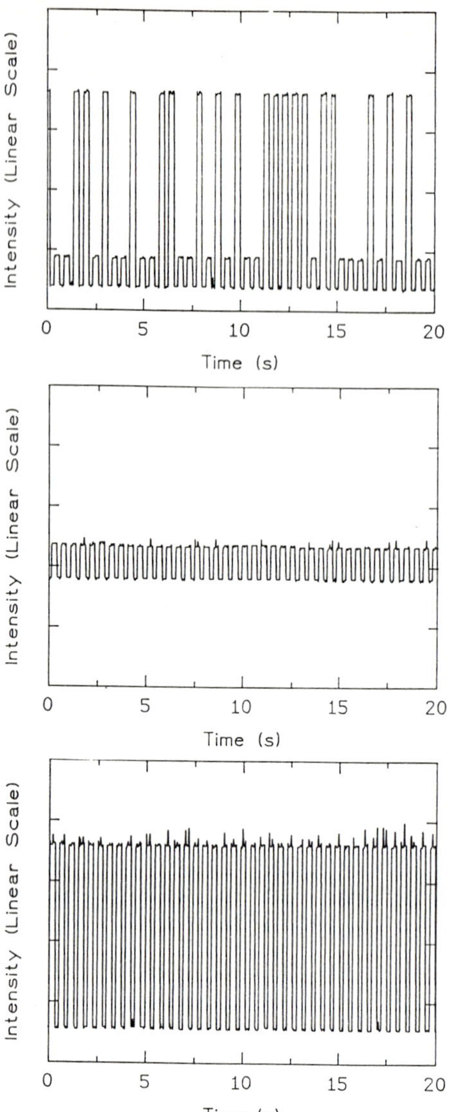

Fig. 17. Measurement of the state of the "recognizer" laser (output from detector D_2 from Fig. 16)
a) no injection
b) right hand helix wave injection
c) left hand helix wave injection

Summary

We have found as prominent features of lasers with transverse degrees of freedom the existence of screw dislocations or vortices of the optical field. As the laser equations can be transformed in the simplest case into a complex Ginzburg-Landau equation, already existing results of the defect interaction [22,25] in the CGLE can be directly applied to the laser. The gradient forces on defects in lasers have been calculated in an approximation and found to agree with an experiment, as well as with mode expansion calculations.

Defects may lie at the root of the pattern recognition capability of lasers. As the appearance of defects is related to a phase symmetry-breaking at the laser threshold, defects will also appear in other systems with such a threshold. Indeed the appearance of vortices in parametric oscillators was recently shown [30].

On a highly speculative level one could - due to the capability of the defects to interact and form "molecules" - expect some form of chemistry and highly complex processes of aggregation of defects in optics.

References

[1] H. Haken, Phys. Lett. 53A, 77 (1975)

[2] C.O. Weiss, A. Godone, A. Olafsson, Phys. Rev. A28, 892 (1983)

[3] Dynamics of Lasers, C.O. Weiss, R. Vilaseca, VCH Verlagsgesellschaft, Weinheim (1991)

[4] L.A. Lugiato, R. Lefever, Phys. Rev. Lett. 50, 220 (1987)

[5] Chr. Tamm, Phys. Rev. A38, 5960 (1988)

[6] Chr. Tamm, C.O. Weiss, Journ. Opt. Soc. Amer. B7, 1034 (1990)

[7] P. Coullet, L. Gil, F. Rocca, Opt. Comm. 73, 403 (1989)

[8] M. Brambilla, F. Battipede, L.A. Lugiato, V. Penna, F. Prati, Chr. Tamm, C.O. Weiss, Phys. Rev. A43, 5090 (1991)

[9] W. Klische, C.O. Weiss, B. Wellegehausen, Phys. Rev. A39, 19 (1989)

[10] C.O. Weiss, Chr. Tamm, P. Coullet, Journ. Mod Opt. 37, 1825 (1989)

[11] These figures have been prepared by R. McDuff

[12] M. Brambilla, C.O. Weiss (unpublished)

[13] M. Brambilla, C. Celtaves, L.A. Lugiato, R. Pirovano, A.B. Coates, C.O. Weiss, "Dynamical Transverse Laser Patterns" (in preparation)

[14] A.G. White, C.P. Smith, N.R. Heckenberg, H. Rubenzstein-Dunlop, R. McDuff, C.O. Weiss, "Interferometric measurements of phase singular ities, Journ. Mod. Opt. (in presss)

[15] F.T. Arecchi, G. Giacomelli, P.L. Ramazza, S. Residori, "Vortices and defect statistics in optical chaos", Phys. Rev. Lett. (submitted)

[16] M. Sargent III, M.O. Scully, and W.E. Lamb, Laser Physics, Addison-Wesley, Reading (1974)
[17] R. Lefever, L.A. Lugiato, Wang Kaige, P. Mandel and N.B. Abraham, Phys. Lett. 135, 254 (1989)
[18] L. Gil, K. Emilsson and G.-L. Oppo "Dynamics of Spiral Waves in a Spatially Inhomogeneous Hopf Bifurcation" (in preparation)
[19] P.S. Hagan SIAM; J. Appl. Math. 42, 762 (1982)
[20] R. Kawasaki, Prog. Theor. Phys. Suppl. 79, 161 (1984)
[21] S. Koga, Prog. Theor. Phys. 67, 164 (1982)
[22] S. Rica and E. Tirapegui, Phys. Rev. Lett. 64, 878 (1990)
[23] S. Rica and E. Tirapegui, Physica D48, 396 (1991)
[24] E. Bodenschatz, A. Weber and L. Kramer "Structure and Dynamics of Spiral Waves and of Defects in Traveling Waves" in "Nonlinear Wave Processes in Excitable Media" A.V. Holden et al., eds., NATO ASI Series B, Plenum Press (1990)
[25] I.S. Aranson, L. Kramer and A. Weber "On the interaction of spiral waves in non-equilibrium media" (to be published in Physica D)
[26] L.D. Landau and E.M. Lifshits; Theoretical Physics, Vol.6, Hydrodynamics, Nauka, Moskau (1988).
[27] R.L. Donnelly; Quantized Vortices in Helium II, Cambridge Univ. Press, Cambridge (1991)
[28] K. Staliunas; Dynamics of Optical Vortices in a Laser Beam, Phys. Rev. Lett. (submitted)
[29] M. Brambilla, L.A. Lugiato, M.V. Penna, F. Prati, P. Pagani, P. Vanotti, M.Y. Li, C.O. Weiss, "The Laser as a nonlinear element of optical associative memory" (in preparation)
[30] K. Staliunas, "Optical vortices in three-wave mixing", Opt. Comm. (in press)

* Alexander von Humboldt fellow, visiting from Vilnius University, 2054 Vilnius, Lithuania

On the Role of Noise in Nonlinear Optics

A. Schenzle

Ludwig-Maximilians-Universität München, Sektion Physik,
Theresienstr. 37, W-8000 München, Fed. Rep. of Germany and
Max-Planck-Institut für Quantenoptik,
Ludwig-Prandtl-Str. 10, W-8046 Garching, Fed. Rep. of Germany

Abstract. The conventional laser, as demonstrated by Haken's single mode quantum theory, generates light with statistical properties analogous to those of a coherent quantum state. This is the most classical state allowed by quantum theory. A generalization of this concept could be obtained by modifying the gain or loss reservoirs, or by including additional nonlinear reversible elements. Such a device emits light with new, possibly non-classical characteristics. A very compact source of squeezed coherent light could consist of a nonlinear optical crystal, doped directly with the laser active material, designed in monolithic form.

1. Introduction

The laser is a paradigm for open systems far from thermal equilibrium which exhibit structural changes in a phase-transition-like behavior. It was Hermann Haken who first demonstrated that there is a general principle behind these non-equilibrium transitions, and the laser is just one - rather typical - example. When a symmetry is broken spontaneously and new structures appear in the time or the space domain, it is only a small number of degrees of freedom that become relevant, dominating the dynamic behavior of the entire system. This is a very general and successful concept of nature. In the early 1960's Hermann Haken developed a quantum statistical theory for the laser which was so general in its scope that it served as a conceptual and mathematical guide line for the understanding of a large variety of nonlinear processes far from equilibrium. This concept remained still valuable in the original field of quantum optics which saw a rapid development following the early days of laser theory. A fundamental aspect in all these problems is the presence of noise - of classical noise, thermal noise or most importantly of quantum noise - that becomes especially visible in the neighborhood of bifurcation points. In the present report I will concentrate only on some recent developments originating basically in Haken's laser theory, but also including the modern concepts of non-classical light.

All macroscopic physical systems, when inspected closely, exhibit randomness, irregularities or noise. Experimentally we can only specify a

Springer Proceedings in Physics, Vol. 69
Evolution of Dynamical Structures in Complex Systems
Editors: R. Friedrich · A. Wunderlin

small number of extensive collective parameters like the volume of the system or its temperature - we have, however, no control over the uncountable number of microscopic degrees of freedom that make up a sizable piece of matter. Another source of randomness emerges from the quantum mechanical description. Mesoscopic systems - in the middle between the microscopic and the macroscopic world - will show both, classical and quantum noise. While the origin of classical statistics lies in a certain ignorance or lack of knowledge about the state of the system, the probabilistic interpretation of quantum mechanics is a matter of principle. It is a widespread belief that noise is only an unpleasant and annoying aspect of real experiments, and the presence of randomness only indicates that the experimental set up should be improved. Therefore noise is not considered as a fundamental or at least interesting phenomenon. In our opinion, this is a too simplistic point of view which entirely misses the point. In reality the ubiquity of noise is just a natural feature that should not be ignored by theory. It adds an essential and interesting dimension to the description of nature. The symmetry, spontaneously broken by a deterministic bifurcation mechanism, is restored by the action of thermal or quantum mechanical noise. Also noise can reveal intrinsic instabilities and can distinguish them from meta-stable stationary states, which are a typical property of nonlinear dynamics. But besides the obvious destabilizing character, it has been demonstrated that under certain conditions noise can also stabilize a state that is deterministically unstable [1].

Noise is entirely inevitable when the microscopic level is reached, and quantum theory takes over. In this regime noise becomes an intrinsic and essential ingredient, required by the consistency of theory. This is the case even in systems over which one has - at least in an idealized sense - total experimental control. All that can be specified about a quantum system is its wave function, which evolves deterministically according to the Schrödinger equation. Nevertheless, the wavefunction only predicts the outcome of an experiment on a pair of canonical variables with certain probability.

2. Classical and Quantum Noise - a Brief Summary

The phenomenon that has marked the beginning of the concept of stochastic processes was the problem of Brownian motion. Another example is a freely suspended pendulum which undergoes random excursions due to the complex motion of the surrounding air. In the Langevin picture the dynamic evolution is described by trajectories which satisfy ordinary differential equations, where the statistical behavior is generated by the addition of a randomly fluctuating force. This idea of a freely diffusing particle, as observed under the microscope, has been applied in the past to many other fields of physics. By generalizing the concept of "random walks" in phase spaces to an arbitrary number of dimensions, driven by internal and external nonlinear forces, for systems far from thermal equilibrium with state dependent noise one arrives at the following stochastic process:

$$d/dt\ x_j(t) = F_j(\{x_l\}) + G_{j,k}(\{x_l\})\ \xi_k(t) \tag{1}$$

j=1,2,3...,D runs over the number of independent degrees of freedom. The random forces $\xi_j(t)$ are not given in explicit form, but are only defined through their statistical properties. It is convenient and sufficient for most cases to assume white noise:

$$\begin{aligned} < \xi_j(t) > &= 0 \\ < \xi_j(t)\ \xi_k(t') > &= \delta_{j,k}\ \delta(t-t') \end{aligned} \tag{2}$$

$F(\{x_j\})$ summarizes the deterministic forces that act on the system. They can be reversible and originate from a Hamiltonian, or can be dissipative and result from the interaction with reservoirs. $G_{j,k}(\{x_l\})$ is a measure of the strength of noise, which can vary in time with the actual position in phase space.

Nonlinear stochastic differential equations - with a few possible exceptions - can only be simulated numerically. Instead of following the fate of an individual trajectory in course of time, one may also characterize a stochastic process by the probability that trajectories will visit a certain neighborhood in phase space. The time evolution of the probability density is determined by the Fokker-Planck equation. While the trajectories satisfy nonlinear ordinary differential equations, the evolution of the probability density is governed by a linear partial differential equation. The two descriptions in terms of trajectories or probabilities are entirely equivalent:

$$\{x_j(t)\} \rightarrow P(\{x_j\},t)$$

The Fokker-Planck equation related to the stochastic differential equation (1) can be written in the form [2]:

$$\begin{aligned} \frac{\partial P(\{x_j\},t)}{\partial t} &= -\frac{\partial}{\partial x_j} F_j P + \frac{1}{2}\frac{\partial}{\partial x_j} G_{jl}(\{x_s\}) \frac{\partial}{\partial x_k} G_{kl} P \\ &\equiv L\ P(\{x_j\},t) \\ &= -\ \partial/\partial x_j\ J_j(\{x_j\},t) \end{aligned} \tag{4}$$

With the definition of the probability current J, the Fokker-Planck equation assumes the form of a continuity equation, which guarantees that in course of time total probability is conserved. The Fokker-Planck equation must be solved subject to appropriate boundary conditions. A random process naturally occurs in a certain domain of phase space, which may be finite or infinite. At the boundaries of that domain, the normal component of the probability current must vanish, otherwise probability would be exchanged between the physically allowed and the forbidden regions. Depending on the details of the problem and the chosen representation, this condition must be imposed explicitly on the solution or may be satisfied automatically by the Fokker-Planck equation [3]. This latter case is rather troublesome when it comes to numerical integration, where the boundary conditions are needed in explicit mathematical form [4].

If there is no external time dependent influence on the system, the process is autonomous and the functions G and F do not dependent on time explicitly. For such a stationary process the Fokker-Planck equation can be cast into the form of an eigenvalue problem:

$$L\, P_{\{n\}} = -\lambda_{\{n\}}\, P_{\{n\}} \tag{5}$$

where {n} stands for a set of s separation parameters or "quantum numbers", which are specified by the boundary conditions. The general time dependent solution can then be written in the form:

$$P(\{x_j\},t) = \sum_{\{n\}} c_{\{n\}}\, P_{\{n\}}\,(\{x_j\})\, \exp(-\lambda_{\{n\}} t) \tag{6}$$

The expansion coefficients $c_{\{n\}}$ are determined uniquely by the choice of the initial distribution $P(\{x_j\},t=0)$. Since L is not a selfadjoint operator the adjoint problem must also be considered:

$$L^+\, W_{\{n\}} = -\lambda^*_{\{n\}}\, W_{\{n\}} \tag{7}$$

The orthonormality condition in terms of the adjoint problems:

$$\int W_{\{n\}}\, P_{\{m\}}\, dx^s = \delta_{\{n\},\{m\}} \tag{8}$$

allows one to determine the expansion coefficients c for any given positive initial condition. In general we expect that there always exists a stationary distribution which is approached in the long time limit $t \rightarrow \infty$:

$$\lambda_{\{0\}} = 0 \quad \text{with} \quad P_{\{0\}}\,(\{x_j\}) \quad \text{and} \quad W_{\{0\}}\,(\{x_j\}) = 1 \tag{9}$$

For the special choice of {n} = 0 the orthogonality relation eq.(7) leads to:

$$\int P_{\{m\}}\, d^D x = \delta_{\{0\},\{m\}} \tag{10}$$

As a consequence:

- the steady state distribution is normalized:

$$\int P_{\{0\}}(\{x_j\})\, d^D x = 1 \tag{11}$$

- the eigenfunctions corresponding to non-trivial eigenvalues $\lambda \neq 0$ must assume negative values, since:

$$\int P_{\{m\}}(\{x_j\})\, d^D x = 0 \quad \text{for } \{m\} \neq 0 \tag{12}$$

and therefore cannot be identified individually with a probability density.

- the general time dependent solution is normalized as well. By integrating over eq.(6) it follows that :

$$c_{\{0\}} = 1 \tag{13}$$

which indicates that the steady state is always included in the expansion of an arbitrary probability density. These conditions are sufficient to guarantee that the solution of the Fokker-Planck equation remains positive for all times. Since the Fokker-Planck equation has a formal similarity to the Schrödinger equation, it may be useful to notice an essential difference. While a quantum system can always be prepared - at least in principle - in an arbitrary eigenstate, this is impossible in case of a diffusion problem. The individual "eigenstates" have no probability interpretation and any normalized state is a superposition that contains the steady state. The eigenvalues in general are complex λ . In order to guarantee relaxation into steady state it is necessary that Re $\lambda_{\{n\}} > 0$ except for $\{n=0\}$.

Returning to our pedagogical example of the suspended pendulum, one may argue that the fluctuations could be reduced by evacuating a glass container surrounding the pendulum. By more and more sophisticated insulation techniques one may reduce and finally get rid of all classical fluctuations, only to find out that there are still fluctuations due to the quantum mechanical nature of the system. Certainly, for a macroscopic mechanical system this randomness is quite impossible to observe, but not impossible in the field of quantum optics. The oscillating pendulum is practically a harmonic oscillator and so is a quantized field mode. For a pair of canonically conjugate variables like x,p even in the ground or vacuum state uncertainties remain, that are expressed quantitatively by the variances:

$$< \Delta^2 x > = < x^2 > - < x >^2 = \frac{1}{2}\left(\frac{\hbar}{m\omega}\right)$$

$$< \Delta^2 p > = < p^2 > - < p >^2 = \frac{1}{2}(\hbar m\omega) \tag{14}$$

On top of these fluctuations classical noise can be present as well and the variances may exceed the vacuum level. This is the case when the considered system interacts with a thermal reservoir at a temperature T, where classical and quantum noise are merely superimposed. In addition, the non-commutability of operators which becomes especially visible in nonlinear dynamic systems introduces noise that has no classical analog and can be characterized only by a rather sophisticated form of random force. The most interesting and puzzling situation occurs, when the quantum properties dominate over the classical ones and non-classical features result.

A realistic model for a quantum optical process must include nonlinear forces, dissipation and some form of gain. The energy flux through the system that drives the process out of equilibrium can either be the result of reversible interactions or due to the coupling with non-thermal reservoirs. The state of the quantum system is then characterized by the statistical operator ρ and the dynamic evolution follows from the master equation:

$$\frac{d}{dt}\rho(t) = \frac{i}{\hbar}[\, H,\rho \,] + \left(\frac{\partial}{\partial t}\rho\right)_{irr} \tag{15}$$

It can be a forbiddingly complex task to solve this equation in a certain representation even for simple physical systems when realistic parameters are used. Instead of treating such a complex high dimensional linear algebra problem, it is in many cases more convenient to reformulate the dynamic equations in terms of quasi-probabilities. Thereby one replaces the bulky matrix equation by a more convenient partial differential equation. In many physical situations this equation has a rather intuitive appearance - it is of the form of a Fokker-Planck equation, even for a purely quantum mechanical problem. However, the essential property that qualifies an equation to be identified with a Fokker-Planck equation is the positive semi-definiteness of the diffusion matrix, and this is by no means a priori guaranteed for an arbitrary quantum process. A quantum process with positive diffusion is practically indistinguishable from a classical one. A master equation which leads to "negative diffusion" does not create a stochastic process, it causes non-classical behavior and cannot be simulated by random trajectories like it is possible for Brownian motion.

3. Quasi-Probabilities

In quantum theory of irreversible processes, the statistical operator ρ contains all the physical information, while classical stochastic processes are described by a probability measure P(x, p, t)dxdp. The formal difference between the classical and the quantum description, is much less obvious, when classical stochastic processes are compared with quantum mechanical ones. In both cases the physical observables are determined only in a probabilistic sense:

$$< b^\dagger b > \;=\; \mathrm{tr}\, \rho(t)\, b^\dagger b \tag{16}$$

and

$$< x\, p > \;=\; \int x\, p\, P(x,p,t)\, dxdp \tag{17}$$

the mathematical concepts for calculating those averages, however, are widely different. It was first demonstrated by Wigner [5] that it is possible to cast quantum mechanics into a probabilistic formalism using a c-number distribution. In connection with quantum optics a similar concept has been developed by Glauber [6], based on the coherent state representation of the field, and which is taylor made for problems in quantum optics. Since these "probabilities" are not necessarily positive, it is safer to use the term quasi-probabilities.

In classical statistics, a useful tool is the characteristic function and the question arises if such a c-number concept exists also for the quantum case. It seems natural to try the following replacement: $P(x.p,t) \rightarrow \rho(t)$, $x,p \rightarrow b^\dagger, b$, and $\int dxdp \rightarrow \mathrm{tr}$ which leads to the following definition [7, 8]:

$$\chi_q(\beta,\beta^*) \;=\; \mathrm{tr}\, \rho(t)\, e^{i\beta^* b^\dagger}\, e^{i\beta b} \tag{18}$$

The quantum averages are now obtained in complete analogy to the classical ones by differentiation:

$$< b^\dagger\, b > \;=\; \frac{\partial}{\partial i\beta^*} \frac{\partial}{\partial i\beta}\, \chi(\beta,\beta^*) \Big|_{\beta=0,\beta^*=0} \tag{19}$$

but now on the basis of a pure c-number formalism. Since the classical characteristic function χ is the Fourier transform of the probability density P, we can always reconstruct the probability by the inverse transformation. Just for curiosity, one could be tempted to use the same idea also in the quantum case, by defining:

$$P(\alpha,\alpha^*,t) \;=\; \int e^{-i\beta^*\alpha^* - i\beta\alpha}\; \chi(\beta^*,\beta,t)\; d\beta d\beta^* \tag{20}$$

While the inverse transformation of χ merely undoes the previous Fourier-transform, it certainly could not return the density operator in the quantum case - since the Fourier transform of a c-number remains a c-number. So what physical meaning is behind this intuitively constructed function $P(\alpha,\alpha^*)$? It seems natural to try:

$$< \alpha^*\alpha > = \int \alpha^*\alpha \, P(\alpha,\alpha^*,t) \, d\alpha d\alpha^* \tag{21}$$

and it is easily verified that this quantity is identical to the quantum mechanical average:

$$< \alpha\alpha^* > \;=\; < b^\dagger b > \tag{22}$$

and more generally $< \alpha^n \alpha^{*m} > = < b^{\dagger m} b^n >$. At this point we have constructed a pure c-number formalism for the description of dissipative quantum dynamics - quite analogous to classical statistical mechanics.

As a matter of principle, a c-number formalism is unable to distinguish between different operator orderings. For such a concept to work, it is necessary to agree on an ordering convention like normal, antinormal or symmetric. The corresponding quasiprobabilities are the Glauber P- , the Wigner W- or the Q-function. The three established functions are merely special cases of the generalized quasiprobability [9]:

$$Z(\alpha,\alpha^*,t,\epsilon) = \int e^{-i\beta^*\alpha^*-i\beta\alpha} \, \chi(\beta,\beta^*,t) \, e^{-\epsilon\beta\beta^*} \tag{23}$$

or, using Glauber's P-function

$$Z(\alpha,\alpha^*,t,\epsilon) \;=\; \frac{1}{\pi\epsilon}\int \exp(-|\alpha-\beta|^2/\epsilon\,) \, P(\beta,\beta^*,t) \, d^2\beta \tag{24}$$

which contains the three familiar distributions as limiting cases :

for $\epsilon = 0$:	$Z(\alpha,\alpha^*)$	=	$P(\alpha,\alpha^*)$
for $\epsilon = 1/2$:	$Z(\alpha,\alpha^*)$	=	$W(\alpha,\alpha^*)$
for $\epsilon = 1$:	$Z(\alpha,\alpha^*)$	=	$Q(\alpha,\alpha^*)$

Before we discuss special physical problems in terms of quasi-probabilities in the next chapter, it may be interesting to emphasize some general features. A classical harmonic oscillator with damping comes to rest, when no external forces are applied. As a result its coordinate and its momentum vanishes as $t \to \infty$. If noise is included, phase sensitive moments still vanish, but phase-invariant ones remain finite. Similarly, for a damped quantum oscillator, all normally ordered moments vanish in the long time limit, since the oscillator relaxes into the vacuum. Therefore, the corresponding Fokker-Planck equation for $P(\alpha,\alpha^*)$ has no diffusion term, and conceptually there is no noise. However, some moments calculated with the Wigner-or the Q- representation must relax toward finite values, since $< b\, b^\dagger >$ is finite in the vacuum. With a dissipative equation nonvanishing asymptotic averages are only obtained if noise is included: consequently, the Fokker-Planck equations for W and Q must always contain a diffusion term.

Deterministic processes are represented by δ-distributions, stochastic ones by regular probability functions. Since the strength of noise increases from P through W to Q, one expects Q to display the smoothest dependence, while P may be rather singular. It can be shown in general [9] that W and Q always exist as regular functions, while $P(\alpha,\alpha^*)$ may contain δ-functions and its derivatives of arbitrary order. Unfortunately, those are just the physical problems that are most interesting, since it is exactly this singular behavior and the non-positivity that distinguishes quantum processes from classical ones.

4. Physical Examples

It may now be the time to illustrate the general ideas discussed in the previous chapters on the basis of a number of physical examples. Each model will be characterized by a Hamiltonian for the modes of the light field only, and will contain dissipative interactions that either cause a random energy flow into the system, or take energy out by means of dissipation. By adding these dissipative terms

$$\left(\frac{\partial\rho}{\partial t}\right)_{irr} = \begin{cases} \text{dissipative loss}: & \Gamma_1\,[\,b, \rho\, b^\dagger\,] + \Gamma_1\,[\,b\,\rho, b^\dagger\,] \\ \text{irreversible gain}: & \Gamma_2\,[\,b^\dagger, \rho\, b\,] + \Gamma_2\,[\,b^\dagger\rho, b\,] \end{cases} \tag{25}$$

to the otherwise reversible evolution equation one obtains an irreversible dynamic quantum process. All linear reservoir interactions can be cast into this general form [10], and the various possible physical mechanisms behind these interactions determine only the parameters Γ_1,Γ_2.

We begin with the simple, but illustrating example of a nonlinear oscillator with damping and gain. It will become obvious that what is called noise in quantum mechanics varies from representation to representation, and it may be interesting to note that the same physical problem can be described by quite different equations. In the paragraph to follow we present the basic quantum optical problem: the laser, but with a generalized absorptive reservoir. This example is chosen to demonstrate the competition of the "quasi-classical" spontaneous emission noise and the non-classical noise of a squeezed reservoir. In the last paragraph we present a model of a laser combined with a nonlinear medium into a self-frequency converting device. The question to be answered is: Can such a system produce non-classical light in the presence of spontaneous emission?

4.1 The Nonlinear Oscillator with Dissipation and Gain

In terms of creation and destruction operators for the harmonic oscillator, the simplest nonlinear model that one can think of is described by the following Hamiltonian:

$$H = \hbar\omega b^\dagger b + \hbar g\, b^\dagger b^\dagger b\, b \tag{26}$$

which certainly has a rather unusual appearance in terms of the canonical variables x,p. The master equation we start from is of the general form of eq.(15) and turns into a partial differential equation, when we use one of the representations i.e. P, W, Q or more conveniently Z. The master equation for ρ turns into the following partial differential equation for $Z(\alpha,\alpha^*)$:

$$\begin{aligned}\frac{\partial Z(\alpha,\alpha^*,t,\epsilon)}{\partial t} = \; & \frac{i\partial}{\partial\alpha}(\omega - 2g\epsilon - i\gamma + id + 2g|\alpha|^2)\alpha\, Z \\ & - \frac{i\partial}{\partial\alpha^*}(\omega - 2g\epsilon + i\gamma - id + 2g|\alpha|^2)\alpha^*\, Z \\ & + 2\,(\epsilon\gamma + (1-\epsilon)d)\,\frac{\partial^2}{\partial\alpha\partial\alpha^*} Z \\ & + ig(2\epsilon - 1)\left[\frac{\partial^2}{\partial\alpha^2}\alpha^2 - \frac{\partial^2}{\partial\alpha^{*2}}\alpha^{*2}\right] Z \\ & + 2ig\epsilon(\epsilon - 1)\left[\frac{\partial^3}{\partial\alpha^2\partial\alpha^*}\alpha^* - \frac{\partial^3}{\partial\alpha^{*2}\partial\alpha}\alpha\right] Z \end{aligned} \tag{27}$$

In complex notation the structure and therefore the interpretation of the second derivatives is not immediately obvious. Therefore we rewrite them in terms of real and imaginary parts: $\alpha = x + iy$, $\alpha^* = x - iy$. While :

$$\partial^2/\partial\alpha\partial\alpha^* = 1/4\,(\partial^2/\partial x^2 + \partial^2/\partial y^2) = 1/4\,\Delta$$

has the form of a classical diffusion term (28)

$$\partial^2/\partial\alpha^2 + \partial^2/\partial\alpha^{*2} = 1/2\,(\partial^2/\partial x^2 - \partial^2/\partial y^2)$$

by itself has no stochastic interpretation.

Glauber's P representation i.e. $\epsilon=0$, involves classical fluctuations only in the form of "gain noise", while loss is not associated with a diffusion term. For $\epsilon=0$ also the third order derivatives vanish. If it were not for the nonlinearity $\propto g$, which is responsible for the second derivatives with a non-positive diffusion matrix, the equation would have the form of a classical stochastic process. Due to the nonlinearity the non-positive diffusion terms will in course of time turn the P function into a highly singular object that is hard to treat analytically, but which could never be handled numerically.

For $\epsilon = 1$ we obtain the evolution equation for the Q-function. Here the classical noise that is generated solely by the interaction with the dissipative reservoir, and which guarantees e.g. that in the long time limit $\langle \alpha^* \alpha \rangle = \langle b b^\dagger \rangle \neq 0$. The third order derivatives also vanish in this case. As a consequence, the only non-classical features come from the same terms as in the Glauber representation. The fact that reservoir noise as well as quantum noise from the nonlinear interaction contribute both to the "diffusion terms" is typical for the P- and Q-representations. This is not the case for the Wigner-distribution.

The equation for the Wigner-distribution is obtained for $\epsilon = 1/2$. Here the second order derivatives represent pure diffusive behavior and contain only the loss and gain noise - in symmetric form. Non-classical behavior in this case follows only from the cubic derivatives. From a mathematical point of view this is a rather unpleasant observation, since it makes the analytical solution of such an equation extremely difficult. On the other hand, when the quantum features dominate, W still exists as a regular function which may be found numerically.

The question may arise what the typical features of quantum noise really are, or why after all do we associate noise with cubic derivatives or the "non-positive diffusion terms". Obviously, they cannot be simulated by random forces. But nevertheless they have something in common with classical noise. For example, let us imagine that the point x=0 is a stationary, unstable point of a deterministic problem. Then in the purely deterministic case $x^n(t) \equiv 0$ for all times if $x(t=0) = 0$ initially. In the presence of classical noise, however, $< x^2(t) >$ will not vanish, and the coordinate will start to evolve even if we start from the same initial state $x(t=0)=0$. This, in the picture of the Fokker-Planck equation, is caused by the presence of second order derivatives. However, the same could be said also if the diffusion matrix would be non-positive, as it is the case for the quantum processes. Therefore, the non-classical terms in the master-equation nevertheless can "kick" a particle off a potential hill in the same way as it is done by classical random forces. The third derivatives in principle have the same effect only in higher orders. What makes these processes differ from the classical Fokker-Planck dynamics is the fact that the P- as well as the W-distributions must not remain positive, even if they started that way initially. A negative probability, however, is impossible in classical statistics and therefore the quantum system can exhibit features that can never be obtained by classical noise.

4.2 Laser with a Squeezed Vacuum

The laser is driven far from thermal equilibrium, by the interaction with two reservoirs. The 'hot' reservoir represents the ensemble of inverted atoms, supplying the energy for the lasing mode, whereas the 'cold' reservoir comprises all dissipative mechanisms - especially the loss through the laser mirrors. In the standard laser model both reservoirs are prepared in simple stationary states. The absorbing 'cold' reservoir is modeled by an ensemble of harmonic oscillators in thermal equilibrium, while the 'hot' reservoir is formally identical, only with creation and annihilation operators interchanged. These reservoirs introduce fluctuations which are practically indistinguishable from classical noise, and this is the reason why the laser can be understood more - or less - in classical terms. In the present context it seems worthwhile to study generalized laser models, where the reservoirs are prepared in a non-classical state. Then we may investigate the question: How are the statistical properties of laser light determined by the properties of the reservoirs. One way of modification would be to prepare the 'cold' thermal reservoir in a squeezed vacuum state, instead of the regular vacuum as in the usual laser theory. It is also conceivable to prepare the atomic ensemble in a non-classical equilibrium state [11]. However, since the latter seems quite difficult to achieve experimentally, we will consider here the first example only. A discussion of the squeezed atomic reservoir will be published separately [12] . Since a squeezed vacuum has modified noise properties depending on the phase of the field, it was argued that this should influence the phase diffusion rate of the laser field and in particular it should reduce the laser linewidth [13].

Here we want to investigate the statistical properties of such a laser in detail. Sect. 4.2.1 presents the modified model along with the corresponding Fokker-Planck equation in P representation. In Sect. 4.2.2 the stationary solution is derived and discussed for various values of the pump and the squeezing parameter. The stationary noise properties are illustrated through a set of 3D plots of the stationary probability density P, below, at and above the laser threshold. These figures demonstrate how the phase invariant laser gain competes with the squeezing tendency of the non-classical reservoir. In Sect. 4.2.3 we investigate the dynamic aspects of the model by solving the eigenvalue problem of the Fokker-Planck process. The non-trivial eigenvalues characterize the transient decay of spontaneous fluctuations and determine the linewidth of the laser. Since phase symmetry is destroyed, the degeneracy of the traditional laser model is lifted and the eigenvectors are no longer related by simple symmetry transformations. In Sect. 4.2.4 the group theoretical aspects of the model are discussed in order to classify the set of eigenfunctions.

4.2.1 The Model

Convenient and systematic methods for deriving a master equation for the density operator of a system coupled to various reservoirs were developed by Haken [7] and by Nakajima and Zwanzig [14]. In Born-Markov approximation the dynamics on the reduced Hilbert space is determined by low order moments of the bath variables. A suitable choice for the reservoir is an ensemble of harmonic oscillators, described by creation and annihilation operators b_k and $b_k^\dagger$. k is the wave number of an individual mode. For the reference state of the bath we now take the squeezed vacuum state. Such a field can be generated experimentally from noise by a parametric amplifier in the unsaturated mode of operation. The state of the reservoir is characterized by the density matrix:

$$R = \Pi\, \rho_k \quad \text{and} \quad \dot{\rho}_k = D_r(k)|0\rangle\langle 0|\, D^{-1}{}_r(k) \tag{29}$$

where the squeezing operator is defined as:

$$D_r(k) = \exp\{\, r\,(-\, b^\dagger_{q+k}\, b^\dagger_{q-k} + b_{q-k}\, b_{q+k}\,)\,\} \tag{30}$$

q is the wave vector of the resonant mode, and the parameter r is a measure of the amount of squeezing imposed. The ensemble averages needed in the derivation of the master equation are:

$$\begin{aligned} < b_k^\dagger\, b_{k'} > &= \delta_{k,k'} \sinh^2(r) \\ < b_k\, b_{k'}^\dagger > &= \delta_{k,k'} \cosh^2(r) \\ < b_k^\dagger\, b_{k'}^\dagger > &= -\, \delta_{k',\, 2q-k} \sinh(r)\cosh(r) \\ < b_k\, b_{k'} > &= -\, \delta_{k',\, 2q-k} \sinh(r)\cosh(r) \end{aligned} \tag{31}$$

After specifying the statistical properties of the reservoir, application of the elimination formalisms is a straight forward task. In the present case it leads to the following irreversible contribution to the master equation:

$$\begin{aligned} \dot{\rho}_{irr} = &-\, C \sinh(r)\cosh(r)\, \{[a^\dagger, \rho a^\dagger\,] + [a^\dagger \rho, a^\dagger\,]\} \\ &-\, C \sinh(r)\cosh(r)\, \{[a, \rho a] + [a\rho, a]\} \\ &+\, C \cosh^2(r)\, \{[a, \rho a^\dagger\,] + [a\rho, a^\dagger\,]\} \\ &+\, C \sinh^2(r)\, \{[a^\dagger, \rho a] + [a^\dagger \rho, a]\} \end{aligned} \tag{32}$$

where $a, a^\dagger$ are the annihilation and creation operator of the lasing mode and C is the coupling constant of the mode and the 'cold' reservoir. The master equation for the reduced density matrix can now be transformed by a systematic and well-known algorithm into a Fokker-Planck equation for the Glauber P-function [6,9]:

$$\frac{\partial P(\alpha,\alpha^*,t)}{\partial t} = -\frac{\partial}{\partial \alpha}\{A - C\cosh^2(r) + C\sinh^2(r) - B|\alpha|^2\}P$$
$$-\frac{\partial}{\partial \alpha^*}\{A - C\cosh^2(r) + C\sinh^2(r) - B|\alpha|^2\}P \tag{33}$$
$$+ C\cosh(r)\sinh(r)\left\{\frac{\partial^2}{\partial \alpha^2} + \frac{\partial^2}{\partial \alpha^{*2}}\right\}P$$
$$+ \{2A + 2C\sinh^2(r)\}\frac{\partial^2}{\partial \alpha \partial \alpha^*}P$$

The terms resulting from the 'hot' reservoir, i.e. the laser gain A and the nonlinear saturation $-B|\alpha|^2$, have just been borrowed from the usual laser theory, since the interaction with the atoms was left unchanged. In order to reduce the number of free parameters and to simplify comparison with the traditional laser model, time and field amplitude will be rescaled. When combining the different terms and decomposing the field into its real and imaginary part:

$$\alpha = x + iy \quad \text{and} \quad \alpha^* = x - iy$$

the Fokker-Planck equation reduces to the following form [15, 16]:

$$\frac{\partial P(x,y,t)}{\partial t} = -\frac{\partial}{\partial x}\{a - (x^2+y^2)\}x\,P$$
$$-\frac{\partial}{\partial y}\{a - (x^2+y^2)\}y\,P \tag{34}$$
$$+\frac{\partial^2}{\partial x^2}\{1 + b\sinh^2(r) + b\sinh(r)\cosh(r)\}P$$
$$+\frac{\partial^2}{\partial y^2}\{1 + b\sinh^2(r) - b\sinh(r)\cosh(r)\}P$$

where the scaled parameters are:

$$a = (A-C)\left(\frac{AB}{2}\right)^{-1/2} \quad \text{and} \quad b = \frac{C}{A}$$

In a realistic experimental situation the parameter B is small compared with A, typically $B/A < 10^{-3}$, and since we are primarily interested in the threshold region $a = 0$ or $A \simeq C$, the additional parameter b is of order one. The numerical solution, however, is not restricted to these parameters, but for the purpose of this paper we will focus only on the realistic parameter range where b is slightly below 1.

4.2.2 Steady State Solutions

Since we are dealing with a stationary stochastic process, the Fokker-Planck equation can be treated as an eigenvalue problem supplemented by suitable boundary conditions. By the following separation ansatz:

$$P(x, y, t) = e^{-\lambda t} P'(x, y)$$

one arrives at an elliptic non-hermitian eigenvalue problem. A stationary stochastic process has a time independent solution which corresponds to the eigenvalue $\lambda = 0$. In case of detailed balancc, this solution can be obtained in closed analytical form. Due to the phase dependence of the noise in the present case, detailed balance is lost and the solution, even for the steady state, must be obtained by numerical methods [16]. An approach, very suitable for this purpose is the method of matrix continued fractions [2]

The results of numerical integration are summarized in Fig. 1, where the stationary P distribution is plotted for various values of the pump parameter and for moderate squeezing $r = 1$. For comparison we have included the same results for the standard laser i.e. $r = 0$ in the same figure for the same values of the pump parameters a. Below threshold ($a = -5$) both distributions are centered at the origin and the influence of squeezing on the shape of the distributions is clearly visible. Note that the squeezing ellipse of the cavity mode is rotated by 90 degrees against the squeezing direction of the 'cold' bath, in agreement with an earlier result obtained by Stenholm. Above threshold there is competition between squeezing and the phase invariant laser gain. While the squeezed fluctuations have the tendency to distribute noise unevenly over the quadratures, spontaneous emission noise attempts to establish phase invariance. But even for high pump parameters ($a = 5$) the influence of squeezing is still seen in a deformation of the distribution. It might be interesting to note that the most probable values of the distribution have again rotated back to the preferred phase of the reservoir. For $r > 0$ the condition of detailed balance is lost, which leads to a non vanishing probability current even in steady state. Since the flow of probability generally has no sources and sinks, this current is divergence free and must have the form of a vortex field. It is a straight forward task to obtain the current from the steady state distribution [16].

4.2.3 Dynamic Properties

The information about the dynamic properties of the model is contained in the non-trivial eigenvalues $\lambda_n \neq 0$ and the corresponding eigenfunctions of the forward and backward Fokker-Planck equation denoted by P_n and Q_n respectively. Provided all λ_n ,P_n and Q_n are known explicitly one can calculate all statistical properties of the process, like the transient decay of statistical moments or of the stationary correlation functions:

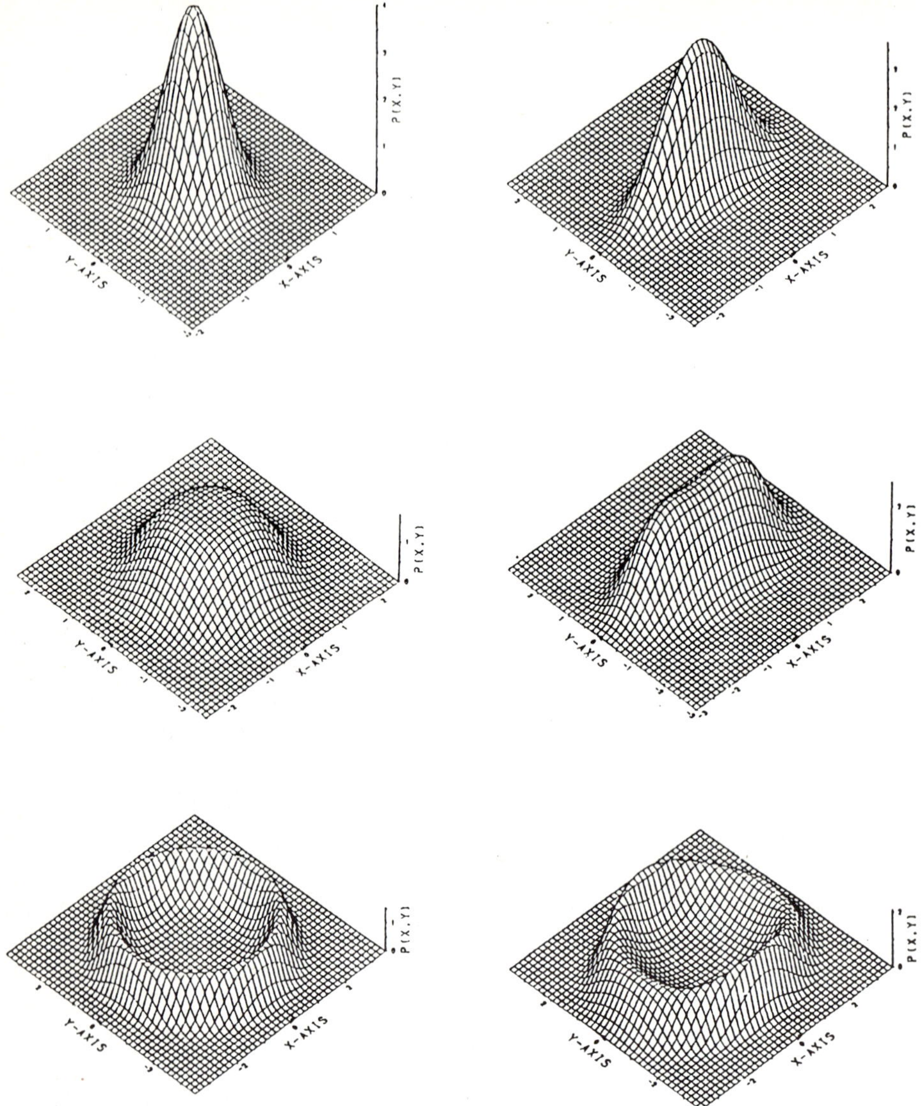

Fig. 1 Steady state distribution for the laser. Left side with regular vacuum (r=o), right side with squeezed vacuum (r=1). The three figures represent three pump conditions: below (a=-4), at (a=0) and above threshold (a=4)

$$< E^{*n}(t)\, E^{m}(t) > =$$

$$\sum_{j=0}^{\infty} \int_{0}^{\infty} (x+iy)^{n}(x-iy)^{m}\, P_j(x,y)\, P_j(x_0,y_0)\, /P_0(x_0,y_0)\, e^{-\lambda_j t}\, dxdy \qquad (35)$$

and

$$G(t) = < E^{*}(t)\, E(t=0) > - < E >^{2} =$$

$$\sum_{j=1}^{\infty} \int\int (x+iy)(x_0-iy_0)\, P_j(x,y) Q_j(x_0,y_0)\, P_0(x_0,y_0)\, e^{-\lambda_j t}\, dxdy\, dx_0 dy_0 \quad (36)$$

Both dynamic results contain basically the same physical information, only in slightly different form. From an experimental point of view the stationary correlation functions are more useful, since they do not require an artificial preparation of the initial state. In leading order, the phase correlation function is given by:

$$G(t) = \int (x+iy)\, P_1(x,y)\, dxdy \int (x-iy)\, Q_1(x,y)\, P_0(x,y)\, dxdy\, \exp(-\lambda_1 t) \quad (37)$$

Its Fourier transform is a Lorentzian centered around the resonance frequency of the cavity. The 'higher' eigenfunctions modify the lineshape around the center, but leave the wings practically unchanged. In the threshold region, the modifications are typically of the order of 10 %. Correlation functions of the field provide the only systematic way for deriving the laser linewidth.

In Fig 2. the first eigenvalue is plotted as a function of the squeezing parameter r. For $r = 0$ the result for the usual laser is obtained. For $r \neq 0$ the degeneracy is lifted and one eigenvalue rises with increasing r, while the other one is slightly decreased, before it also rises with increased squeezing. But even the slight reduction in one eigenvalue does not indicate necessarily a slight reduction in the linewidth as well, since this eigenvalue carries only part of the statistical weight.

For small r the eigenvalue which rises from the beginning carries statistical weight of similar order. In any case, the effect is so minute that it is of hardly any practical relevance. Besides the region of moderate squeezing, the linewidth generally increases with increasing r.

The corresponding eigenfunctions are shown in Fig.3. For $r = 0$ the spectrum is degenerated and there exist two linearly independent eigenfunctions, rotated by an angle of 90 degrees. The first excited state of the laser is plotted in Fig.3a for r=0. When squeezing is applied, the rotational symmetry is broken and the eigenfunctions are no longer related by a simple symmetry transformation. For $r = 1$ the results are shown in

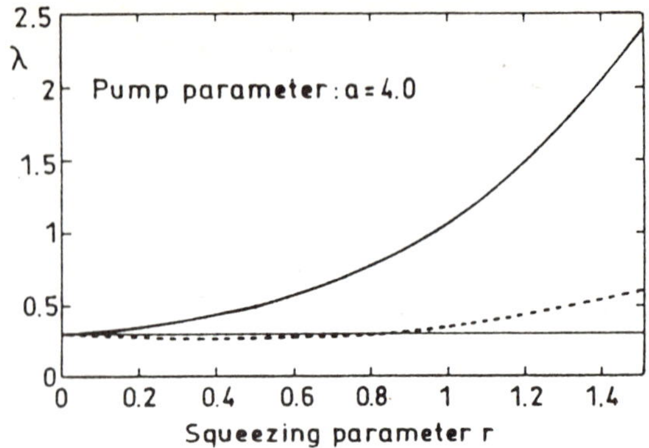

Fig.2 The smallest eigenvalues as a function of the squeezing paramenter r, well above threshold a=4. These rates determine the linewidth of the laser.

Fig.3b. The eigenfunctions corresponding to the next pair of eigenvalues are plotted in Fig.4b for r=1 and compared with the degenerate pair of functions for the usual laser in Fig.4a. In the case without squeezing, the pairs of degenerate eigenfunctions are related by simple symmetry transformations: $\lambda_1 \rightarrow$ rotation by 90 degrees, $\lambda_2 \rightarrow$ rotation by 45 degrees. This is no longer the case with squeezing. A group theoretical argument illustrates this property.

4.2.4 Symmetry

Due to the rotational invariance of the standard laser model, the eigenvalues are doubly degenerated. The symmetry transformations that leave the laser equation invariant form the continuous group $C_{\infty\nu}$ which contains rotations by an arbitrary angle, especially infinitesimal angles and inflections. The character table of this group is well known. From there we learn that this group has indefinitely many two-dimensional representations which are not equivalent. The different representations are labeled by an index k. The first two degenerated eigenfunctions plotted Fig.3a belong to the representation k = 1 . When squeezing is applied the symmetry of the model reduces to the symmetry of a rectangle. This group is isomorphic to the Klein-Group (KG) [16]. This group consists of four elements: the identity, reflection about the x-axis, reflection about the y-axis and a rotation about 180 degrees. The character table of this group is well known and shows that there are only four one dimensional representations. When the value of r is changed from zero to a finite

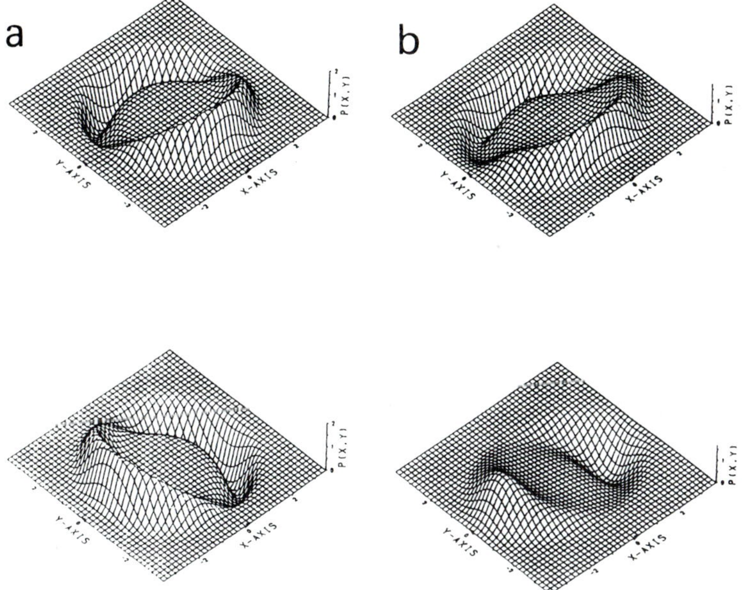

Fig.3 Degenerate pair of eigenfunctions (a, b) for the first nontrivial eigenvalue for r=0 λ=0.22 and for the laser with squeezing r=1, λ=0.25 and 0.81

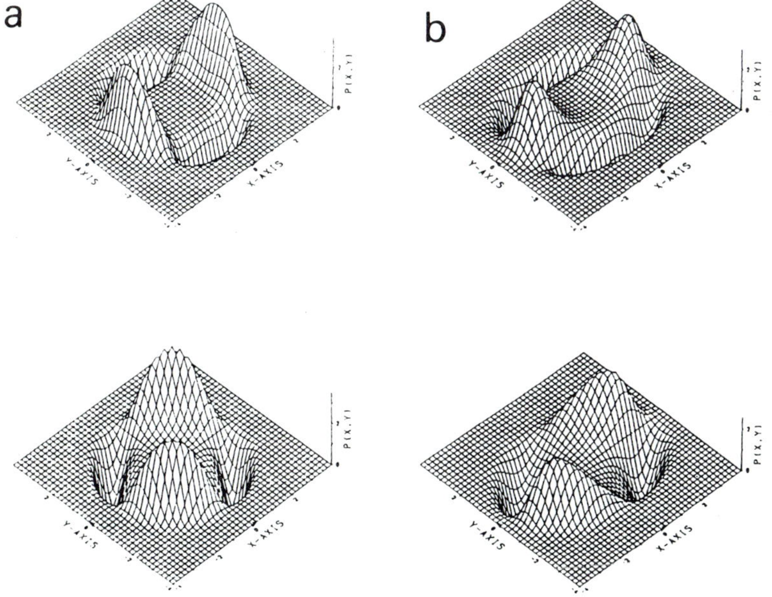

Fig.4 Degenerate pair of eigenfunctions (a, b) for the second eigenvalue for r=0 λ=0.87 and for the laser with squeezing r=1, λ=1.52 and 1.79

value, the two dimensional irreducible representations of $C_{\infty\nu}$ become reducible representation of the Klein-Group. With the help of the orthorgonality relations among the group characters it is possible to calculate in which irreducible representation of the Klein-Group a given representation of $C_{\infty\nu}$ decays. As an example, Fig.4a shows the degenerated eigenfunctions corresponding to the second eigenvalue of the standard model ($r = 0$), which belong to $k = 2$. For $r = 1$ the degeneracy is lifted again, and the two functions now belong to two different representations of the Klein-Group. The function in Fig.4b(top) belongs to D_1 and Fig.4b(bottom) belongs to D_2. Since all representations are one-dimensional it is easy to see if the eigenfunctions have the correct symmetry. The group theoretical arguments therefore provide an independent way to check the consistency of the numerical results.

4.3 The Self-Frequency-Converting Laser

Recently various optical systems have been suggested and demonstrated experimentally which should have the potential of creating non-classical light [17-20]. A squeezed vacuum state can be produced by a parametric oscillator near threshold [21]. A squeezed coherent state can be generated by frequency doubling in a χ^2 nonlinear crystal [22]. A very compact device could combine the lasing material as well as the nonlinear crystal in one doubly resonant cavity. Doping of the nonlinear crystal directly with the lasing material could produce a compact monolithic device which is capable to emit non-classical light. Such a model is also of interest from the theoretical point of view. By including the lasing process into the dynamical system, the laser noise is consistently included. This is in contrast to the usual models in nonlinear optics, where the driving field is simulated by a classical source. With respect to the nonclassical aspects this is an important improvement, since spontaneous emission noise from the laser source competes with the squeezing tendency of the parametric process. It is not a priori clear which noise source will dominate. The assumption of a classical deterministic light source breaks phase invariance. Including the lasing mechanism into the dynamics eliminates this nonrealistic artifact. In addition, the laser in this model is not a rigid or passive device. At the onset of parametric conversion the laser field providing the energy is depleted. This leads conceptually to an interesting nonlinear dynamic problem. Both processes, frequency up- and down-conversion, produce squeezed light in a certain range of operation [23,24]. Experimentally the two cases are quite different, but for demonstrating the basic ideas it is quite sufficient here to focus on only one of the processes.

4.3.1 The Self-Down-Converting Laser

Since the laser model in quasi-probability representation has already been introduced in the previous chapter, we construct the model directly in this representation. The lasing mode is α_2 while the down-converted mode is characterized by α_1. In combining the Haken Laser-Fokker-Planck equation with the quasi-probability equation of the parametric process [25-27] we arrive at the following equation of motion for $P(\alpha_1,\alpha_1^*,\alpha_2,\alpha_2^*,t)$ in scaled variables:

$$\begin{aligned}\frac{\partial P}{\partial t} = \; & -\left\{ \frac{\partial}{\partial \beta_1}[\, -\gamma_1\beta_1 + \chi\beta_1^*\beta_2\,] + \text{c.c.} \right\} P \\ & -\left\{ \frac{\partial}{\partial \beta_2}[\, (g-\gamma_2)\beta_2 - b|\beta_2|^2\beta_2 - \chi/2\; \beta_1^2\,] + \text{c.c.} \right\} P \\ & +\left\{ 2g \frac{\partial^2}{\partial\beta_2\partial\beta_2^*} + \chi/2 \left[\frac{\partial^2}{\partial\beta_1^2}\beta_2 + \text{c.c.} \right] \right\} P \end{aligned} \tag{38}$$

time was measured in units of the inverse damping rate of the lasing mode and the fields were scaled as

$$\alpha_i = \beta_i\,(b/\gamma_1)^{1/2}$$

b is the parameter of nonlinear saturation. The evolution equation contains nonlinear deterministic force in the drift term and two contributions with second order derivatives: a conventional noise term that represents spontaneous emission noise from the laser process and non-classical "noise" terms which originate from the nonlinear Hamiltonian of the parametric process. These two sources of noise are in competition and we expect that for small pump rates and weak fields the classical source will dominate, while far above threshold there is a chance that the non-classical fluctuations may win.

4.3.2 The Classical Dynamics

The classical evolution equations for vanishing noise are determined by the drift term of the "Fokker-Planck equation" (38) :

$$\begin{aligned} \alpha_1 &= -\alpha_1 + \kappa\alpha_1^*\,\alpha_2 \\ \alpha_2 &= f(|\alpha_2|^2)\alpha_2 - 1/2\;\alpha_1^2 \end{aligned} \tag{39}$$

where $f(z) = (\,g - \Gamma - z\,)$ and $\Gamma = \gamma_1/\gamma_2$

These equations of motion are invariant under the transformation

$$(\alpha_1,\alpha_2) \rightarrow (\alpha_1\exp(i\phi),\alpha_2\exp(2i\phi))$$

for an arbitrary phase angle ϕ. This phase invariance suggests the definition of a reduced set of variables:

$$\begin{aligned}
x_1 &= |\alpha_1|^2 \quad \text{or} \quad \alpha_1 = \sqrt{x_1}\,\exp(i\Phi_1) \\
x_2 &= |\alpha_2|^2 \quad \text{or} \quad \alpha_2 = \sqrt{x_2}\,\exp(i\Phi_2) \\
x_3 &= \mathrm{Re}(\alpha_1{}^2\alpha_2{}^*) \; = \; x_1\sqrt{x_2}\,\cos(2\Phi_1-\Phi_2)
\end{aligned} \tag{40}$$

Thereby the two equations for the complex fields α_1,α_2 reduce to three coupled equations for the real variables x_i, plus an equation for the phases Φ_1 or Φ_2 which decouples. The dynamics in the classical regime has the following fixed point structure [28]:

(i)	$g \leq \Gamma$	$x_1=x_2=x_3=0$
(ii)	$\Gamma \leq g \leq g_1$	$x_1=0,\ f(x_2) = 0,\ \ x_3 = 0,\quad g_1 = \Gamma + 1/\kappa^2$
(iii)	$g_1 \leq g \leq g_2$	$x_1 = x_2f(x_2),\ x_2 = 1/\kappa^2,\ x_3 = x_1/\kappa$
(iv)	$g \geq g_3$	$x_1 = 4x_2,\ f(x_2) = 4,\ x_3 = 4x_2/\kappa$

The two critical values g_2 and g_3 can be determined analytically, but are rather involved - they should be taken from the original paper [28]. Depending on the actual parameters of the system, g_2 can be either smaller or larger than g_3. In the latter case, the two regimes melt into one, while in the first case a gap opens between the two stability regions. In this gap no stationary solution exists, but the system is attracted by a limit cycle.

4.3.3 The Noise Spectrum

Since we are interested in noise properties on the level of vacuum noise, it is not sufficient to solve the quantum dynamics inside the cavity. In addition, one has to relate the properties inside the cavity to the outside world by the input-output concept of Collet and Gardiner [29,30]. This can be done in a straight forward manner as long as the dynamics is linear. In the full nonlinear domain this is a rather difficult task [31]. We will obtain a first insight into the problem by linearizing the dynamics about the classical fixed points. The equation of motion for $P(\alpha_1,\alpha_2,t)$ then is of the general form:

$$\frac{\partial P}{\partial t} = - \Sigma A_{j,k} \frac{\partial}{\partial q_j}(q_k P) + \frac{b}{2\gamma_1} \Sigma D_{j,k} \frac{\partial^2}{\partial q_j}\partial q_k\ P \quad (41)$$

where the drift and the diffusion matrices are taken at the fixed point values (10) of α_1 and α_2. The explicit expressions are quite complex and can be found in ref.[10]. The complex fields are separated in real and imaginary parts: $\alpha_1 = q_1 + iq_2$, $\alpha_2 = q_3 + iq_4$.

Since the outside world is not quantized in cavity modes, it is more realistic and closer to experiment to determine the spectrum of the emission field [29,32]. The spectral matrix is found directly from the coefficients of the "Fokker-Planck" equation:

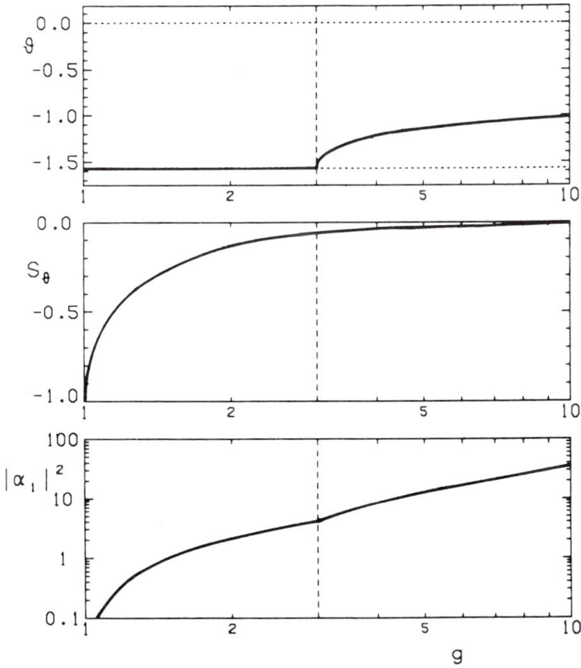

Fig.5 Characteristic laser parameters in the region at and above the parametric threshold as a function of the laser gain. From top to bottom: quadrature angle of maximum squeezing, squeezing at the optimal angle (100% squeezing at threshold), intensity of the subharmonic field.

$$S(\omega) = (\ A + i\omega 1\ )^{-1}\ D\ (\ A^T - i\omega 1\ )^{-1} \quad (42)$$

The matrix combines in an efficient way the two modes of the field as well as the two quadrature components. The spectrum of the down converted field with the quadrature phase angle θ is :

$$S_\theta(\omega) = 8\,\Gamma\,\{\cos^2\theta\,S_{11} + \sin^2\theta\,S_{22} + 2\sin\theta\cos\theta\,(S_{12} + S_{21})\} \tag{43}$$

while the noise in the lasing mode is determined by:

$$S_\theta(\omega) = 8\quad\{\cos^2\theta\,S_{33} + \sin^2\theta\,S_{44} + 2\sin\theta\cos\theta\,(S_{34} + S_{43})\} \tag{44}$$

These spectra are normalized such that $S_\theta = 0$ is the shot noise level and $S_\theta = -1$ corresponds to perfect squeezing.

In Fig.5 we show the properties which are related to squeezing: The quadrature angle θ for which optimal squeezing is realized, the amount of squeezing and the intensity of the sub-harmonic field, plotted as a function of the pump parameter g. The squeezing spectrum is shown in Fig.6. At threshold of sub-harmonic generation perfect squeezing is obtained at resonance. With increasing pump rate a coherent amplitude is established, but squeezing is reduced and maximum noise reduction occurs off resonance. However, for moderate brightness, large noise suppression is still achieved. In contrast to the case of second harmonic, nothing is gained by pushing the pump rate into the regime of the last stable fixed point.

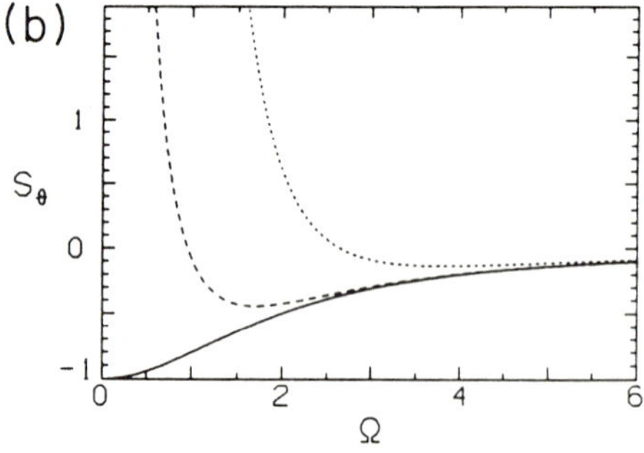

Fig 6. Noise spectrum for different pump rates g=1.0001 (close to threshold, solid line), g=1.2 (dashed line), g=2 (doted line). The quadrature angle was chosen for optimal squeezing.

5. Conclusions

While in many regions of physics randomness or fluctuations play a rather unimportant role, this is different in quantum optics, where, due to the extreme experimental sensitivity, noise is present in almost every measurement. In sophisticated experiments one will attempt to eliminate technical noise as much as possible. Thermal fluctuations play no role in the optical domain but are an important source of randomness in micro-wave experiments. Lowering the temperature into the mK regime will also eliminate the thermal photons. In open nonlinear systems, driven far from equilibrium, irreversible gain and loss mechanisms become important which are inevitably accompanied by noise. Spontaneous emission noise in lasers is a wellknown example, which is mathematically indistinguishable from thermal fluctuations. Nonlinear quantum systems exhibit, on top of all the different classical fluctuations, additional randomness which is of pure quantum origin. These fluctuations differ in their mathematical properties from classical noise. They can not be simulated by Langevin type random forces and can therefore not be associated with a classical diffusion process. Nevertheless, in common with the classical statistical forces, these fluctuations can probe the stability of a system and can trigger the decay of unstable states. Experimentally, the signature of quantum noise is detectable, since quantum fluctuations violate inequalities of classical statistics. The competition of classical and quantum noise in nonlinear dissipative systems is one of the most fascinating aspects of quantum optics.

References

1. R.Graham and A.Schenzle, Phys.Rev.A26, 1676 (1982)
2. H.Risken, "The Fokker-Planck Equation", (Springer Verlag Berlin 1984)
3. W.Feller, "An Introduction to Probability Theory and its Applications", (John Wiley, N.Y. 1968)
4. Ch.Ginzel and A.Schenzle, to be published.
5. E.P.Wigner, Phys.Rev.40, 749 (1932)
6. R.J.Glauber, Phys.Rev.Lett. 10, 84 (1963), Phys.Rev.131, 2766 (1963)
7. H.Haken in "Encyclopedia of Physics", XXV/2c, ed. by L.Genzel (Springer Verlag, Berlin 1969)
8. M.Hillery, R.F.O'Connel, M.O.Scully, and E.P.Wigner, Phys. Rep. 106, 121 (1984)
9. K.E.Cahill and R.J.Glauber, Phys.Rev.177, 1882 (1969)
10. P.N.Argyres and P.L.Kelley, Phys.Rev.A134,98 (1964)
11. M.A.Dupertuis, S.M.Barnett and S.Stenholm, J.Opt.Soc.Am.B,4, 1102 (1987) B,4 1124 (1987)
12. Ch.Ginzel and A.Schenzle, to be published
13. J.Gea-Banacloche, Phys.Rev.Lett. 59, 543 (1987)
14. R. Zwanzig, Physica 30, 1109 (1964)
15. Ch.Ginzel,J.Gea-Banacloche and A.Schenzle, Acta Phys. Pol.A78, 123 (1990)

16. Ch.Ginzel, R.Schack and A.Schenzle, J.Opt.Soc.Am.B8, 1704 (1991)
17. Y.Yamamoto, S.Machida and O.Nilson, Phys.Rev.A34, 4025 (1986)
18. R.E.Slusher, L.W.Hollberg, B,Yurke, J.C.Mertz and J.F.Valley, Phys.Rev.Lett.55, 2409 (1985)
19. T.Debuisschert, S.Reynaud, A.Heidmann, E.Giacobino and C.Fabre, Quantum.Opt.1, 3, (1989)
20. J.Mertz, A.Heidmann, C.Fabre, E.Giacobino and S.Reynaud, Phys.Rev.Lett.64, 2897 (1990)
21. L.A.Wu, H.J.Kimble, J.L.Hall and H.Wu, Phys.Rev.Lett.57, 2520 (1986)
22. A.Sizmann, R.J.Horowicz,G.Wagner and G.Leuchs, Opt.Comm.80, 138 (1990)
23. A.Sizmann, R.Schack and A.Schenzle, Europhys.Lett.13, 109 (1990)
24. D.F.Walls, M.J.Collet and A.S.Lane, Phys.Rev.A42, 4366 (1990)
25. P.D.Drummond, K.J.McNeil and D.F.Walls, Optica Acta, 27, 321 (1980)
26. P.D.Drummond, K.J.McNeil and D.F.Walls, Optica Acta, 28, 211 (1981)
27. M.Dörfle and A.Schenzle, Z.Phys.B 65, 113 (1986)
28. R.Schack, A.Sizmann and A.Schenzle, Pys.Rev.A43, 6303 (1991)
29. C.W.Gardiner and M.J.Collet, Phys.Rev.A31, 3761 (1985)
30. C.W.Gardiner, Quantum Noise in Springer Series of Synergetics Vol.56,1991
31. F.Kärtner, T.Langer, Ch.Ginzel and A.Schenzle, to appear in Phys.Rev.A
32. M.J.Collet and D.F.Walls, Phys.Rev.A32, 2887 (1985)

Part III

Fluid Dynamics and Solid State Physics

Pattern Formation in Fluids – Variational and Non-Variational Models

M. Bestehorn

Institute of Theoretical Physics and Synergetics,
University of Stuttgart, Pfaffenwaldring 57/4,
W-7000 Stuttgart 80, Fed. Rep. of Germany

Abstract. The synergetic theory of pattern formation allows the reduction of basic equations of motion to unified and simplified order parameter equations. For the particular case of a thermal instability of a fluid, these concepts are extremely fruitful and yield results that may be directly compared to experiments. In this article, we discuss pattern formation in several extended systems far from thermal equilibrium. We thereby distinguish between processes that are mainly ruled by a gradient dynamics of the order parameters and those that may show time dependence even in the long time limit. Special emphasis will be placed on oscillatory instabilities, pattern formation of fluids with low viscosity as well as the occurrence of convective structures in a rotating fluid layer. All these results show good agreement with recent experiments.

1. Synergetics and Pattern Formation

Pattern formation in non-equilibrium systems is surely one of the most active and fascinating fields of research in our time. Spatial patterns emerge as a result of spontaneous self-organization; the systems under consideration undergo certain phase transitions and bifurcations. Well defined control parameters allow external access and may drive the systems through critical points where qualitatively new patterns arise.

The synergetic theory of pattern formation [1-4] opens the door to a unified description far from thermal equilibrium, applying to systems in quite different fields, such as physics, chemistry, and biology.

The problems considered have the common feature, quite often of being composed of many subsystems and also possess the ability of self-organization on a macroscopic scale. This may lead to the evolution of patterns with a high degree of order in space and/or in time. The spatio-temporal behaviour of a system in the vicinity of a phase transition can very often be described by a few state variables which in analogy to Ginzburg-Landau theory of equilibrium thermodynamics are called *order parameters*. The large number of state variables assigned to the dynamics of the subsystems can be expressed by the order parameters in a unique way; the order parameters enslave the many degrees of freedom of the subsystems. The mathematical concept of synergetics allows a drastic reduction of the complexity of the systems under consideration by

Springer Proceedings in Physics, Vol. 69
Evolution of Dynamical Structures in Complex Systems
Editors: R. Friedrich · A. Wunderlin

elimination of the enslaved modes [5]. It thereby turns out that the resulting equations for the spatial and temporal behaviour of the order parameters have a similar form to the phenomenologically derived Ginzburg-Landau equations for phase transitions in thermal equilibrium. Beyond that, these *generalized Ginzburg-Landau equations* (GGLE) for phase transitions in open systems can be derived from basic physical laws, such as the Navier-Stokes equations [1, 2] or the Maxwell equations [1, 6] for hydrodynamic problems and for laser instabilities, respectively.

The present article is concerned with pattern formation in fluids forced by an externally applied, vertical temperature gradient, the so-called convection instability or Bénard problem (for a review see e.g. [7] and references therein). The temporal behaviour of pattern formation of this particular case may be divided roughly into two distinct groups:

1. Relaxational temporal evolution. After a certain initial transient phase a final, time independent state is reached asymptotically. These transients can take a very long time compared to the natural time scale of the convective system, the horizontal diffusion time of heat.

2. Fully time-dependent patterns. The pattern stays time dependent and reaches, after initial transients, a periodic, quasi-periodic, or even chaotic state.

It is an obvious step to use the unifying character of generalized Ginzburg-Landau equations to find an appropriate mathematical description of these two phenomena. The underlying basic hydrodynamic equations are not variational, i.e. may not be found by variation of a potential. On the other hand, when deriving the order parameter equations (OPE) it turns out that some of these may originate from a potential, at least under some approximations. If these approximations are violated or if non-potential forces determine the spatio-temporal evolution of the system, we may find time-dependent solutions. As examples we mention the inclusion of Coriolis forces as in the rotating convection cell [8] or the occurrence of a horizontal mean flow in convection of fluids with a low Prandtl number [9]. Both effects violate the variational character of the governing equations and additional, non-variational terms become necessary. It thereby turns out that spatial dislocations of the patterns play a crucial role and can mediate the non-relaxational behaviour.

To demonstrate the features of variational and non-variational terms in the generalized Ginzburg-Landau equations, the article is organized as follows: First we review some different experimental setups to study convection. A large number of experiments have been done in all of these fields in recent years. Then we give a description in terms of a potential theory, explaining pattern formation and selection of different symmetries (rolls, squares, hexagons). In the last part we shall consider non-variational effects. Special emphasis will be given to convection in binary mixtures, the effect of rotation and the inclusion of a mean flow for fluids with low viscosity. All these results will be compared with recent experimental findings.

2. Several Experiments Showing Convective Patterns

2.1 Convection of a pure fluid

Pattern formation in the field of hydrodynamic instabilities has attracted great experimental and theoretical interest since the experimental work of H.Bénard [10] on thermal convection in fluids at the beginning of the century (For a review see e.g. [11-13] and references therein).

The Bénard instability of a pure fluid (fig.1) is concerned with a homogeneous fluid layer contained between two horizontal plates and a uniform vertical temperature gradient. The external heating, which drives the system away from thermal equilibrium, induces an unstable density distribution of the fluid. At

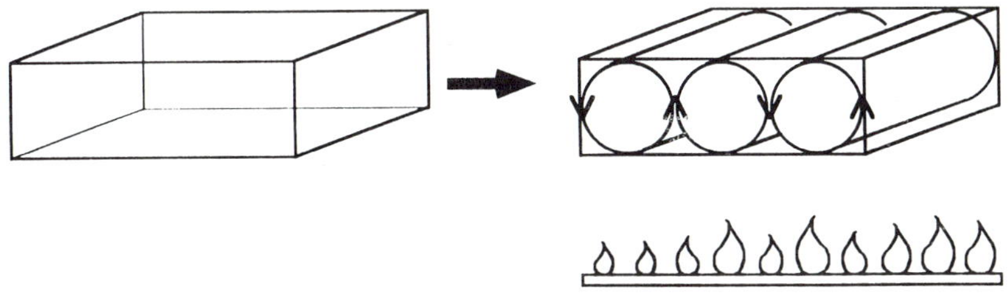

Figure 1: The Bénard problem. A fluid in a rectangular container is heated from below. The motionless heat conducting state (left) becomes unstable at a certain temperature gradient and a quite regular pattern of fluid motion occurs.

a certain critical temperature gradient, convection sets in in various forms of ordered regular patterns. Quite often, a regular pattern of rolls emerge. For poorly heat conducting boundaries a pattern of quadratic convective cells is observed.

A fluid with density ρ, viscosity η, and thermal conductivity κ is described by the velocity field $\mathbf{v}(\mathbf{r},t)$, the temperature field $T(\mathbf{r},t)$, the pressure $p(\mathbf{r},t)$ as well as a state equation for the density. The conservation laws for an incompressible fluid under the influence of an externally applied gravitational acceleration g read:

$$\begin{aligned}
\rho(\mathbf{r},t)\{\partial_t \mathbf{v}(\mathbf{r},t)+\mathbf{v}(\mathbf{r},t)\cdot\nabla\mathbf{v}(\mathbf{r},t)\} &= \rho(\mathbf{r},t) g \mathbf{z}_0 \\
&\quad -\nabla p(\mathbf{r},t)+\eta\Delta\mathbf{v}(\mathbf{r},t) \\
\nabla\cdot\mathbf{v}(\mathbf{r},t) &= 0 \\
\partial_t T(\mathbf{r},t)+\mathbf{v}(\mathbf{r},t)\cdot\nabla T(\mathbf{r},t) &= \kappa\Delta T(\mathbf{r},t),
\end{aligned} \tag{1}$$

where $\mathbf{z}_0$ is the unit vector in vertical direction. In the Boussinesq approximation the variation of the density ρ is neglected except for the external force term, where it results in buoyancy effects. In this term a linear variation of density with the temperature is assumed:

$$\rho(T) = \rho_0[1 - \alpha(T(\mathbf{r},t) - T_0)], \tag{2}$$

where T_0 is the temperature at the bottom plate and the thermal expansion coefficient is denoted by α.

For the velocity we make the usual decomposition into a toroidal and a poloidal part, represented by two scalar fields:

$$\mathbf{v}(\mathbf{r},t) = \nabla \times \{\phi(\mathbf{r},t)\mathbf{z}_0\} + \nabla \times \nabla \times \{\psi(\mathbf{r},t)\mathbf{z}_0\}. \tag{3}$$

Introducing the variation $\Theta(\mathbf{r},t)$ of the temperature from the basic linear temperature profile and eliminating the pressure by forming the curl and twice the curl of the Navier-Stokes equations one arrives at the following set of evolution equations:

$$\begin{aligned}
\{\Delta - \frac{1}{Pr}\partial_t\}\Delta\Delta_2\psi(\mathbf{r},t) &= -R\Delta_2\Theta(\mathbf{r},t) \\
&\quad -\frac{1}{Pr}\{\nabla \times \nabla \times (\mathbf{v}(\mathbf{r},t)\cdot\nabla\mathbf{v}(\mathbf{r},t))\}_z \\
\{\Delta - \frac{1}{Pr}\partial_t\}\Delta_2\phi(\mathbf{r},t) &= -\frac{1}{Pr}\{\nabla \times (\mathbf{v}(\mathbf{r},t)\cdot\nabla\mathbf{v}(\mathbf{r},t))\}_z \\
\{\Delta - \partial_t\}\Theta(\mathbf{r},t) &= \Delta_2\psi(\mathbf{r},t) + \mathbf{v}(\mathbf{r},t)\cdot\nabla\Theta(\mathbf{r},t),
\end{aligned} \tag{4}$$

with the Prandtl number $Pr = \eta/\rho_0\kappa$ and the horizontal Laplacian $\Delta_2 = \partial_{xx} + \partial_{yy}$. Time and length are scaled by the vertical diffusion time κ/d^2 and the layer depth d, respectively, and all quantities are dimensionless. The control-parameter R, the Rayleigh number, is given by:

$$R = -\frac{\rho_0 g\alpha\beta d^4}{\kappa\eta}.$$

R is proportional to the temperature gradient β. Assuming vanishing velocity components on the boundaries (rigid boundary conditions), (3) leads to

$$\phi(\mathbf{r},t) = \psi(\mathbf{r},t) = \partial_{\mathbf{n}}\psi(\mathbf{r},t) = 0 \tag{5}$$

for $\mathbf{r}$ on the horizontal and vertical walls and $\mathbf{n}$ perpendicular. On the lateral walls, we have in addition

$$\partial_{\mathbf{n}}\phi(\mathbf{r},t) = 0, \qquad \Delta_2\psi(\mathbf{r},t) = 0. \tag{6}$$

General boundary conditions for the temperature field can be expressed in the form

$$\partial_z\Theta(\mathbf{r},t) = \pm Bi\Theta(\mathbf{r},t) \qquad \text{for} \quad z = 0,1,$$

where Bi is another dimensionless parameter, the so-called Biot number, standing for the ratio of the thermal conductivity of the corresponding wall and to those of the fluid. A perfectly conducting boundary corresponds to $Bi \to \infty$, a poorly heat conducting boundary to $Bi << 1$.

2.2 Non-Bousssinesqian and surface tension effects

Hexagonal cells may be obtained if fluids with so-called non-Boussinesqian material properties are studied. These properties can become important if one uses a very thin layer, and, as a consequence, a large temperature gradient along the cell. Then the viscosity and thermal conductivity also vary with the temperature and higher order contributions to (2) have to be taken into account (see e.g. [14]):

$$\rho(T) = \rho_0[1 - \alpha(T(\mathbf{r},t) - T_0) - \alpha'(T(\mathbf{r},t) - T_0)^2]. \tag{7}$$

Hexagonal convective patterns are also formed if the instability is not dominated by buoyancy but by the temperature dependence of the surface tension of the fluid at a free boundary [11]. This so-called Marangoni effect dominates the instability in thin fluid layers or in microgravity experiments. Surface tension effects are usually included by a fourth dimensionless number, the Marangoni number:

$$M = \frac{\gamma \alpha d^2}{\rho \nu \kappa},$$

where the coefficient γ describes the linear dependence of the surface tension on the temperature along the surface [15]. On the upper surface one obtains a boundary condition that links the velocity field to the temperature (see e.g. [15, 16]). In terms of the variables Θ, ψ, and ϕ they read:

$$\partial_z \phi(\mathbf{r},t) = 0, \qquad \psi(\mathbf{r},t) = 0, \qquad \partial_z^2 \psi(\mathbf{r},t) = M\Theta(\mathbf{r},t) \tag{8}$$

at $z = 1$.

2.3 Binary mixtures

The instability of the heat conductive state in the Bénard experiment and the Bénard-Marangoni instability has a non-oscillatory character: a slight displacement of a fluid particle simply tends to increase in the course of time. A convective instability with an oscillatory character can be observed in the case of a fluid consisting of two miscible components like for instance a water-ethanol mixture. An oscillatory behaviour of the instability is expected to occur if instead of a simple increase of an initial displacement a temporal oscillation around the initial position of the fluid particle sets in. In a binary fluid mixture this may happen due to the Soret effect. Experiments on the onset of convection in binary fluid mixtures reveal an astonishingly complex spatio-temporal behaviour of the flow already close to onset [17]. The fluid motion consists of convection rolls which move in horizontal direction forming traveling waves. Usually there exist several wave trains which interact and behave chaotically in time and simultaneously exhibit irregular spatial patterns [18]. In another parameter region, the instability is again non-oscillatory; nevertheless the non-linear mechanism of pattern selection may tend to an oscillation of two perpendicular sets of rolls, reflecting the symmetry of the layer, e.g. in a square or circular vessel [19-21].

If the fluid is a mixture of two miscible components the relative concentration $C(\mathbf{r},t)$ of the two components has to be taken into account as an additional state variable. Density variations due to inhomogeneous concentrations can be regarded as a second mechanism for instability. Consequently, (2) has to be extended to the form

$$\rho(T,C) = \rho_0[1 - \alpha(T(\mathbf{r},t) - T_0) - \tilde{\alpha}(C(\mathbf{r},t) - C_0)]. \tag{9}$$

An additional transport equation for the concentration is given by:

$$\{L\Delta - \partial_t\}C(\mathbf{r},t) = -\Delta_2\psi(\mathbf{r},t) - L\Delta\Theta(\mathbf{r},t) + \mathbf{v}(\mathbf{r},t)\cdot\nabla C(\mathbf{r},t), \tag{10}$$

with the dimensionless number L (Lewis number) being the ratio between thermal and mass diffusion. Due to (9), the first equation of (4) has to be extended on the right hand side by

$$RS\Delta_2 C(\mathbf{r},t)$$

with the so-called separation ratio S. S gives the ratio between the two destabilizing effects of buoyancy forces due to temperature and concentration gradients. The Hopf bifurcation occurs if $S < -L^2$ (for more details see e.g. [22]).

2.4 Convection in a rotating layer

If a layer containing a pure liquid is rotated around its vertical axis, the equations of motion have to be augmented by terms induced by the Coriolis and the centrifugal forces [23]. The centrifugal force has a potential and may be easily included into the pressure. Qualitatively new effects arise from the Coriolis terms, entering the right-hand side of the first equation (4) as

$$-Ta\partial_z\Delta_2\phi(\mathbf{r},t),$$

and of the second one as

$$-Ta\partial_z\Delta_2\psi(\mathbf{r},t),$$

where Ta is the Taylor number and proportional to the angular velocity Ω of the layer:

$$Ta = \frac{2\Omega d^2\rho}{\eta}. \tag{11}$$

For large Pr, the linear system is only changed quantitatively, the critical Rayleigh number and the critical wave length of the unstable patterns increase with increasing Ta. A new feature is obtained for a sufficiently large Taylor number above a certain critical value: The set of parallel rolls already known from the simple Bénard problem, becomes unstable due to a second set of rolls with an angle of about 60^0. After saturation this new set becomes in turn unstable to a third set with again 60^0 relative orientation, and so on, resulting in a pattern of rolls that alters periodically. The sequence of alteration ($\pm 60^0$) is thereby well defined by the sense of rotation. In contrast to the three cases discussed before, this instability is a secondary one and based exclusively on non-linear effects; it is called the Küppers-Lortz instability [8].

Furthermore, if Ta is below the critical value, a once formed stable pattern shows rigid rotation. The angular velocity thereby is proportional to Ta and to the number of defects of the rigidly rotating pattern. This can be understood in terms of OPEs including non-variational expressions [24] and will be discussed in more detail in sect.4.

3. Variational Models

In this and in the following section, we wish to describe pattern formation in the weakly non-linear regime, i.e. in the vicinity of the critical points defining phase transitions. We therefore project the solutions of the basic equations onto a certain number of Galerkin modes, and obtain a large system of coupled ordinary differential equations. The slaving principle of synergetics then allows a drastic reduction of dependent variables in mode space. It turns out that the resulting mode or amplitude equations ca be derived from a potential under certain assumptions. To include also the influence of lateral boundary conditions and to describe the formation of dislocations and grain boundaries in the bulk of the layer, we derive extended Swift-Hohenberg models. In contrast to amplitude equations, these models have the great advantage that they may be formulated in a rotationally invariant way in the horizontal plane. Numerical solutions are presented for several physical systems.

3.1 Amplitude equations

Since the treatment of low viscosity fluids, binary mixtures, or rotating layers leads immediately to non-variational problems, we restrict ourselves in this section to the case of a pure, non-rotating fluid with $1/Pr = 0$. Eq.(4) reads

$$\Delta\Delta_2\phi(\mathbf{r},t) = 0 \tag{12}$$

and, together with (5,6):

$$\phi(\mathbf{r},t) = 0 \tag{13}$$

in the whole layer.

The order parameter equations in Fourier space. As Galerkin modes we take the solutions of the linearized problem (4) for an infinite lateral layer:

$$\begin{aligned} (d_z^2 - k^2)^2 q_{1\ell} + R q_{2\ell} &= 0 \\ (d_z^2 - k^2 - \lambda_\ell(k^2)) q_{2\ell} + k^2 q_{1\ell} &= 0. \end{aligned} \tag{14}$$

The condition

$$\lambda_\ell(R, k^2) = 0$$

defines for a given k the smallest Rayleigh number necessary for the onset of convection [23]. The critical wave number minimizes R and is given by

$$\frac{dR}{dk^2} = 0$$

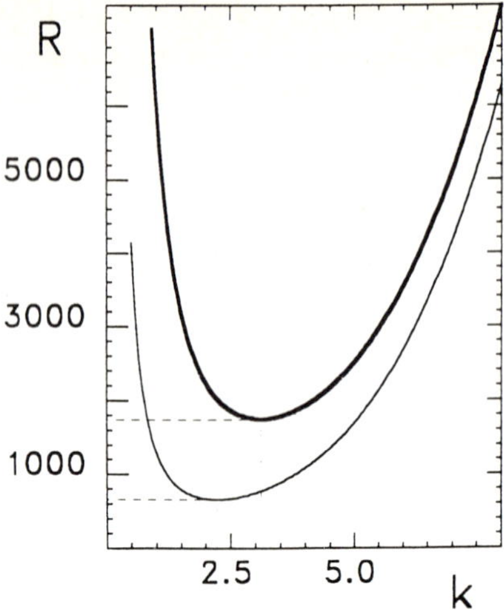

Figure 2: The minimum Rayleigh number for the onset of convection as a function of the absolute value of the wave vector. Cells having the same lateral extent as the thickness of the fluid layer become unstable first. Solid line: free-free boundary conditions at the upper and lower plate, thin line: rigid-rigid boundary conditions.

(compare fig.2). The projection

$$\begin{bmatrix} \psi(\mathbf{r},t) \\ \Theta(\mathbf{r},t) \end{bmatrix} = \sum_{\ell} \int_{-\infty}^{\infty} d^2\mathbf{k} \quad \xi_\ell(\mathbf{k},t)\mathbf{q}_\ell(\mathbf{k},z)e^{i\mathbf{kx}} + c.c. \tag{15}$$

with

$$\mathbf{k} = (k_x, k_y), \qquad \mathbf{x} = (x, y)$$

leads to the amplitude equations:

$$\begin{aligned} \dot{\xi}_\ell(\mathbf{k},t) &= \lambda_\ell(\mathbf{k})\xi_\ell(\mathbf{k},t) + \sum_{\ell'\ell''} \int_{-\infty}^{\infty} d^2\mathbf{k}' d^2\mathbf{k}'' \quad c_{kk'k''}^{\ell\ell'\ell''} \xi_{\ell'}(\mathbf{k}',t)\xi_{\ell''}(\mathbf{k}'',t) \\ &\quad \times\delta(\mathbf{k}-\mathbf{k}'-\mathbf{k}''), \end{aligned} \tag{16}$$

where the coefficients c are matrix elements formed by $\mathbf{q}_\ell$ [2]. Here we are still at the same level of complexity; the infinitely many degrees of freedom intrinsic in the basic partial differential equations are expressed by an infinite number of mode amplitudes $\xi_\ell(\mathbf{k},t)$.

Along the lines of synergetics, we divide the eigenmodes into two groups, according to their eigenvalues:

$\lambda_u \approx 0, \Longrightarrow \xi_u(\mathbf{k},t)$, u=unstable, order parameter

λ_l ↗ ↘

$\lambda_s \ll 0, \Longrightarrow \xi_s(\mathbf{k},t)$, s=stable, enslaved modes.

The slaving principle allows us to express the amplitudes of the enslaved modes as a functional of those of the order parameters in a unique way. In other words, we may eliminate the enslaved mode amplitudes in (16). The simplest way to do this is invoking the adiabatic elimination [1, 2]. In this case, the dynamics of the enslaved modes is neglected; they instantaneously follow the order parameters. The remaining equations for the order parameters ξ_u read (here and in the following we suppress the index "u"):

$$\begin{aligned}\dot{\xi}(\mathbf{k},t) &= \lambda(k^2)\xi(\mathbf{k},t) + \int d\mathbf{k}_1 d\mathbf{k}_2 d\mathbf{k}_3 \Gamma(\mathbf{k}_1,\mathbf{k}_2,\mathbf{k}_3)\xi(\mathbf{k}_1,t)\xi(\mathbf{k}_2,t)\xi(\mathbf{k}_3,t) \\ &\quad \times\delta(\mathbf{k}-\mathbf{k}_1-\mathbf{k}_2-\mathbf{k}_3). \qquad (17)\end{aligned}$$

We note that we have still a system describing the motion of an infinite number of order parameters that denote the amplitudes of plane waves with the two-dimensional wave vectors $\mathbf{k}$. If the system is isotropic in real space, the linear part of the OPE may only depend on k^2, i.e. the unstable modes lie on a ring in Fourier space with radius k_c. We assumed inversion symmetry for the order parameter which results in only odd powers of ξ in (17). This symmetry is violated for non-Boussinesqian or surface tension effects leading to the inclusion of a quadratic term in ξ.

We now make the basic assumption that the mode amplitudes ξ near threshold are excited essentially only in a narrow circular band in 2D k-space, i.e. on a ring with radius k_c and width Δk. Introducing polar coordinates k and φ we may express the two-dimensional vector $\mathbf{k}$ solely by its orientation φ. Then the delta function under the integral in (17) requires a coupling of the four wave vectors $\mathbf{k}, \mathbf{k}_1, \mathbf{k}_2$, $\mathbf{k}_3$, forming a rhombus. Therefore the cubic part in (17) may be written as

$$(k_c\Delta k)^3 \int_0^{2\pi} d\varphi' f(\varphi-\varphi')\xi(\varphi)|\xi(\varphi')|^2, \qquad (18)$$

where

$$f(\varphi-\varphi') = \Gamma(\varphi',\varphi,\varphi'+\pi) + \Gamma(\varphi',\varphi'+\pi,\varphi) + \Gamma(\varphi,\varphi',\varphi'+\pi). \qquad (19)$$

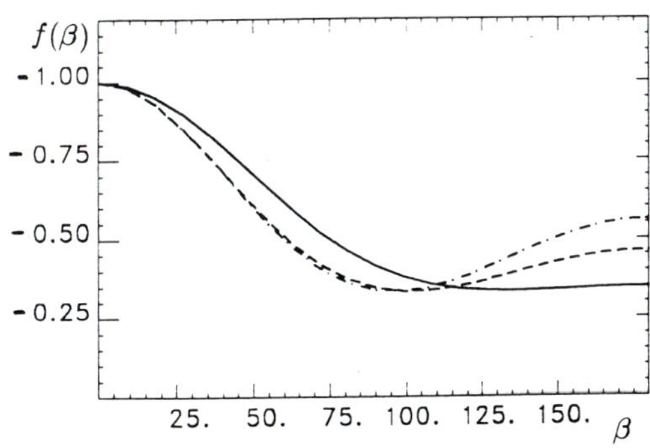

Figure 3: Coupling coefficient f as a function of the coupling angle between two modes for several Biot numbers. Solid: $Bi = 1000$, dashed: $Bi = 1$, short-long dashed: $Bi = 0.1$. If $Bi < 1$ squares are the preferred structure.

Fig.3 shows computed values for f for a pure fluid and different Biot numbers. Due to the condition of isotropy in real space, f may only depend on the relative angle between $\mathbf{k}$ and $\mathbf{k}'$. If we assume that Γ contains scalar products of its arguments $\mathbf{k}_i$, we may approximate f by a Taylor series with respect to $\cos(\varphi - \varphi')$ along

$$f(\beta) = \sum_{n=0}^{N/2} a_{2n}(\cos\beta)^{2n}. \tag{20}$$

Because of the symmetry with respect to $\varphi' \to \varphi' + \pi$ in the expression under the integral (18), odd powers of $\cos\beta$ cancel and only even powers have to be taken into account.

Perfect patterns. In the following, we shall denote patterns described by a composition of wave vectors with the same wavelength but arbitrary directions as *perfect patterns*. They have no dislocations or grain boundaries. The OPEs may be derived by a potential if the coupling coefficients obey the relation:

$$f(\beta) = f(-\beta) \tag{21}$$

in accordance with (20). Then the potential reads:

$$V[\xi] = -\int_0^{2\pi} d\varphi \left\{ \lambda(k_c^2)|\xi(\varphi)|^2 + \frac{1}{2}\int_0^{2\pi} d\varphi' f(\varphi - \varphi')|\xi(\varphi)|^2|\xi(\varphi')|^2 \right\} \tag{22}$$

and

$$\dot{\xi}(\varphi) = -\frac{\delta V}{\delta \xi^*(\varphi)}. \tag{23}$$

We note that the inclusion of a dependence on $\sin(n\beta)$ in (20) as well as the extension to a wave band in the non-linearity may lead to non-potential expressions that cannot be cast into the form (23). We shall encounter the first case in a rotating convection cell, the latter in fluids with low viscosity (see sect.4).

Here we wish to discuss briefly the stability of several perfect patterns as stable solutions of (23). If we treat patterns on a lattice having N-fold rotation symmetry, the OPEs may be written as

$$\dot{\xi}_i = \lambda(k_c^2)\xi_i + \sum_{j=1}^{N} f_{ij}\xi_i|\xi_j|^2, \tag{24}$$

where the index i denotes the several directors of the lattice. A perfect pattern where all N modes are excited equally with

$$|\xi_i|^2 = \frac{\lambda}{\sum_j^N f_{ij}}$$

Figure 4: Perfect patterns having 10-fold (left) and 12-fold (right) rotation symmetry.

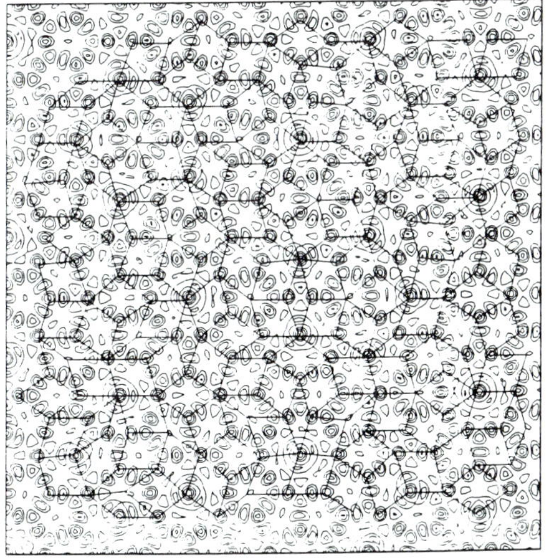

Figure 5: Connecting certain maxima and minima of a ten-fold symmetric pattern, a Penrose tiling occurs composed of two different rhombuses.

is stable if all disturbances have negative growth rates σ_i. σ_i are the eigenvalues of f_{ij}. If all nondiagonal elements of f are equal, the perfect pattern is stable as long as the self-coupling coefficient f_{ii} has a larger absolute value than the interaction coefficients f_{ij}. Fig.4 shows perfect 10 and 12-mode patterns. A so-called Penrose tiling may be constructed inside the 10-mode pattern (fig.5), see e.g. [25].

The inclusion of a quadratic term originating from non-Boussinesqian or Marangoni effects in (17) leads to an expression of the form

$$B\xi_{k_c'}\xi_{k_c''}$$

in (24), where the vectors $\mathbf{k}'_c, \mathbf{k}''_c$, and $\mathbf{k}_c$ have to form an equilateral triangle. The square term causes a stabilization of groups of three plane waves with relative angles of 60^0. If only one such group is excited, the pattern consists of regular hexagons. The case of two groups (6 plane waves) leads also to a pattern as shown in fig.4, right frame.

Finite band width excitation. If the system is laterally limited, the emerging patterns may no longer be described by a single wavelength but by a wave packet. The mode amplitudes belonging to a plane wave with wave vector k_c become slowly space dependent. The non-linear part is still regarded locally in real space. This is the concept of *finite band width excitation* and means nothing other than the reintroduction of low order spatial derivatives in (24). These derivatives are placed only in the linear part and stem from a Taylor expansion of k around k_c, substituting

$$k^2 \rightarrow (\mathbf{k}_c(\varphi_i) - i\nabla_2)^2,$$

where ∇_2 is the horizontal gradient. We now have for each direction in real space one partial differential equation to solve. One arrives immediately at the famous Newell-Whitehead equation [26] if one reduces the treatment only to one direction, describing the formation of rolls. The neglect of higher derivatives leads to the particular form of (24):

$$\dot{\xi}(\mathbf{x},t) = \left[\varepsilon - \zeta^2(2ik_c\partial_x + \partial_y^2)^2\right]\xi(\mathbf{x},t) - |\xi(\mathbf{x},t)|^2\xi(\mathbf{x},t). \tag{25}$$

The bifurcation parameter ε denotes the distance from threshold $\varepsilon = (R - R_c)/R_c$, and ζ^2 is proportional to the curvature of the largest eigenvalue around k_c^2. Note that (25) can still be derived from a potential.

3.2 Extended Swift-Hohenberg models

The decisive drawback of amplitude equations is the lack of a rotationally invariant formulation in 2D real space. This would require an infinite number of coupled equations since there are infinitely many directions. A quite elegant method is the construction of a fast varying wave function that contains the whole spatial dependence of the order parameter [2]. Therefore, we sum up over all directions in real space according to

$$\Psi(\mathbf{x},t) = \int_0^{2\pi} d\varphi \xi_\varphi(\mathbf{x},t) e^{i\mathbf{k}_c(\varphi)\mathbf{x}}. \tag{26}$$

Going back all the way to (15) and (3), we may express the hydrodynamic variables $\mathbf{v}$ and Θ by that wave function. The relations read in lowest order of the 2D order parameter ($\ell = u$)

$$\begin{aligned} \mathbf{v}(\mathbf{r},t) &= \nabla \times \nabla \times \{q_{1u}(\Delta_2, z)\Psi(\mathbf{x},t)\mathbf{z}_0\} + O(\Psi^2) \\ \Theta(\mathbf{r},t) &= q_{2u}(\Delta_2, z)\Psi(\mathbf{x},t) + O(\Psi^2). \end{aligned} \tag{27}$$

Due to its k^2-dependence, the eigenvector $\mathbf{q}_u$ is now a differential operator. Keeping in mind that Ψ is mainly excited on a ring with radius k_c, we may substitute the Laplacians in the eigenvectors by $-k_c^2$. Then the eigenvectors are simple functions of z. In this approximation, the relations to the basic variables read:

$$\begin{aligned}
\mathbf{v}(\mathbf{r},t) &= \begin{bmatrix} d_z f(z)\partial_x \Psi(\mathbf{x},t) \\ d_z f(z)\partial_y \Psi(\mathbf{x},t) \\ -f(z)\Delta_2 \Psi(\mathbf{x},t) \end{bmatrix} + O(\Psi^2) \\
\Theta(\mathbf{r},t) &= g(z)\Psi(\mathbf{x},t) + O(\Psi^2).
\end{aligned} \tag{28}$$

From this last equation, we derive immediately boundary conditions for Ψ on the lateral walls. Assuming vanishing horizontal velocity and temperature field, they are:

$$\partial_{\mathbf{n}}\Psi(\mathbf{x},t) = \Psi(\mathbf{x},t) = 0 \tag{29}$$

for $\mathbf{x}$ on and $\mathbf{n}$ perpendicular to the lateral walls.

The order parameter equation in real space. Now we give a short derivation of the evolution equation of (26) that we shall call in the following 2D-OPE. In the linear part we allow for a variation of the wave vector in a band centered at k_c. It can be approximated in real space by

$$\lambda(\Delta_2) \approx \varepsilon - g^2(1+\Delta_2)^2. \tag{30}$$

Here and in the following we rescale the spatial coordinates to have $k_c^2 = 1$. The angular dependence of the third order term in (24) may be approximated by spatial derivatives, or, with regard to a rotationally invariant formulation of the problem, by powers of the 2D-Laplacian. If, for the cubic term, we consider the general expression

$$\Psi \sum_{n=0}^{N/2} A_{2n} \left(\frac{\Delta_2}{2}+1\right)^{2n} \Psi^2 \tag{31}$$

and if we assume again a single excited wavelength but arbitrarily many excited directions of the mode amplitudes, the relation between A_n and a_n of the formulas (20) and (31) may be established as follows:

$$\begin{aligned}
a_0 &= 3A_0 + \sum_{n=1}^{N/2} A_{2n} \\
a_{2n} &= 2A_{2n}, \qquad n = 1,2,3 \ldots N/2.
\end{aligned} \tag{32}$$

(32) enables us to express the fully angular dependence intrinsic in (17) or (24) of the mode coupling between plane waves with given arbitrary orientation via the local form (31) including only rotationally invariant expressions of spatial derivatives. The only restriction in our derivation is the reduction of the excited band to its critical value k_c, therefore our model can describe only qualitatively side-band instabilities where the instability mechanism is due to the growth of a disturbance having a different absolute value of the wave vector and not of the orientation in real space.

Numerical results: rolls, squares, quasi-periodic structures. Now we wish to discuss briefly the properties of several truncations with respect to N of (30) and (31). The crudest approximation belongs to $N = 0$, leading to the well-known Swift-Hohenberg equation [27] that was numerically treated in two spatial dimensions in [28] as a model of the convection instability:

$$\dot{\Psi}(\mathbf{x},t) = [\varepsilon - g^2(1+\Delta_2)^2]\Psi(\mathbf{x},t) + A_0\Psi^3(\mathbf{x},t) \tag{33}$$

and $A_0 < 0$. In terms of (20), no angular dependence of the coupling coefficients can be expressed. This leads to the preference of parallel rolls as stationary patterns and corresponds quite well to the properties of a high Prandtl number fluid having perfectly heat conducting horizontal boundaries [29]. Numerical solutions are presented in fig.6.

Figure 6: Stable solution of Swift-Hohenberg equation (33). As initial condition a perfect parallel roll pattern was used. After a long integration time in units of (33), the pattern matches the boundary conditions and rolls are oriented perpendicular to the walls.

The next level is $N = 2$. Now we may describe a coupling with one extremum in the interval $[0, \pi]$ of coupling angles. Due to the conditions discussed in sect.3.1, we may find an exchange of stability between rolls and squares. A simplified equation reads:

$$\dot{\Psi}(\mathbf{x},t) = [\varepsilon - g^2(1+\Delta_2)^2]\Psi(\mathbf{x},t) + A\Psi^3(\mathbf{x},t) + B\Psi(\mathbf{x},t)\Delta_2^2\Psi^2(\mathbf{x},t) \tag{34}$$

and $B < 0$. If A changes from negative to positive values, rolls lose stability and give way to squares. Fig.7 shows the evolution of a random dot pattern to a regular square structure. The coefficients in (34) were calculated numerically for poor thermal conductors on the bottom and the top of the layer. Experimental evidence may be found in [30] and in a binary mixture in [19].

In contrast to (33), (34) cannot longer be derived from a potential. Nevertheless we saw in sect.3.1 that all sorts of coupling coefficients are potential.

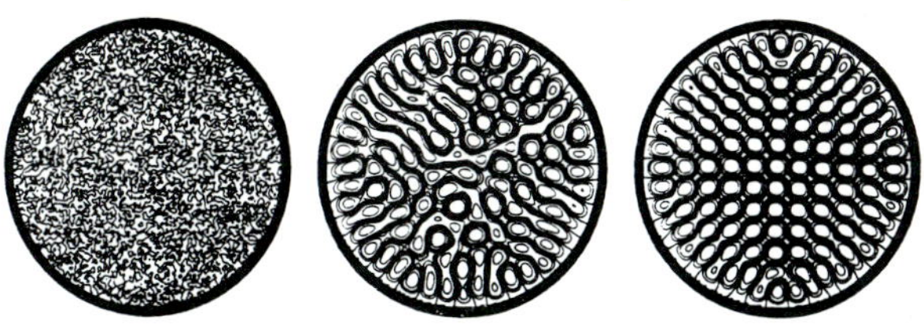

Figure 7: Temporal evolution of a random initial state according to (34). The coefficients were computed for poor thermal conductors at the bottom and the top of the fluid. Frames at $t = 0, 50, 1500$.

if they fulfil the condition (21) and if the patterns consist mainly of modes with the same wavelength. We conjecture that this is the reason for the relaxational behavior encountered in all numerical investigations in the literature of equations having the general form (30) and (31).

Including even higher derivatives, one can model hypothetic couplings with more minima. In that way a hexagonal pattern can be stabilized with $N = 6$. Here we wish to show the formation and stabilization of a pattern having 10-fold symmetry in a very large aspect ratio system. The dependence of the coupling coefficients on the angle is shown in fig.8, for f we take ($N = 10$):

$$\begin{aligned} f(\beta) &= -3 + 56\cos^2\beta - 392\cos^4\beta + 1120\cos^6\beta \\ &\quad -1280\cos^8\beta + 512\cos^{10}\beta. \end{aligned} \tag{35}$$

This fixes the values of A_n according to (32). Fig.9 shows the formation of a quasi crystalline structure in 2D-space. The initial condition was chosen randomly.

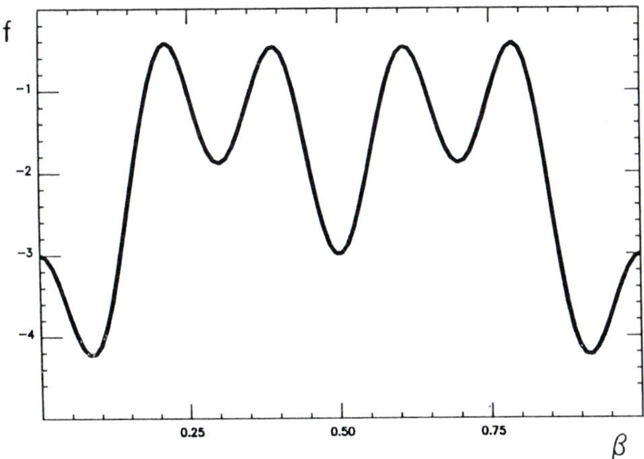

Figure 8: Mode coupling f as a function of coupling angle β. If the coupling has five equidistant minima, a quasi-periodic pattern having ten-fold rotation symmetry may be selected as a stable solution of the order parameter equation.

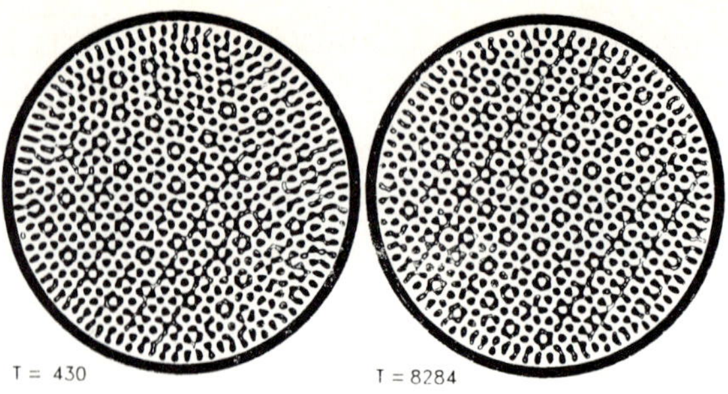

Figure 9: Two states of temporal evolution determined by a mode coupling shown in fig.8. Eventually, a quasi-periodic 2D structure is stabilized.

Even order non-linearities. Until now we have considered only cubic non-linearities. If the mirror symmetry with respect to the vertical midplane is violated, terms with even powers in the order parameter occur. The simplest one is quadratic, as already mentioned in sect.3.1.. In the form without spatial derivatives, it retains the variational property of (33). Adding the expression

$$\delta \Psi^2(\mathbf{x}, t)$$

to the right-hand side of (33), we arrive at an equation first derived by Haken [2] and numerically investigated in [31]. Stability analysis shows that hexagons are stable if

$$\varepsilon < \frac{4}{3A_0}\delta^2$$

(see fig.10). δ corresponds directly to non-Boussinesqian or surface tension effects and can be calculated as a function of the Marangoni number. A detailed study of the stability of rolls and hexagons in the case of Bénard-Marangoni convection is given in [32]. Recent experiments are reported in [33].

4. Non-variational Models

Now we shall treat several examples of convection instability that may lead to pattern formation exhibiting non-variational dynamics. We consider in particular the occurrence of an oscillatory instability in binary mixtures, the influence of low viscosity, and finally the thermal instability of a rotating fluid layer.

4.1 Oscillatory instabilities

The order parameter equation describing pattern formation in binary mixtures is derived in the way outlined in sect.3. The basic equations have to be extended by (9) and (10). Linear theory [22] shows that a Hopf bifurcation occurs if

$$S < -\frac{L^2}{L^2 + L + 1}. \tag{36}$$

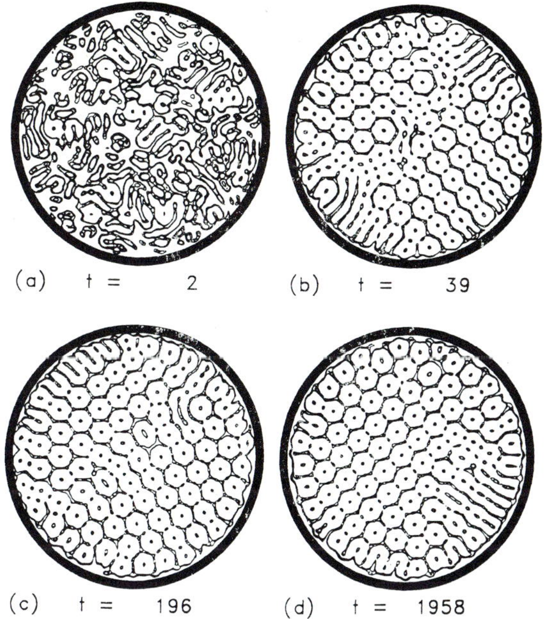

Figure 10: The inclusion of a quadratic term in the order parameter in (33) allows for the selection of hexagons. This term is proportional to non-Boussinesqian or surface tension effects. The lines of vanishing horizontal velocity of the fluid are plotted.

A pair of complex conjugate eigenvalues crosses the real axis if

$$R = \frac{27}{4}\pi^4 \frac{L+1}{S+1} \tag{37}$$

with the imaginary parts

$$\omega_c = \pm\frac{3}{2}\pi^2\sqrt{-L^2 - \frac{S(L+1)}{S+1}}. \tag{38}$$

The relations (36), (37), and (38) are valid for so-called perfect (free) boundary conditions in vertical direction. For physical boundary conditions, a first Galerkin approximation using polynomials leads to the values ($L = 0$):

$$R = \frac{1707}{S+0.96}, \qquad \omega_c = \pm 19.7\sqrt{-\frac{S}{S+0.96}}. \tag{39}$$

The order parameter is now a complex wave function whose solutions consist of traveling or standing wave trains. The 2D-OPE is the complex generalization of (30) and (31). The linear expression (30) has to be completed by dispersion:

$$\lambda(\Delta_2) \approx \varepsilon + i\omega_c - g^2(1+\Delta_2)^2 - i\gamma(1+\Delta_2). \tag{40}$$

The simplest non-linear model that may differentiate between the interaction of left and right traveling waves then reads [34]:

$$\dot{\Psi}(\mathbf{x},t) = [\varepsilon - (1+\Delta_2)^2 + i(\omega_c - \gamma(1+\Delta_2))]\Psi(\mathbf{x},t) \tag{41}$$
$$+A\Psi(\mathbf{x},t)|\Psi(\mathbf{x},t)|^2 - B\Psi(\mathbf{x},t)|\nabla_2\Psi(\mathbf{x},t)|^2 + C|\Psi(\mathbf{x},t)|^4\Psi(\mathbf{x},t).$$

Here we include 5th order terms to give the possibility of describing a subcritical bifurcation, as is the usual case for a binary mixture. In one spatial dimension, traveling waves are selected if $\mathrm{Re}(A-B) > 2\mathrm{Re}(A)$ if $C = 0$, $A, B < 0$. This can be seen most easily by decomposing the order parameter Ψ into its right- and left-traveling components, according to:

$$\Psi(x,t) = \eta(x,t)e^{ix} + \xi(x,t)e^{-ix}.$$

The one-dimensional coupled equations for the amplitudes [35] η and ζ read:

$$\begin{aligned}\dot{\eta} &= [\varepsilon + v\partial_x + D\partial_{xx} + A_s|\eta|^2 + A_c|\zeta|^2 + C(6|\zeta|^2|\eta|^2 + 3|\zeta|^4 + |\eta|^4)]\eta \\ \dot{\zeta} &= [\varepsilon - v\partial_x + D\partial_{xx} + A_s|\zeta|^2 + A_c|\eta|^2 + C(6|\zeta|^2|\eta|^2 + 3|\eta|^4 + |\zeta|^4)]\zeta,\end{aligned}$$

where $A_s = A - B, A_c = 2A$. We performed a numerical treatment of (41) with the boundary conditions (5) in a circular as well as in a rectangular layer using a pseudo-spectral method and a semi-implicit time integration scheme.

Fronts and spirals. First we examine the case where all non-linear coefficients in (41) are real-valued and $C = -1$. As initial condition we use patterns where only small (singular) centers of the layer have a non-vanishing wave function whereas Ψ is set to zero on the rest of the layer. Due to the rotation symmetry of (41), the pattern development starts with circular or spiral rings, depending on the initial distribution in the centers. These rings have the wavelength $\lambda_c = 2\pi$ and travel with the phase velocity $v_{ph} = \omega_c$. The envelope of each center thereby propagates radially until it encounters an obstacle like the wall or another wave front. We studied this evolution in both sub- as well as supercritical regimes [34]. The cross-coupling term A becomes important at the moment where two wavetrains encounter one another in any part of the layer. Here we demonstrate the influence of the self-coupling term $A - B$ for the case of a small negative cross coupling. Fig.11 shows the situation for a negative self coupling of the same magnitude as the cross coupling. The waves can penetrate after collision and no formation or stabilization of a front between counter-propagating wave trains can be seen. This changes dramatically if the self coupling expression becomes positive (fig.12). Now a front is created between the centres. However the front is not stable and moves towards that region with the initially somewhat smaller amplitude. The same behaviour is obtained for a cubic order parameter equation having stabilizing (negative) self- as well as cross-coupling coefficients. If A is increased further, the emerging front has a larger slope and turns out to be stable with respect to fluctuations or non-symmetric initial conditions (see fig.13). Next we performed time series for the

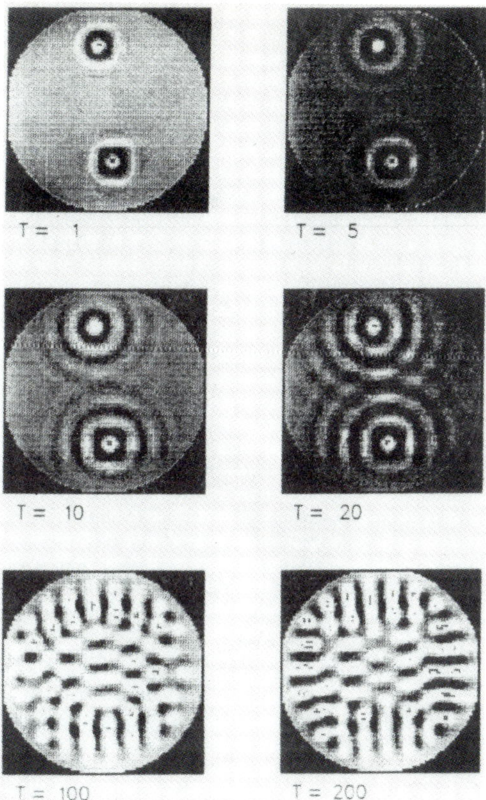

Figure 11: Numerical solutions of (41) for real non-linear coefficients and a negative self-coupling A_s. The initial condition consists of two domains with non-zero values for Ψ and $\Psi = 0$ for the rest of the plane. The circles spread out in form of propagating rings. Where they collide, they penetrate each other.

same parameter values for very large aspect ratios, starting now with different kinds of centres (fig.14). The "ring type" has azimuthal symmetry, the "spiral type" has an angular dependence according to $\exp \pm i\varphi$ in the initial states. The time series presented in fig.14 shows the emergence and stabilization of fronts between several domains of radially traveling waves. Note that there is no essential difference between the dynamics of circular rings or spirals, both types expand with about the same velocity and have the same properties with respect to their ability of forming stable regions and fronts between them. Qualitatively the same patterns are well-known from non-equilibrium chemical systems, e.g. the famous Belousov-Zhabotinsky reaction [36].

We finally mention that the same type of equations (41) may be derived from non-linear reaction diffusion equations, such as the Brusselator [37] or the Oregonator [38].

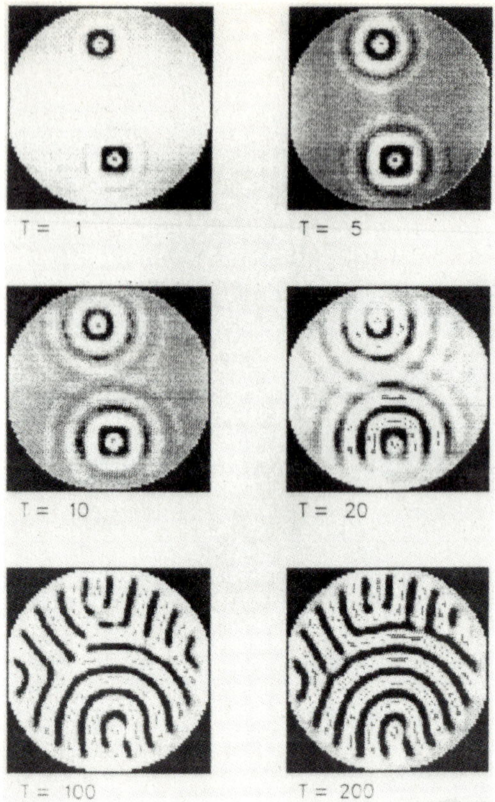

Figure 12: Same as fig.11, but for slightly positive values of A_s. Now fronts are formed in the colliding regions which are not stable but move towards the centre with a somewhat smaller initial amplitude.

Pulses. Several experiments have recently revealed the existence of pulses of traveling waves (TW) in an annulus very near threshold of convection in a binary fluid mixture [39]. If these pulses have a small amplitude, they travel with a certain group velocity, according to the linear theory. But as soon as non-linear effects become important they begin to slow down. Eventually, the pulses are stationary or at least very slow [40] in the laboratory frame. Theoretical as well as numerical investigations based on a one-dimensional amplitude equation show indeed pulses of confined traveling waves (CTW) [41]. These pulses have a vanishing group velocity only for singular values of the control parameters and not for a wide parameter region as observed experimentally. The inclusion of non-linear terms with spatial derivatives may decrease that velocity but can never eliminate a constant shift of the envelope function [42].

To model pulse creation and propagation in an annular narrow channel, we shall analyse solutions of (41) using rigid boundary conditions in y-direction,

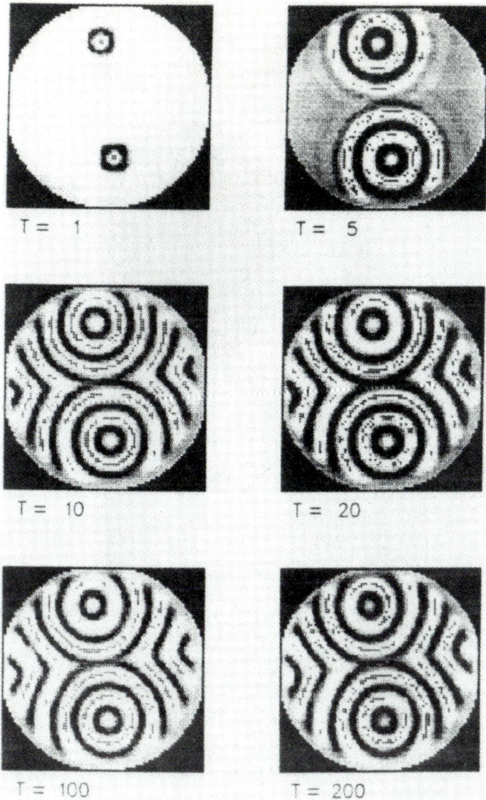

Figure 13: Same as fig.11, but for large positive A_s. The front between the centres is stabilized.

periodic ones in x-direction. The aspect ratios are $\Gamma_x = 9\pi, \Gamma_y = 3/4\pi$, the coefficients are assumed to be complex valued: $A = 3 + i, B = 0, C = -2.75 + i$. The small aspect ratio perpendicular to the channel direction shifts the value for the onset of convection to $\varepsilon_L = 0.93$ [43].

Numerical examination of (41) in the subcritical domain $\varepsilon < \varepsilon_L$ leads to the following new results: Besides the trivial solution (conducting state), a confined state of traveling waves (CTW), already known from amplitude equations, is recovered. The qualitatively new feature is the occurrence of a third kind of stable solution that consists of a source of symmetrical left and right traveling waves. The convection is confined to a narrow pulse of width about $1.5\lambda_c$ which is more or less half the extension of the pulse enveloping the CTW-solution. In the following, we shall refer to it as SST (symmetric source type). Its phase as a function of x is symmetric to the midpoint of the pulse and decreases to the outside, whereas a CTW shows a monotonically increasing or decreasing phase with x.

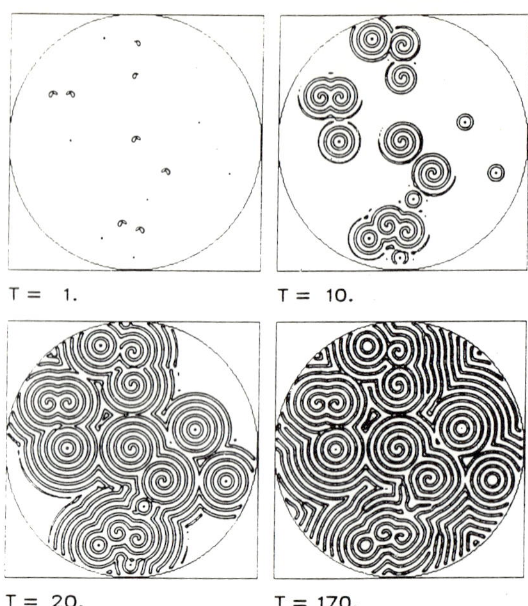

Figure 14: Same as fig.13, but for a spatially more extended system. The initial conditions are now rings and spirals with azimuthal dependence exp $\pm i\varphi$. Smaller domains are unstable with respect to larger ones. The domain walls at $T = 170$ are stable. All domains consist of circular or rotating spiral waves traveling towards the lateral wall.

For different confined initial conditions, such as CTW or localized standing waves, SST is formed in the subcritical region. The pulse moves along the channel but slows down and reaches finally zero velocity. The dispersion γ influences only the width of the pulse and not the temporal behaviour.

By increasing ε ($\varepsilon = 0.7$), the branch of CTW is reached (fig.15). By lowering ε again, this solution persists, showing hysteretic behaviour. Clearly, CTW, SST, and the conducting state are simultaneously stable.

To demonstrate that SST has a wide range of attraction and that there also exist time dependent mixed states of CTW and SST, a collision of a CTW state with SST was performed (fig.15). At the beginning a rather long phase of transients emerges where either a TW state to the left or to the right determines the pattern for a time and leads to an oscillation of the center of the confined state with respect to x. The intermediate CTW has a longer extent compared to the pure CTW. The counterpropagating wave train formed at the leading edge may reverse the direction of propagation of the whole pulse. It may also happen that a stable CTW is emitted, leaving a pure SST behind ($T = 380$). After one full x-cycle, the CTW collides with the SST from the other side and the interaction begins again. Now a temporally oscillating solution of a mixed state occurs, turning periodically from a pattern formed mainly by left TW (e.g. $T = 1080$) followed by the reverse ($T = 1260$), etc. This confined state

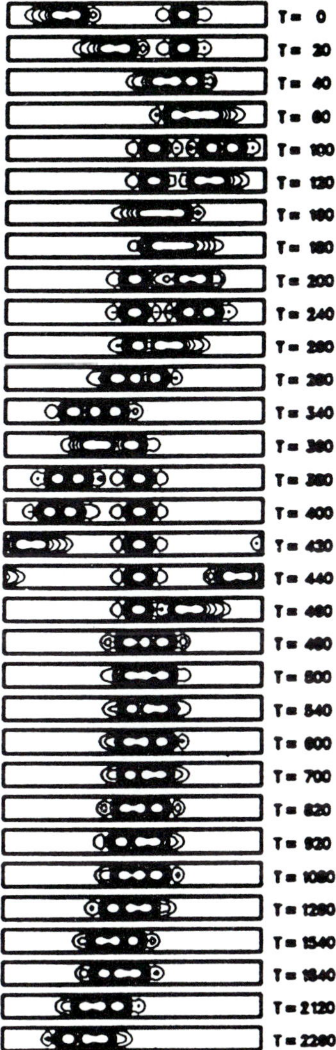

Figure 15: Temporal evolution of an initial state combined of CTW and SST, produced by a former run for $\varepsilon = 0.8$, and $\varepsilon = 0.6$, respectively. For this series, ε was set at $\varepsilon = 0.7$, where both particular solutions are stable. After a long phase of transitions, the system eventually reaches a mixed state of CTW (left and right) and SST, that shows no tendency to settle down to a stationary solution during the computations.

is present for the rest of the numerical computation. Another unstable CTW-state is shown in fig.16, for a slightly larger value of ε. After the collision, a CTW survives with a clearly longer spatial extent than the stable CTW. After almost two cycles of traveling, this state suddenly transforms to a SST

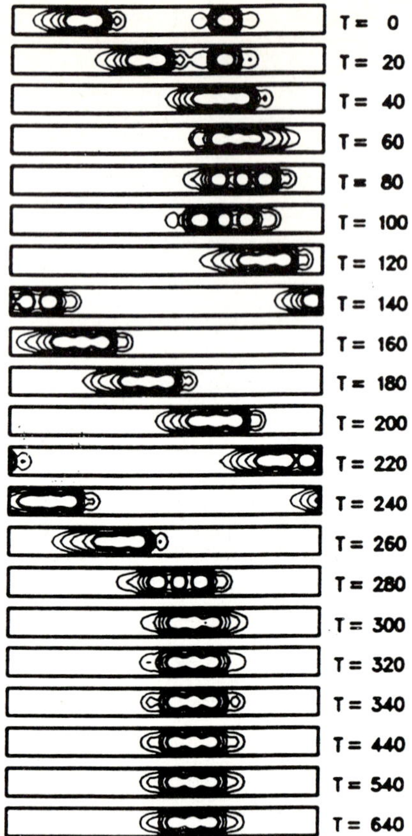

Figure 16: Same as fig.15, but for $\varepsilon = 0.8$. In the first part of the transition, a spatially more extended, "excited" CTW is created during the collision. However, this pulse is not time stable but evolves into an extended, but stable SST.

($T = 280$) with three times the extent of that of a stable SST. This state is completely stable for the rest of the computation and can be thought of a composition of three adjacent pure SSTs.

A collision of pure states may lead to a sort of "excited states" that may persist for long times and that are time dependent. These mixed states have a longer extent than the pure states and may emit from time to time a pure state, like in the run of fig.15. For a long evolution time they can either form a temporally stable symmetric state or continue to oscillate, showing an alternation of left and right TW. Therefore it seems natural to distinguish between the pure SST and CTW states that are quantized in their spatial extent and propagation speed, and excited states.

Despite the fact that SST is stable in the laboratory frame and confined to a narrow pulse of the same size than the critical wave length, it differs from the experimentally observed pulses found in binary mixtures which have no reflection symmetry and are usually of the CTW-type. However, quite recent experiments [44] show stabilization of a SST state after the collision of two counter-propagating CTWs.

4.2 Fluids with low viscosity – Mean flow effects

Experiments in fluids having a low Prandtl number show time dependent behaviour just at onset [45]. This became theoretically understandable with the work of Siggia et al. [9]. They considered the horizontal mean flow field ϕ (see eq.(3)) as a second order parameter.

If the value of Pr is low, the order parameter Ψ may generate a vorticity or mean flow field via the non-linearities of the second equation (4). The mean flow is of order ε and couples to the order parameter by the non-linear terms of the two remaining equations, leading finally to a cubic order in Ψ.

The OPE for moderate and low Prandtl number. The OPEs are derived in the same way as in sect.3; the left-hand-side of (15) has to be extended to

$$\begin{bmatrix} \psi(\mathbf{r},t) \\ \Theta(\mathbf{r},t) \\ \phi(\mathbf{r},t) \end{bmatrix} , \tag{42}$$

with $\mathbf{q}_\ell$ being now the eigenmodes of the complete linearized system (4). After an adiabatic elimination, the OPEs (17) acquire additional terms of the form

$$\frac{1}{Pr}\int d\mathbf{k}_1 d\mathbf{k}_2 d\mathbf{k}_3 \Gamma^{(1)}(\mathbf{k}_1,\mathbf{k}_2,\mathbf{k}_3)\xi(\mathbf{k}_1,t)\xi(\mathbf{k}_2,t)\xi(\mathbf{k}_3,t) \tag{43}$$
$$[(\mathbf{k}_2\times\mathbf{k}_1)+(\mathbf{k}_3\times\mathbf{k}_1)]\,(\mathbf{k}_2\times\mathbf{k}_3)\delta(\mathbf{k}-\mathbf{k}_1-\mathbf{k}_2-\mathbf{k}_3),$$

where $\times$ denotes the cross product. Due to the antisymmetry of the cross product, this term vanishes if $|\mathbf{k}| = k_c$ is assumed for all k-vectors. Clearly (43) violates the variational character.

Since one group of eigenvalues is proportional to Pr, the adiabatic elimination breaks down for extremely low Pr. Then the order parameter ξ or Ψ may again represent only the temperature and the poloidal velocity field. The spatio-temporal description of the mean flow is covered by a second order parameter field $\Phi(\mathbf{x},t)$, introduced as:

$$\phi(\mathbf{r},t) = \Phi(\mathbf{x},t)h(z), \tag{44}$$

where $h(z)$ has to fulfil the boundary conditions (5).

The 2D-OPEs for low Prandtl number. Based on these ideas, which are discussed in more detail in [46], we examined the two coupled evolution equations Ψ and Φ in the two horizontal dimensions [47]:

$$\begin{aligned} \partial_t\Psi(\mathbf{x},t) &= [\varepsilon-(1+\Delta_2)^2]\Psi(\mathbf{x},t)-\Psi^3(\mathbf{x},t)-g\mathbf{V}_H(\mathbf{x},t)\cdot\nabla_2\Psi(\mathbf{x},t) \\ \partial_t\Delta_2\Phi(\mathbf{x},t) &= Pr(\Delta_2-c^2)\Delta_2\Phi(\mathbf{x},t)+[\nabla_2\Psi(\mathbf{x},t)\times\nabla_2\Delta_2\Psi(\mathbf{x},t)]_z \\ \mathbf{V}_H(\mathbf{x},t) &= (-\partial_y\Phi(\mathbf{x},t),\partial_x\Phi(\mathbf{x},t)). \end{aligned} \tag{45}$$

The constant c depends on the vertical profile $h(z)$ (for rigid-rigid conditions (5), $h(z)$ may be approximated as a Poiseuille flow leading to $c^2 = 10$; for free-free conditions, h=const and $c^2 = 0$).

In a recent paper Bodenschatz et al. [48] report on convection in a cylindrical large aspect ratio system with pressurized CO_2. They discovered n-armed spiral patterns formed by convective rolls. These spirals are stationary in a *rotating* frame of reference. To model this experimentally detected behavior, we solved the system (45) numerically for a circular geometry with a large aspect ratio. The boundary conditions at the circular lateral wall read:

$$\begin{aligned} \Psi(\mathbf{x},t) &= d, \qquad & \partial_n \Psi(\mathbf{x},t) &= 0, \\ \Phi(\mathbf{x},t) &= 0, \qquad & \partial_n \Phi(\mathbf{x},t) &= 0. \end{aligned} \tag{46}$$

The parameter d measures lateral heating on the sidewall that plays a major role in the experiments in [48, 49]. Patterns with stable n-armed spirals can be obtained already for the Swift-Hohenberg equation forced by lateral heating ($d \neq 0$). To this end we started from an initial condition with a one-armed spiral with its tip located at the centre of the cylindrical container. The pattern evolves towards a stable one-armed spiral that matches the concentric rolls near the sidewall due to the lateral forcing. In the matching zone, a defect is created (see fig.17).

If non-variational terms are included ($g \neq 0$) the spiral begins to rotate rigidly. It is seen that the mean flow is essentially created by the defect, where the horizontal drift velocity is maximal. In turn, the mean flow acts on this dislocation and rotates the whole spiral. We note that the angular velocity is proportional to the strength of non-potential effects. There is no threshold value for Pr for the onset of rotation and the occurrence of non-variational behaviour. This is due to the fact that the continuous rotation symmetry in the cylindrical layer is broken spontaneously by the convection pattern. The disturbances due to any pattern that represent infinite rotation have zero growth rate. These disturbances are excited by the vorticity field. In contrast, the numerical results for rectangular cells [46] showed a significant influence of the mean flow on pattern formation only for very small Pr.

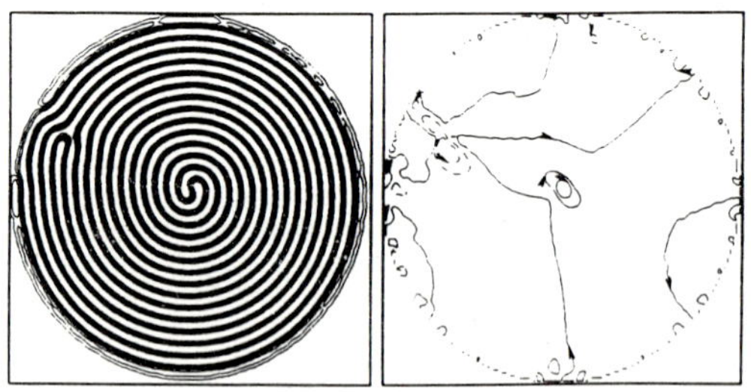

Figure 17: Rigidly rotating pattern found by numerical integration of (45). The spiral rotates counterclockwise (left) due to the mean flow (right). The mean flow is a horizontal drift that is created mainly by the defects. In turn the mean flow acts on these defects and rotates the spiral with a constant angular velocity. $\varepsilon = 0.7, g = 20, Pr = 1, c^2 = 10, d = 0.15$.

The angular velocity of the rigidly rotating spiral can be estimated by the expression (see next sect., eq.(52)):

$$\Omega = -g \frac{\int r dr d\varphi \frac{\partial}{\partial\varphi}\Psi(r,\varphi)\mathbf{V}_H(r,\varphi)\cdot\nabla_2\Psi(r,\varphi)}{\int r dr d\varphi \left[\frac{\partial}{\partial\varphi}\Psi^0(r,\varphi)\right]^2}, \tag{47}$$

where r and φ are cylindrical coordinates.

4.3 Convection in a rotating cell

The influence of rotation changes the linear part of the basic equations and couples the poloidal and toroidal components of the velocity representation due to Coriolis forces. In this section, we follow the work of Fantz et al. [24].

For infinite Prandtl number, the OPEs in Fourier space have the same form as (17). The crucial difference is the functional dependence of the coupling coefficients Γ on the orientation of the wave vectors. Due to the structure of the expressions reflecting the Coriolis force in a rotating frame of reference, Γ now also depends on cross products of the wave vectors. This introduces a sine dependence of the coupling on the coupling angle β in (20), according to:

$$f(\beta) = \sum_{n=0}^{N/2} a_{2n}(\cos\beta)^{2n} + \sum_{n=1}^{M} b_n(Ta)\sin(2n\beta), \tag{48}$$

where $b_n(Ta)$ vanish with the Taylor number Ta. The latter term in (48) is non-variational since it violates the condition (21).

In 2D-real space, this structure of coupling may be approximated by expressions of the form:

$$\Psi\left[\nabla_2\Psi \times \nabla_2\Delta_2^n\Psi\right]_z. \tag{49}$$

If Ta is not too large, $n = 1$ is a good approximation (see fig.18).

We finally present numerical solutions of the 2D-OPE that describes convection in a rotating layer and has the form:

$$\begin{aligned}\dot{\Psi}(\mathbf{x},t) &= [\varepsilon - (1+\Delta_2)^2]\Psi(\mathbf{x},t) - \Psi^3(\mathbf{x},t) \\ &+C(Ta)\Psi(\mathbf{x},t)\left[\nabla_2\Psi(\mathbf{x},t)\times\nabla_2\Delta_2\Psi(\mathbf{x},t)\right]_z, +D(Ta)\Psi(\mathbf{x},t)\Delta_2^2\Psi^2(\mathbf{x},t)\end{aligned} \tag{50}$$

where $C(Ta)$ and $D(Ta)$ can be calculated explicitly from the basic equations [24].

If Ta is small, the non-variational terms in (50) can be regarded as small disturbances of the Swift-Hohenberg equations. Let $\Psi_0(r,\varphi)$ be a solution of the unperturbed Swift-Hohenberg equation, then a rotated function

$$\Psi(r,\varphi) = \mathbf{e}^{\alpha L_z}\Psi_0(r,\varphi) = \Psi_0(r,\varphi+\alpha) \tag{51}$$

is also a solution for a cylindrical geometry. L_z is the z-component of the operator assigned to angular momentum $L_z = \partial_\varphi$. Inserting (51) into (50) and neglecting quantities that reflect a deformation of the rigidly rotating pattern, gives a first approximation for the value of the angular velocity:

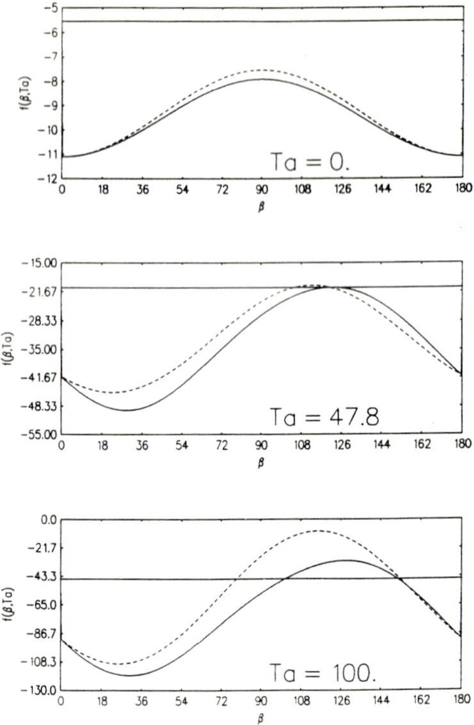

Figure 18: Mode coupling $f(\beta)$ as a function of coupling angle for several Taylor numbers and perfect boundary condition at the bottom and top of the fluid. The Küppers-Lortz instability sets in if $f(0)/2 < f(\beta)$ for any β. This is the case for $Ta = Ta_c = 47.8$ (middle frame). Solid line: Exact values found by computation of the coupling coefficients in Fourier space. Dotted line: Approximation according to (50).

$$\dot{\alpha} = C(Ta)\frac{\int rdrd\varphi L_z\Psi_0(r,\varphi)\left[\nabla_2\Psi_0(r,\varphi)\times\nabla_2\Delta_2\Psi_0(r,\varphi)\right]_z}{\int rdrd\varphi\left[L_z\Psi_0(r,\varphi)\right]^2}. \tag{52}$$

Note that (52) describes an additional rotation observed in the rotating frame of references.

Evaluation of the right-hand side of (52) shows that the angular velocity of a pattern is linked to the existence of dislocations and grain boundaries. Perfect patterns in the sense of sect.3.1 do not rotate. Fig.19 shows a sequence of a rigidly rotating structure for a low Taylor number.

If Ta is increased further, the rotating patterns become deformed more and more and finally get unstable at a certain critical Ta_c. In a moderate or small aspect ratio geometry, this behavior is well known as Küppers-Lortz instability [8]. A set of rolls becomes unstable due to a second set with a relative orientation of 58^0. This second set becomes unstable with respect to a third one and so on, giving a time dependent pattern formation at onset. For a larger box, this mechanism is accompanied by defect motion and may lead to chaotic behavior. Fig.20 shows the Küppers-Lortz instability for a Taylor number slightly above Ta_c.

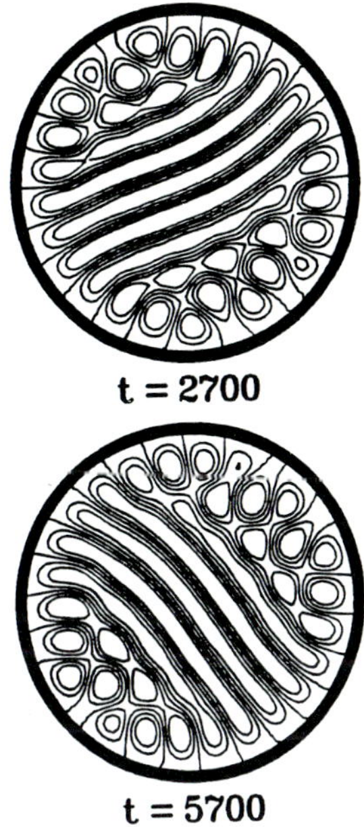

Figure 19: Rigidly rotating pattern with respect to the rotating frame of references. The Taylor number is well below the critical value, $Ta = 20$. The rotation is driven by the dislocation of the pattern.

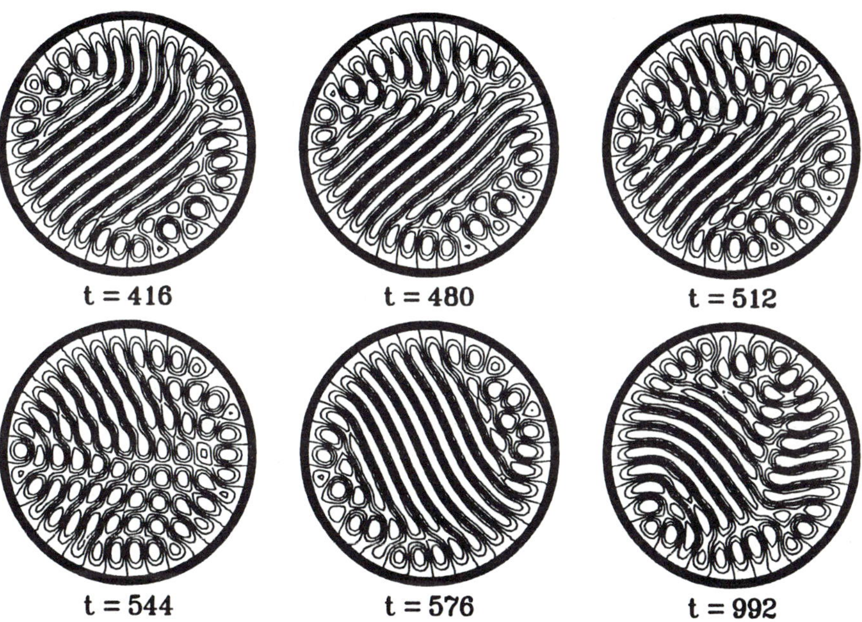

Figure 20: Küppers-Lortz instability for a $Ta = 50$, slightly above Ta_c. Rolls are unstable due to a second set of rolls with a relative orientation of about 60^0 (see also fig.18). The pattern remains time dependent.

Acknowledgments

It is a pleasure to thank R.Friedrich and M.Fantz for interesting and helpful discussions as well as A. Greiner for helpful suggestions during the preparation of the manuscript. I am grateful to M.Fantz for leaving me figures 18-20 and W.A.Lorenz for figure 1.

References

[1] H.Haken, *Synergetics. An Introduction*, Springer Berlin 3rd Ed. (1983)

[2] H.Haken, *Advanced Synergetics*, Springer Berlin, 2nd printing (1987)

[3] H.Haken, Rep. Prog. Phys. **52**, 515 (1989)

[4] H.Haken, *Information and Self-Organization*, Springer Berlin, (1988)

[5] A.Wunderlin and H.Haken, Z. Phys. **44**, 135 (1981)

[6] H.Haken in *Encyclopedia of Physics*, Vol XXV/2c, Springer Berlin (1970)

[7] R.Friedrich, M.Bestehorn, and H.Haken, Int. J. Mod. Phys. **B4**, 365 (1990)

[8] G.Küppers and D.Lortz, J.Fluid.Mech. **35**, 609 (1969); F.H.Busse and K.E.Heikes, Science **208**, 173 (1980)

[9] D.Siggia and A.Zippelius, Phys. Rev. Lett. **47**, 835 (1981)

[10] H.Bénard, Rev. Gen. Sci. Pur. Appl. **11**, 1261 (1900)

[11] C.Normand, Y.Pomeau, and M.Velarde, Rev. Mod. Phys. **49**, 581 (1977)

[12] F.H.Busse, Rep. Prog. Phys. **41/II**, 1931 (1978)

[13] R.P.Behringer, Rev. Mod. Phys. **57**, 657 (1985)

[14] F.H.Busse, J. Fluid. Mech. **30**, 625 (1967)

[15] C.Yih, *Fluid Mechanics*, Mc Graw-Hill Book Company (1969)

[16] J.K.Platten and J.C.Legros, *Convection in Liquids*, Springer Verlag Berlin (1984)

[17] V.Steinberg, E.Moses, and J.Fineberg, Nucl. Phys (proc. sup.) **B2**, 109 (1987);
D.Bensimon, P.Kolodner, C.M.Surko, H.Williams, and V.Croquette, J. Fluid Mech. **217**, 441 (1990)

[18] M.Bestehorn, R.Friedrich, and H.Haken, Z. Phys. **B75**, 265 (1989);
M.Bestehorn, R.Friedrich, and H.Haken, Z. Phys. **B77**, 151 (1989);
M.Bestehorn, R.Friedrich, and H.Haken, Physica **D37**, 295 (1989)

[19] E.Moses and V.Steinberg, Phys. Rev. **A43**, 707 (1991); Pys. Rev. Lett. **57**, 2018 (1986)

[20] H.W.Müller and M.Lücke, Phys. Rev. **A38**, 2965 (1988)

[21] M.Bestehorn, Physica D, accepted (1992)

[22] D.Gutkowicz-Krusin, M.A.Collins, and J.Ross, Phys. Fluids **22**, 1443, 1457 (1979)

[23] S.Chandrasekhar, *Hydrodynamic and Hydromagnetic Stability*, Clarendon, Oxford (1961)

[24] M.Fantz, Diplom thesis Universität Stuttgart, (1991);
M.Fantz, R.Friedrich, M.Bestehorn, and H.Haken, to be published

[25] G.M.Zaslavsky, R.Z.Sagdeev, D.A.Usikov, and A.A.Chernikov *Weak chaos and quasi-regular patterns*, Cambridge University Press (1991)

[26] A.C.Newell and J.A.Whitehead, J. Fluid. Mech. **38**, 279 (1969);
A.C.Newell in *Propagation in Systems far from Equilibrium*, ed. by J.E.Wesfreid, H.R.Brand, P.Manneville, G.Albinet, N.Boccara, Springer Series in Synergetics **Vol 41**, (Springer Berlin 1988)

[27] J.Swift and P.C.Hohenberg, Phys. Rev. **A15**, 319 (1977)

[28] H.S.Greenside, W.M.Coughran Jr., and N.L.Schreyer, Phys. Rev. Lett. **49**, 729 (1982);
H.S.Greenside and W.M.Coughran Jr., Phys. Rev. **A30**,398 (1984)

[29] P.Bergé and M.Dubois, Contemp. Phys. **25**, 535 (1984)

[30] P.Le Gal, A.Pocheau, and V.Croquette, Phys. Rev. Lett. **54**, 2501 (1985);
P.Le Gal and V.Croquette, Phys. Fluids **31**, 3440 (1988)

[31] M.Bestehorn and H.Haken, Phys. Lett. **A99**, 265 (1983)

[32] M.Bestehorn and C.Pérez-García, Europhys. Lett. **4**, 1365 (1987)

[33] S.Ciliberto, E.Pampaloni, and C.Pérez-García, J. Stat. Phys. **64**, 1045 (1991)

[34] M.Bestehorn and H.Haken, Phys. Rev. **A42**, 7195 (1990)

[35] M.C.Cross, Phsy. Rev. Lett. **57**, 2935 (1986)

[36] A.T.Winfree, Scient. Americ. **6**, 82 (1974);
A.N.Zaikin and A.M.Zhabotinsky, Nature **225**, 535 (1970);
Z.Nagy-Ungvarai, S.C.Müller: In *Propagation in Systems Far from Equilibrium*, J.E.Wesfreid, H.R.Brand, P.Manneville, G.Albinet, N.Boccara (eds.), Springer Series in Synergetics **Vol 41**, Springer Berlin (1988)

[37] I.Prigogine and G.Nicolis, J. Chem. Phys. **46**, 3542 (1967);
I.Prigogine and R.Levever, J. Chem. Phys. **48**, 1695 (1968)

[38] R.J.Field and R.M.Noyes, J. Chem. Phys **160**, 1877 (1974)

[39] J.J.Niemala, G.Ahlers, and D.S.Cannell, Phys. Rev. Lett. **64**, 1365 (1990); K.E.Anderson and R.P.Behringer, Phys. Lett. **145A**, 323 (1990); J.A.Glazier and P.Kolodner, subm. to Phys. Rev. A, (August 1990)

[40] P.Kolodner, Phys. Rev. Lett. **66**, 1165 (1991)

[41] O.Thual and S.Fauve, J. Phys. (France) **49**, 1829 (1988); S.Fauve and O.Thual, Phys. Rev. Lett. **64**, 282 (1990); W.van Saarlos and P.C.Hohenberg, Phys. Rev. Lett. **64**, 749 (1990)

[42] R.J.Deissler and H.R.Brand, Phys. Lett. **130A**, 293 (1988)

[43] M.Bestehorn, Europhys. Lett. **15**, 473 (1991)

[44] P.Kolodner, subm. to Phys. Rev. A (1991)

[45] G.Ahlers and R.P.Behringer, Phys. Rev. Lett. **40**, 712 (1978); V.Steinberg, G.Ahlers and D.S.Cannell, Phys. Scripta **T9**, 97 (1985); M.S.Heutmarker and J.P.Gollub, Phys. Rev. **A35**, 242 (1987)

[46] P.Manneville, J. Phys. (Paris) **44**, 759, L-903 (1983).

[47] M.Bestehorn, M.Fantz, R.Friedrich, H.Haken, and C.Pérez-García, preprint

[48] E.Bodenschatz, J.R. de Bruyn, G.Ahlers and D.S.Cannell, Phys. Rev. Lett. **67**, 3078 (1991)

[49] G.Ahlers, at the NATO-ARW in Estella, Spain, September 1991, to appear in *New Trends in Non-Linear Dynamics: Non-Variational Aspects*, Physica D

Dynamic Theory of Planetary Magnetism and Laboratory Experiments

F.H. Busse

Institute of Physics, University of Bayreuth,
W-8580 Bayreuth, Fed. Rep. of Germany

Abstract

The problem of the origin of planetary magnetism is formulated as a bifurcation problem and some recent theoretical work on the generation of magnetic fields by buoyancy–driven convection in rotating spherical shells is briefly reviewed. Since the lack of laboratory experiments has hampered the theoretical progress, a possible configuration for a laboratory apparatus is proposed.

1. Introduction

The Earth's magnetism has played an enigmatic role for a long time in the history of mankind. The ancient Chinese are usually credited with being the first people who were aware of the strange force that acts on lodestones floating, for instance, on wooden boards in vessels of water. But claims have been made [1] that the Olmecs in Mexico knew about the mysterious properties of the lodestones even earlier around 1000 B.C. The compass needle and its role in navigation is well known to everybody, but little knowledge seems to exist about the origin of the force causing the orientation of the needle. The answer that my daughter got in elementary school, namely that there is a huge pile of magnetized ore under northern Canada and Greenland, appears to be typical of the views held by many people.

This answer is not all that far from what scientists thought until about 100 years ago. The scientific study of the phenomenon of geomagnetism goes back several centuries and the first treatise on this subject, "De Magnete" by W. Gilbert which appeared in 1600, is regarded as the first monograph in the modern sense in the field of physics. After measuring the direction of the field on the surface of a spherical magnet and comparing it with the observations on the Earth, Gilbert came to the conclusion that "Magnus magnes ipse est globus terrestris". With his far reaching conclusions based on experiments, Gilbert exerted a significant influence on the development of the sciences in the 17th century.

The idea that the Earth's magnetic field is caused by the remanent magnetism of material in the Earth's interior was challenged only after it became obvious that the Curie temperature beyond which remanent magnetisation vanishes is reached at a depth of about 30 km. The idea of Larmor [2] that magnetism can be produced by motions in an essentially homogeneous electri-

Springer Proceedings in Physics, Vol. 69
Evolution of Dynamical Structures in Complex Systems
Editors: R. Friedrich · A. Wunderlin

cally conducting fluid was first proposed as an explanation for the magnetic field of sunspots. But it was soon adopted as a possibility for the explanation of the origin of geomagnetism. There were few theoretical efforts to pursue this possibility in the first half of this century and the two volume monograph "Geomagnetism" by Chapman and Bartels [3] all but ignores the question of the origin. The famous theorem of Cowling [4] stating that axisymmetric magnetic fields cannot be produced by the dynamo mechanism proposed by Larmor certainly discouraged studies of the dynamo hypothesis for a long time. Today this difficulty has been overcome and many theoretical solutions are known which exhibit deviations from axisymmetry which are much lower than those of the Earth's magnetic field.

Similarly to the transition from laminar to turbulent fluid flow induced by hydrodynamic instability, the dynamo mechanism basically describes the transition of a state of motion without magnetic field in an electrically conducting fluid to a state with magnetic field. In the latter state a new sink in the form of ohmic dissipation is opened for the mechanical energy just as additional viscous dissipation by the fluctuating component of motion characterizes the turbulent state of fluid motion. In contrast to the hydrodynamical problem for which laboratory experiments have provided large amounts of useful data, there hardly exists a field of experimental dynamos. As a consequence dynamo theory has become a mathematically oriented discipline and a number of concepts have been developed with only rather limited checks provided by physical reality.

This unsatisfactory state of affairs from the physical point of view led to early attempts to construct laboratory dynamos. In order to overcome the high magnetic diffusivity of liquid metals Lowes and Wilkinson [5,6] used iron cylinders rotating in a solid block of iron with the gaps filled by mercury. But the nonlinear nature of the magnetisation of iron and the effects of hysteresis led to a very complex behavior of this laboratory dynamo and no quantitative theory for the explanation of the data has ever been published. Clearly, for theoretical as well as geophysical reasons, it is desirable to have laboratory dynamos which do not rely on the high magnetic permeability of iron. One purpose of the present paper is to draw attention to a proposal for such a dynamo made some years ago [7].

In the following we shall first outline a theoretical approach to the problem of dynamos in planetary cores and then consider a simple dynamo which can serve as the basis for the design of a laboratory dynamo experiment.

2. Dynamos in Rotating Spheres Heated from Within

Since typical time scales involved in the generation of magnetic fields by fluid motions are long in comparison with the propagation of electromagnetic waves across the fluid domain, there is no need to consider the full Maxwell's equations. By neglecting the displacement current we eliminate the possibility of electromagnetic waves just as it is usually convenient to neglect compressibility in the analysis of a turbulent fluid in order to eliminate the possibility of acoustic waves — except that this kind of approximation turns out to be much better justified in the magnetic case. Using Ohm's law for an electrically conducting fluid, $\vec{j} = \sigma(\vec{E} + \vec{u} \times \vec{B})$, after eliminating the electric field $\vec{E}$ and the current $\vec{j}$, we arrive at the following equation for the magnetic flux density $\vec{B}$,

$$(\frac{\partial}{\partial t} + \vec{u} \cdot \nabla)\vec{B} - \lambda\nabla^2\vec{B} = \vec{B} \cdot \nabla\vec{u} \tag{1}$$

where λ is the magnetic diffusivity, $\lambda = (\mu_o\sigma)^{-1}$, and where an incompressible fluid, $\nabla \cdot \vec{u} = 0$, has been assumed.

Any solution for $\vec{B}$ of equation (1) that grows in time for a given velocity field $\vec{u}$ is called a dynamo. There are no sufficient conditions which guarantee dynamo solutions for general velocity fields. But a number of necessary conditions are known for which we refer to recent monographs [8,9]. Equation (1) can be compared with the stability equation for the infinitesimal disturbance $\vec{\tilde{u}}$ of a steady flow $\vec{u}$,

$$(\frac{\partial}{\partial t} + \vec{u} \cdot \nabla)\vec{\tilde{u}} - \nu\nabla^2\vec{\tilde{u}} - \nabla\tilde{p} = -\vec{\tilde{u}} \cdot \nabla\vec{u} \tag{2}$$

which determines the point of bifurcation where a secondary solution may bifurcate from the basic solution $\vec{u}$. In order to follow the bifurcating solution, the nonlinear term $\vec{\tilde{u}} \cdot \nabla\vec{\tilde{u}}$ must be included in the analysis. In the case of equation (1) the nonlinearity appears through the Lorentz force $\vec{j} \times \vec{B}$ in the equation of motion. In the full magnetohydrodynamic dynamo problem the Navier–Stokes equation of motion coupled through the Lorentz force with the dynamo equation (1) must be considered. The coupling of the two equations through the Lorentz force has the consequence that for every dynamo with $\vec{B}$ there exists a dynamo with $-\vec{B}$. This statement cannot be made, in general, for the evolving disturbances $\vec{\tilde{u}}$.

In the case of convection–driven dynamos in rotating spherical shells further symmetry properties can be used. Since the preferred form of convection driven by a spherically symmetric distribution of heat sources in a spherically symmetric gravity field exhibits a symmetry with respect to the equator, the solutions $\vec{B}$ of equation (1) can be separated into two classes, the quadrupolar class which exhibits the same symmetry as the convection velocity field and the dipolar class which exhibits the opposite symmetry with respect to the equator. In Fig. 1 the two classes of magnetic fields are sketched through the use

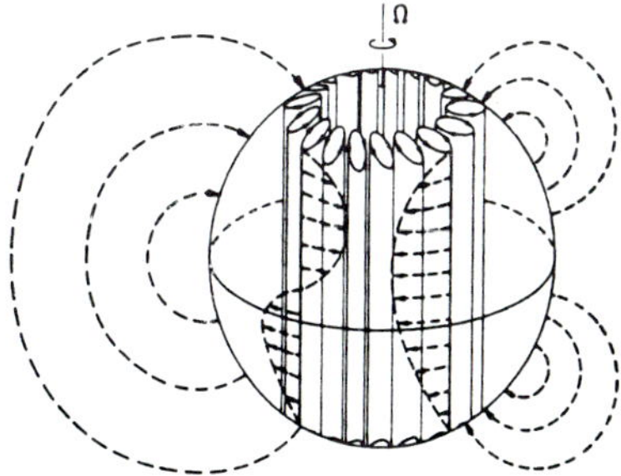

Figure 1: Sketch of axisymmetric components of dipolar (left) and quadrupolar (right) classes of magnetic fields

of the axisymmetric poloidal and toroidal components. Clearly this symmetry property is preserved when the Lorentz force is included in the analysis. Since the convection velocity field that is realized when the spherically symmetric static state becomes unstable assumes the form of azimuthally periodic waves propagating in the prograde direction, some further symmetries can be used in studying different classes of magnetic fields bifurcating from the non–magnetic state. These symmetries are listed in the diagram of Fig. 2. Only the case in which the magnetic field exhibits the same azimuthal periodicity as the convection flow yields axisymmetric components of $\vec{B}$ at the point of bifurcation. Furthermore the geomagnetic field is steady in a first approximation because its reversals occur on a much longer time scale than the typical magnetic time

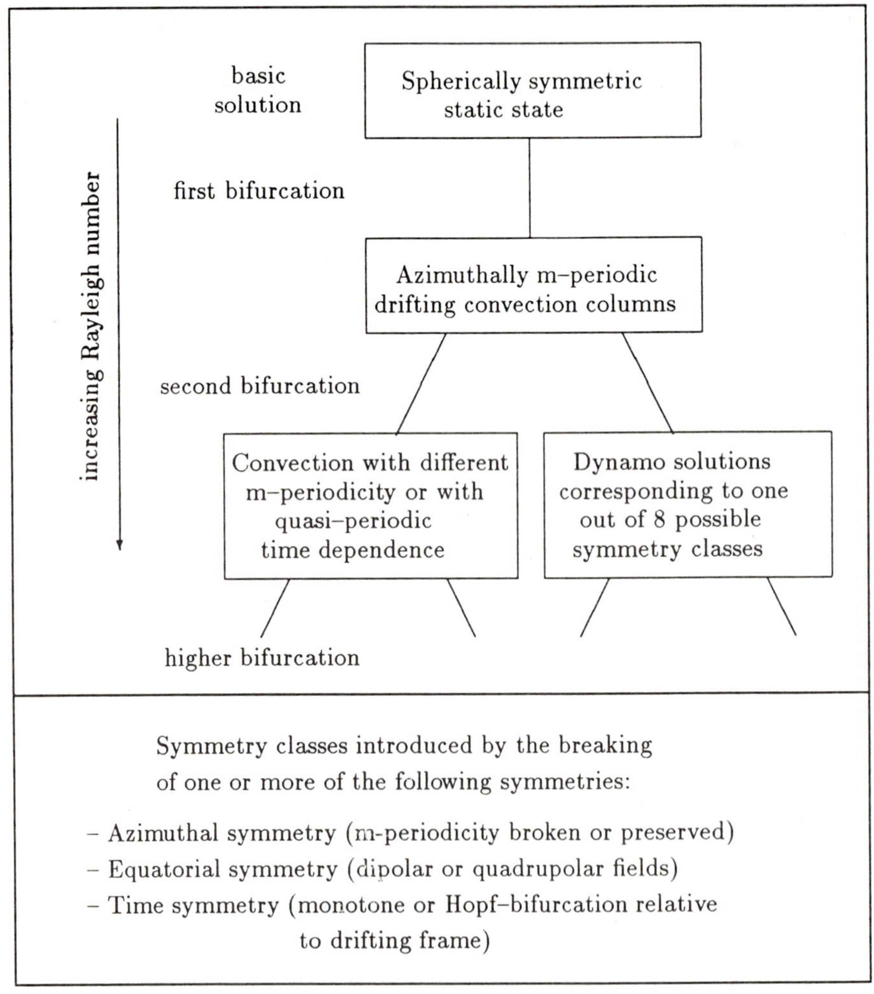

Figure 2: Bifurcation Diagram

scale r_0^2/λ of the Earth's core where r_0 denotes its radius. For these reasons and for reasons of numerical economy only steady, m–periodic dynamos have been investigated by Zhang and Busse in a recent series of papers [10,11,12].

Although the geomagnetic field is primarily dipolar even if extrapolated to the Earth's core as shown in Fig. 3, the substantial deviation from dipolar symmetry raises some questions. Assuming spherically symmetric physical conditions in the Earth's core and at its boundary, one could look for higher bifurcations in which the dipolar–quadrupolar symmetry is broken. While the magnetic Reynolds number, $R_m = Vr_0/\lambda$ based on a typical velocity V in the core is only of the order 10^2 to 10^3, numerous possibilities for bifurcations exist and the core is undoubtedly in a turbulent state. For reasons of symmetry, however, there should always be two solutions of the mixed dipolar–quadrupolar kind with equal probability for realisation in the turbulent state. In the long time average their contributions to the quadrupolar component are expected to cancel such that the magnetic field of the averaged turbulent state exhibits a dipolar symmetry. Paleomagnetic measurements seem to disagree with this expectation since the coefficient of the axisymmetric quadrupole field always amounts to a finite fraction of dipolar field strength of definite sign. Alternatively, the dipolar–quadrupolar symmetry may be broken by deviations from the North–South symmetry in the physical properties of the Earth's core or of the lower mantle which, through its finite electrical conductivity, exerts a considerable influence on the geodynamo. Indeed, the presence of large scale convection in the Earth's mantle, which manifests itself through the process of plate tectonics at the Earth's surface, is likely to cause major thermal and chemical inhomogeneities near the core–mantle boundary. Strong evidence for the influence of lateral inhomogeneities on the dynamo process has recently been obtained from paleomagnetic observations of the path of the "virtual" magnetic pole during reversals of the geomagnetic field. These observations suggest that the field does remain approximately dipolar during reversals with the north- and south magnetic poles progressing in the neighborhood of the 90^0 W and 90^0 E longitudes [13]. The implications for dynamo theory of this important discovery can not yet be fully comprehended. But the influence of lateral inhomogeneities can hardly be overestimated.

It is evident from our brief remarks, that the theory of planetary dynamos and of the geodynamo in particular is still in its early stages of evolution. The lack of knowledge about physical conditions in the Earth's core and at its boundary and the lack of opportunities for comparisons between theory and laboratory experiments continue to be severe handicaps in the development of a theory capable of making quantitative predictions.

3. An Approximate Solution for a Simple Dynamo

In this section we consider the possibility of growing solutions of equation (1) for the case of an electrically conducting fluid with constant diffusivity λ contained within a cylindrical box of height d and with the radius s_0. The velocity field $\vec{u}$

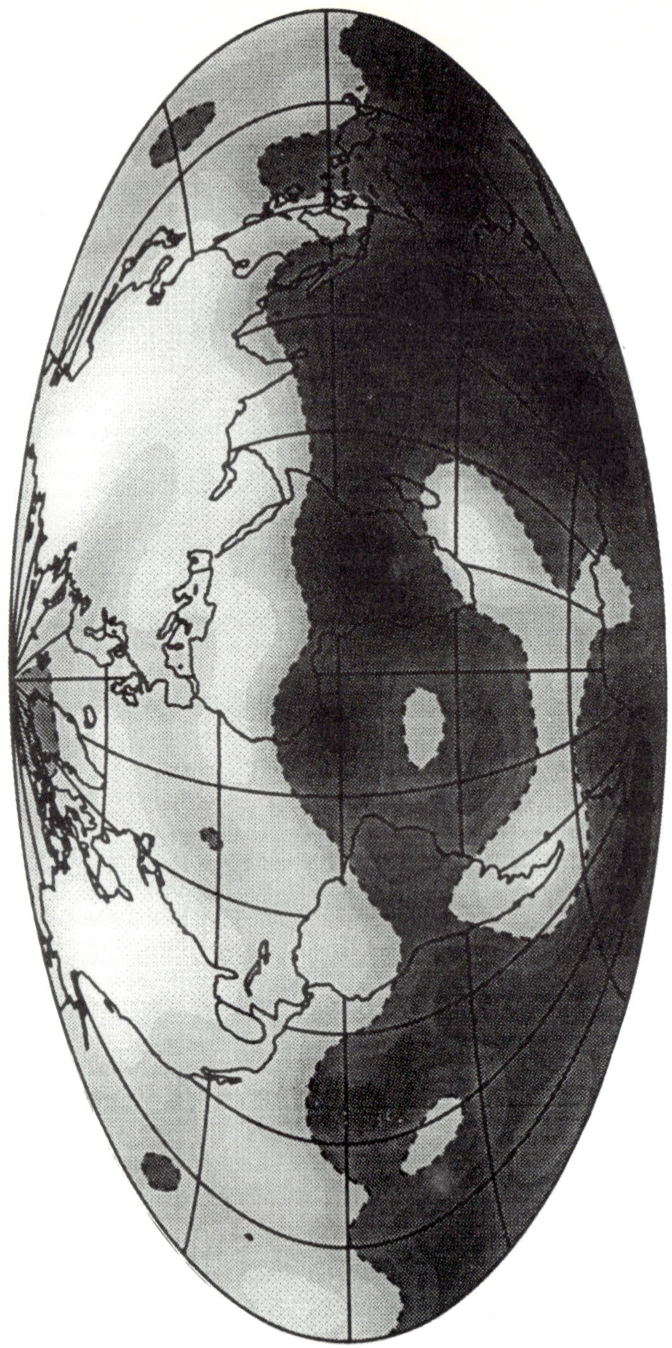

Figure 3: Lines of constant magnetic flux density intersecting the core–mantle boundary of the Earth. Regions of positive (dark) and negative (light) radial components $\vec{r} \cdot \vec{B}$ are separated by dashed lines.

of the fluid is given by

$$\vec{u} = \nabla \times \vec{k}\psi + \vec{k}w \tag{3a}$$

$$\text{with} \quad \psi \equiv wA/C = A \sin \alpha x \sin \alpha y \tag{3b}$$

where $\vec{k}$ is the unit vector in the z-direction of a cylindrical system of coordinates (s, φ, z) and where the definitions

$$x = s \cos \varphi, \quad y = s \sin \varphi$$

have been used. We shall assume the limit

$$\alpha \gg \max(s_0^{-1}, d^{-1}) \tag{4}$$

such that there is no need to satisfy boundary conditions for $\vec{u}$ on the surface of the cylindrical box. As is already suggested by our choice of the small scale periodic velocity field (3), we are trying to adopt one of the periodic dynamos of G.O. Roberts [14] for a finite fluid domain. We thus look for solutions of equation (1) in the form

$$\vec{B} = (\vec{\bar{B}} + \vec{\check{B}}) \exp\{pt\} \tag{5}$$

where the bar indicates the average over square intervals with length $2\pi/\alpha$ in the x, y–plane. We thus find

$$\bar{\psi} = \bar{w} = \overline{\vec{\check{B}}} = 0.$$

The average over equation (1) yields

$$(p - \lambda\nabla^2)\vec{\bar{B}} = \nabla \times \overline{(\vec{u} \times \vec{\check{B}})}. \tag{6}$$

After substracting (6) from (1) we find for the fluctuating component of the magnetic field

$$(p - \lambda\nabla^2)\vec{\check{B}} = \vec{\bar{B}} \cdot \nabla\vec{u} - \vec{u} \cdot \nabla\vec{\bar{B}} + \nabla \times (\vec{u} \times \vec{\check{B}} - \overline{\vec{u} \times \vec{\check{B}}}). \tag{7}$$

The limit (4) implies

$$| \vec{\check{B}} | \ll | \vec{\bar{B}} | \tag{8}$$

as will be demonstrated at the end. This condition allows us to neglect the last term in equation (7). For the same reason we may neglect the second term on the right hand side of (7) in comparison with the first. Anticipating that the growth rate p will satisfy the relationship $p \ll \lambda a^2$, we obtain as solution of (7)

$$\vec{\check{B}} \approx \vec{\bar{B}} \cdot \nabla\vec{u}/2\alpha^2\lambda \tag{9}$$

where again boundary conditions on the surface of the cylinder have been disregarded because of the small scale of $\vec{\check{B}}$. Inserting (9) into equation (6) we find

$$(p - \lambda\nabla^2)\vec{\bar{B}} = \nabla \times (\vec{\bar{B}} - \vec{k}\vec{k} \cdot \vec{\bar{B}})AC/4\lambda. \tag{10}$$

Using the general representation

$$\vec{B} = \nabla \times (\nabla \times \vec{k}h) + \nabla \times \vec{k}g \tag{11}$$

for the solenoidal vector field $\vec{B}$, we can derive two scalar equations for h and g by taking the z–components of equation (10) and of its curl,

$$(p - \lambda\nabla^2)\Delta_2 h = \frac{AC}{4\lambda}\Delta_2 g \tag{12a}$$

$$(p - \lambda\nabla^2)\Delta_2 g = \frac{AC}{4\lambda}(\nabla^2 - \Delta_2)\Delta_2 h \tag{12b}$$

where Δ_2 denotes the two–dimensional Laplacian

$$\Delta_2 \equiv \frac{1}{s}\frac{\partial}{\partial s}s\frac{\partial}{\partial s} + s^{-2}\frac{\partial^2}{\partial\varphi^2}.$$

Equations (12) can be solved if the boundary conditions at the surface of the cylinder are reduced to the requirement that the normal component of the current vanishes

$$\vec{j}\cdot\vec{n} \equiv \vec{n}\cdot\nabla\times(\nabla\times\vec{k}g) - \vec{n}\cdot\nabla\times\vec{k}\nabla^2 h = 0 \quad \text{at} \quad z = \pm\frac{d}{2} \quad \text{and at} \quad s = s_0. \tag{13a}$$

This condition is not sufficient to insure the continuity of the magnetic field with a potential field outside, but it is difficult to satisfy the latter condition in the present problem. Restricting the attention to the axisymmetric case, we obtain from (13) the condition for g

$$\frac{\partial}{\partial s}g = 0 \quad \text{at} \quad s = s_0 \quad \text{and} \quad g = 0 \quad \text{at} \quad z = \pm\frac{d}{2} \tag{13b}$$

while no condition is imposed on h since it corresponds to a purely azimuthal current density. After eliminating h we obtain the equation

$$(p - \lambda\nabla^2)^2\Delta_2 g = -\left(\frac{AC}{4\lambda}\right)^2\frac{\partial^2}{\partial z^2}\Delta_2 g. \tag{14}$$

The highest value of the real part of p is attained for the solution

$$g = J_0(\beta s)\cos(\pi z/d) \tag{15}$$

which satisfies all conditions (13b) if βs_0 is chosen equal to the first zero of the Bessel function $J_1(r)$,

$$J_1(r) = \frac{\partial}{\partial r}J_0(r) = 0 \quad \text{for} \quad r = \alpha_{11} \equiv 3.83\ldots \quad \text{or} \quad \beta = \alpha_{11}/s_0. \tag{16}$$

Since the growth rate p satisfies the relationship

$$p = -\lambda\left(\beta^2 + \left(\frac{\pi}{d}\right)^2\right) + \pi AC/4\lambda d \tag{17}$$

the condition for dynamo action can be written in the form

$$\frac{ACd}{\lambda^2} > 4\pi(1 + (\frac{\alpha_{11}d}{\pi s_0})^2). \tag{18}$$

The solution derived above is exact for the problem in which many cylinders are stacked upon each other and enclosed by cylindrical rings whose radii s_n correspond to the zeros α_{1n} of the Bessel function $J_1(r)$ through the relationship $s_n = \alpha_{1n} s_0 \alpha_{11}^{-1}$. Since insulating interfaces between these domains of electrically conducting fluids are assumed, the continuation of solution (15) and corresponding expression for h satisfy the appropriate boundary conditions exactly, with alternating signs of the magnetic field in adjacent domains. The fact that the poloidal field component described by h is more tightly constrained in this case than in the case of the single cylindrical box with insulating exterior suggests that the exact value of the right hand side in inequality (18) will be somewhat less than the approximate value that we have derived. Finally we observe that because of assumption (4) condition (8) is well satisfied when inequality (18) holds.

4. A Possible Laboratory Experiment

Using magnetic Reynolds numbers R_A and R_C based on the maximum amplitudes of the two components of the velocity field (3),

$$R_A \equiv A\alpha s_0/\lambda \quad , \quad R_C \equiv Cd/\lambda \tag{19}$$

the criterion (18) for dynamo action can be written in the form

$$R_A R_C > \alpha s_0 4\pi(1 + (\frac{\alpha_{11}d}{\pi s_0})^2). \tag{20}$$

Since magnetic diffusivities for liquid metals are typically of the order of 0.1 m^2 sec^{-1} and since the dimensions of a laboratory apparatus should not exceed the order of 1 m, criterion (20) implies that velocities of 1 m sec^{-1} will be required. Thus it is virtually impossible to satisfy condition (20) by convection flows driven by thermal buoyancy even though the form of the motion (14) is rather similar to that realized in a convection experiment rotating about a vertical axis. Instead flows of sufficient strength must be generated by external pumps. The appropriate technology for this purpose already exists in the form of coolant circuits in modern fast breeder reactors in which molten sodium is pumped in large volumes of hundreds of cubic meters. Indeed, the possibility for dynamo action has been considered in connection with those circuits [15], but it has been found that the geometry and the form of the flow field are not conducive to the onset of dynamo action [16]. The helical character of the flow field (3) is responsible for its nearly optimal property as dynamo as is evident from a comparison with other periodic flows considered by G.O. Roberts [14]. We therefore have proposed [7] that a flow (3) can be realized by placing a sufficient number of helical conduits in a cylindrical vessel filled with liquid

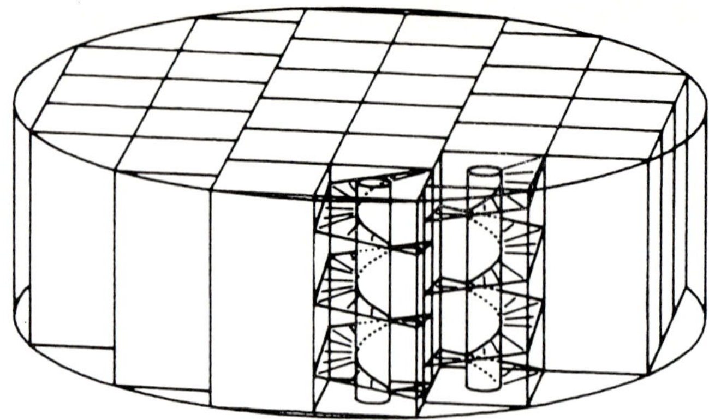

Figure 4: Sketch of a laboratory apparatus for a dynamo experiment. Two conduits for liquid sodium flow are combined. One conduit consists of the vertical pipes with alternating up– and down–flow. The other conduit consists of helical channels around the pipes. Again, the vertical direction of the flow changes sign in neighboring channels, but the helical sense remains the same. Connections of the pipes and channels at the top and bottom are not shown.

sodium with a diameter of about 2 m and a height of 1 m. A sketch of the proposed configuration is shown in Fig 4. This is not the appropriate place to discuss in detail the technical aspects of the experiment. The main point is the long circuit within the vessel of nearly uniform conductivity which minimizes the effects of advection to and from the pumps. Using the capacity of 100 m^3/h at a pressure head of 6 bars of currently available pumps and a magnetic diffusivity of liquid sodium of about 0. 1 m^2/sec we find that the condition (20) for dynamo action can be exceeded by at least a factor 10. Such a large margin should be sufficient to account for uncertainties introduced by the turbulent flow and the only approximate validity of the theory given in the preceding section.

5. Concluding Remarks

Dynamo theory is concerned with the self–excited generation of magnetic energy through a bifurcation from a non–magnetic state of mechanical motion. Although the process is of a fundamental nature, it has found little mentioning in textbooks of physics in contrast to the technical dynamo which relies on remanent magnetisation. Even the simplest dynamo in the form of the disk dynamo is not often found in books on electromagnetism. To geophysicists and astrophysicists the dynamo process is of special interest in the form of the homogeneous dynamo operating in a singly connected domain of a fluid of essentially uniform electrical conductivity. The development of the theory of homogenous dynamos has been carried out by theoretical geophysicists and astrophysicists and by applied mathematicians. The difficulty in realizing laboratory experiments is probably responsible for the lack of interest in the problem among physicists until recently.

An attempt to create a dynamo of the Ponomarenko type [17] has been undertaken by Gailitis and coworkers [18]. Their apparatus consists basically of two coaxial cylindrical pipes which are traversed by a helical flow of liquid sodium in opposite directions. The conditions for self-excitation have not been reached so far and it remains questionable whether the growth rate of the magnetic field can be made sufficiently strong before decay sets in as the magnetic field is advected out of the apparatus of only 3 m length. Obviously the conditions for self-excitation cannot be improved by an increase of the flow rate beyond a certain limit.

The main purpose of any experimental dynamo would be the determination of the equilibrium magnetic field amplitude and its dependence on the parameters of the system. In the proposed configuration shown in Fig. 3, for instance, the flow rate in the inner vertical cylinders and the outer helical manifolds can be varied independently. Of primary interest will be the dependence of the magnetic energy on the flow rate in the experiment which can be compared with yet to be developed nonlinear theories.

Acknowledgement: The author is indebted to Professor U. Müller, Karlsruhe, for discussions on the possibility of laboratory dynamos.

References

[1] Carlson, J. B., *Lodestone Compass: Chinese or Olmec Primacy?*, Science **189**, 753–760, 1975

[2] Larmor, J., *How could a rotating body such as the sun become a magnet?*, Brit. Ass. Advan. Sci. Rep. 159–160, 1919

[3] Chapman, S., and Bartels, J., *Geomagnetism, Vols 1 and 2*, Oxford University Press, 1940

[4] Cowling, T. G., *The magnetic field of sun spots*, Monthly Not. Roy. Astr. Soc. **94**, 39-48, 1934

[5] Lowes, F. J., and Wilkinson, I., *Geomagnetic dynamo: A laboratory model*, Nature **198**, 1158–1160, 1963

[6] Lowes, F. J., and Wilkinson, I., *Geomagnetic dynamo: An improved laboratory model*, Nature **219**, 717–718, 1968

[7] Busse, F. H., *Definition und Entwurf zweier magnetohydrodynamischer Experimente*, Report. IRB, Kernforschungszentrum Karlsruhe, pp. 1–20, 1979

[8] Moffat, H. K., *Magnetic Field Generation in Electrically Conducting Fluids*, Cambridge University Press, 1978

[9] Fearn, D., Roberts, P. H., and Soward, A. M., *Convection, stability and the dynamo*, pp. 60–324 in "Energy stability and convection, G. P. Galdi and B. Straughan, eds. Pitman Research Notes in Mathematics, vol. **168**, 1988

[10] Zhang, K.-K., and Busse, F. H., *Finite amplitude convection and magnetic field generation in a rotating spherical shell*, Geophys. Astrophys. Fluid Dyn. **44**, 33–53, 1988

[11] Zhang K.-K., and Busse, F. H., *Convection Driven Magnetohydrodynamic Dynamos in Rotating Spherical Shells*, Geophys. Astrophys. Fluid Dyn. **49**, 97–116, 1989

[12] Zhang K.-K., and Busse, F. H., *Generation of Magnetic Fields by Convection in a Rotating Spherical Fluid Shell of Infinite Prandtl Number*, Phys. Earth Planet. Int. **59**, 208–222, 1990

[13] Laj, C., Mazaud, A., Weeks, R., Fuller, M., and Herrero-Bervera, E., *Geomagnetic reversal paths*, Nature **351**, 447, 1991

[14] Roberts, G. O., *Dynamo action of fluid motions with two–dimensional periodicities*, Phil. Trans. Roy. Soc. London **A271**, 411–454, 1972

[15] Bevir, M. K., *Possibility of electromagnetic self–excitation in liquid metal flows in fast reactors*, Brit. J. Nucl. Energy **12**, 455–458, 1973

[16] Pierson, E. S., *Electromagnetic Self–Excitation in the Liquid–Metal Fast Breeder Reactor*, Nuclear Sci. Eng. **57**, 155–163, 1975

[17] Ponomarenko, Y. B., *On the theory of the hydromagnetic dynamo*, Zh. Prikl. Mech. Tech. Fiz. (USSR) **6**, 49–51, 1973

[18] Gailitis, A., *The Helical MHD Dynamo*, pp. 147–156 in "Topological Fluid Mechanics", H. K. Moffatt and A. Tsinober, eds., Cambridge University Press, 1990

Influence of Colored Noise on Energy Transport and Optical Line Shapes in Dimers

Ch. Warns and P. Reineker

Abteilung Theoretische Physik, Universität Ulm,
Albert-Einstein-Allee 11, W-7900 Ulm, Fed. Rep. of Germany

Abstract. The excitation energy transport between the molecules of a dimer and its optical absorption line shape are investigated using a model in which the influence of the phonons is modeled by a dichotomic stochastic process with colored noise giving rise to modulations of the molecular excitation energies. Equations of motion for the density matrix of the system and for correlation functions describing optical absorption are derived and solved algebraically on a computer. Transport properties, optical lineshapes and the connection between both are discussed.

1.Introduction

The dynamics of electronic excitations in condensed matter is of importance in such phenomena as exciton transport in molecular crystals, sensitized luminescence, formation of excimers, energy transfer between antenna molecules and the reaction centers in photosynthetic systems, etc. In these phenomena both the electronic interaction between neighboring molecules and the interaction between electronic and vibrational degrees of freedom are essential [1–6].

These interactions are difficult to treat simultaneously and therefore various approximations have been applied. One of these approaches is to describe the influence of the phonons on the electronic degrees of freedom by a stochastic process [1–4]. In this description the Hamiltonian consists of a time independent and a stochastically time dependent part:

$$H = H_0 + H_1(t) \tag{1}$$

$$H_0 = \sum_m \epsilon\, a_m^+ a_m + \sum_{m \neq n} J_{mn} a_m^+ a_n \tag{2}$$

$$H_1(t) = \sum_m h_{mm}(t)\, a_m^+ a_m + \sum_{m \neq n} h_{mn}(t)\, a_m^+ a_n \,. \tag{3}$$

In (1–3) a_m^+ and a_m are creation and annihilation operators for an excitation at site m, ϵ is the local excitation energy and J_{mn} the electronic transfer matrix element between sites m and n. $H_1(t)$ in (3) describes the influence of the phonons via fluctuations of the excitation energy ($h_{mm}(t)$) and of the transfer matrix element ($h_{mn}(t)$). In [1–4] it is assumed that the fluctuations are

Springer Proceedings in Physics, Vol. 69
Evolution of Dynamical Structures in Complex Systems
Editors: R. Friedrich · A. Wunderlin

described by a δ-correlated Gaussian Markov process with disappearing mean value (Haken-Strobl model) :

$$\langle h_{mn}(t)\rangle = 0 \tag{4}$$

$$\langle h_{mn}(t)\, h_{m'n'}(t')\rangle = 2\Lambda(m,n,m',n')\,\delta(t-t') \tag{5}$$

$$\Lambda(m,n,m',n') = \gamma_{|m-n|}\delta_{mn'}\delta_{nm'} + \bar{\gamma}_{|m-n|}\delta_{mm'}\delta_{nn'}(1-\delta_{mn})\,. \tag{6}$$

To simplify the discussion, in the following we shall take into account only energy fluctuations and restrict the consideration to a dimer system consisting of two molecules situated at sites 1 and 2, respectively. The diagonal elements of the density operator ρ_{11} and ρ_{22} describe the probabilities of finding the exciton at the two sites, and the non-diagonal elements ρ_{12} and ρ_{21} describe phase relations between them. Taking into account the properties of the stochastic process, we arrive at the following equation of motion for the stochastically averaged density matrix :

$$\frac{d}{dt}\begin{pmatrix}\rho_{11}\\ \rho_{22}\\ \rho_{12}\\ \rho_{21}\end{pmatrix} = \begin{pmatrix}0 & 0 & +iJ & -iJ\\ 0 & 0 & -iJ & +iJ\\ +iJ & -iJ & -2\gamma_0 & 0\\ -iJ & +iJ & 0 & -2\gamma_0\end{pmatrix}\begin{pmatrix}\rho_{11}\\ \rho_{22}\\ \rho_{12}\\ \rho_{21}\end{pmatrix}\,. \tag{7}$$

With the ansatz $\rho(t) = \exp(Rt)\,\rho$, (7) transforms to a non-hermitean eigenvalue problem which can be solved analytically. The eigenvalues are given by

$$R_1 = 0\,, \quad R_2 = -2\gamma_0\,, \quad R_{3,4} = -\gamma_0 \mp \sqrt{\gamma_0^2 - 4J^2}\,. \tag{8}$$

Assuming that the excitation is initially sitting at site 1, from the general solution we get the probability of finding the exitation there at time t as

$$\rho_{11}(t) = \frac{1}{2} + \frac{\sqrt{\gamma_0^2-4J^2}-\gamma_0}{4\sqrt{\gamma_0^2-4J^2}}\,e^{R_3 t} + \frac{\sqrt{\gamma_0^2-4J^2}+\gamma_0}{4\sqrt{\gamma_0^2-4J^2}}\,e^{R_4 t}\,. \tag{9}$$

From (8) and (9) we immediately see that the occupation probability at site 1 decays in a purely exponential manner as long as $\gamma_0 > 2J$. For $\gamma_0 < 2J$ the square root becomes imaginary and the occupation probability shows damped oscillations. In the first case, Haken and Strobl denoted the exciton motion as incoherent, in the second one as coherent.

To show the connection of the model parameters with experimentally accessible quantities, e.g. the optical absorption line shape can be calculated. To that end in the Hilbert space in addition to the excited states also the ground state of the dimer has to be taken into account. The equation of motion for the density operator can be derived in the same way as (7). The corresponding eigenvalue problem decays into two problems, i.e. the optical absorption is described by a set of eigensolutions which is independent from the one describing the transport. For a dimer with differently oriented dipole moments at the two molecules, we get two perpendicularly polarized line shapes centered at $\epsilon - J$ and $\epsilon + J$ and the half halfwidth is given by γ_0:

$$\chi'' = \frac{\gamma_0}{[\omega - (\epsilon \pm J)]^2 + \gamma_0^2} . \tag{10}$$

Whereas the eigenvalue problems for the optical absorption and the transport problem are separate, there is anyway a close connection with respect to the physical parameters. The transition from coherent to incoherent transport occurs for the same value of γ_0, namely $\gamma_0 = 2J$, for which the half halfwidth of the optical absorption line equals the splitting of the two absorption lines (Davydov–splitting).

2. Colored Noise: Energy Tranport

Recently a generalization of the model [7–10] was introduced by replacing the white noise by colored dichotomic and Gaussian noise. Within this model the mean square displacement for a quantum particle has been calculated in [7–10] using various approximation procedures. The application of the dichotomic colored noise model to the dimer problem allows the exact calculation of the transport between the molecules and of the optical absorption line shape. The results were first obtained numerically [11,12] and just recently in an analytical way [13] using computer algebra [14].

For simplicity we again neglect fluctuations of the transfer integrals and denote the energy fluctuations by $\epsilon_m(t) = h_{mm}(t)$. In a dichotomic colored noise process the energy fluctuations assume only two values, $(\epsilon_m(t) = \pm\Delta)$, with a switching rate λ. The process is then characterized in the following way:

$$\langle \epsilon_m(t) \rangle = 0 \tag{11}$$

$$\langle \epsilon_m(t)\, \epsilon_n(t) \rangle = \delta_{nm}\, \Delta^2\, e^{-\lambda|t-t'|} \tag{12}$$

$$\langle \epsilon(t_1)\, \epsilon(t_2) \cdots \epsilon(t_{2\mu}) \rangle = \langle \epsilon(t_1)\, \epsilon(t_2) \rangle \cdots \langle \epsilon(t_{2\mu-1})\, \epsilon(t_{2\mu}) \rangle \tag{13}$$

$$t_1 \geq t_2 \geq \cdots \geq t_{2\mu} ,$$

i.e. the average value of the fluctuations vanishes, two-times correlation functions decay exponentially, and multi-time correlation fuctions are calculated according to (13). The white noise limit is obtained for large values of λ. Integrating (5) and (12) with respect to t, we see that the limit has to be carried out in the following way: $\lambda \to \infty$, $\Delta \to \infty$ such that $\Delta^2/\lambda = \gamma_0$ remains finite.

We again derive the equation of motion for the density operator of the system. In contrast to the white noise case the equations of motion cannot be closed using the four averaged matrix elements of the density operator alone. In addition to the equation of motion for $\langle \rho_{11} \rangle$ we have to derive equations of motion for $\langle \epsilon_1 \rho_{11} \rangle$, $\langle \epsilon_2 \rho_{11} \rangle$, $\langle \epsilon_1 \epsilon_2 \rho_{11} \rangle$, and corresponding equations for the other elements of the density operator. These equations are most conveniently derived using a theorem by Shapiro and Loginov [15] :

$$\frac{d}{dt}\langle \epsilon(t)\, \Phi_t[\epsilon] \rangle = \langle \epsilon(t)\, \frac{d}{dt}\Phi_t[\epsilon] \rangle - \lambda \langle \epsilon(t)\, \Phi_t[\epsilon] \rangle \,. \tag{14}$$

In this way we arrive at a set of 16 coupled differential equations of the form

$$\dot{\rho} = L\,\rho\,, \tag{15}$$

where ρ is a 16-dimensional column vector and L a 16-dimensional non-hermitean matrix. The structure of this matrix is such that its eigensolutions could be obtained using computer algebra. We introduce the scaled model parameters and eigenvalues by

$$R_{sn} = \frac{R_n}{2J}\,, \quad \lambda_s = \frac{\lambda}{2J}\,, \quad \Delta_s = \frac{\Delta}{2J}\,. \tag{16}$$

The eigenvalues are then given by

$$\begin{aligned}
&R_{s1} = 0 \\
&R_{s2} = -2\lambda_s \\
&R_{s3} \ldots R_{s6} = -\lambda_s \\
&R_{s7,8} = -\lambda_s \pm i \\
&R_{s9} \ldots R_{s12} = -\lambda_s \pm \frac{1}{\sqrt{2}} \\
&\qquad \times \sqrt{\lambda_s^2 - 4\Delta_s^2 - 1 \pm \sqrt{16\Delta_s^4 - 8\lambda_s^2\Delta_s^2 + 8\Delta_s^2 + \lambda_s^4 + 2\lambda_s^2 + 1}} \\
&R_{s13} \ldots R_{s16} = -\lambda_s \pm \sqrt{\lambda_s^2 - 1 - 2\Delta_s^2 \pm 2\sqrt{\Delta_s^4 - \lambda_s^2}}\,.
\end{aligned} \tag{17}$$

A closer inspection of the structure of the eigenvectors shows that $\langle\rho_{11}(t)\rangle$, the probability of finding the excitation at the first molecule, is only determined by the eigenvalues R_{s1}, R_{s3}, which are purely real, and by R_{s13} ... R_{s16}, which may be real or complex depending on the magnitudes of λ_s and Δ_s. Fig. 1 shows the areas in the λ_s^{-1}-Δ_s^{-2} plane, where four, two and none of the four eigenvalues R_{s13} ... R_{s16} are complex (indicated by the numbers in the brackets).

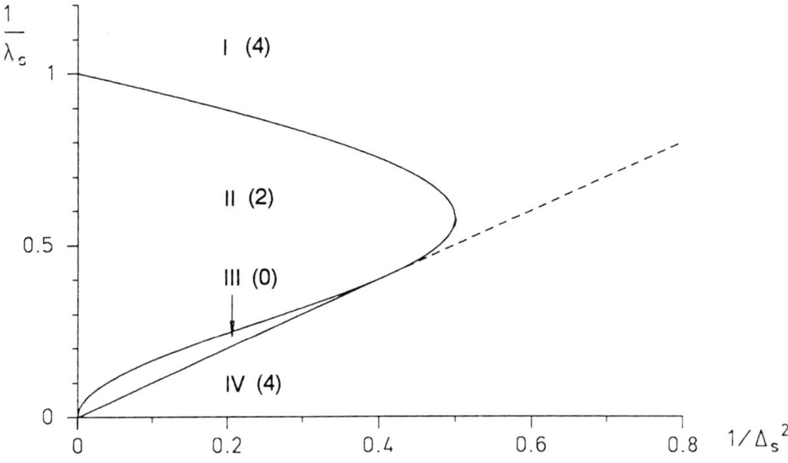

Figure 1: Number of complex eigenvalues $R_{s13}, \ldots, R_{s16}$ in the $1/\lambda_s - 1/\Delta_s$-plane

With the notation $y = \lambda_s^{-1}$, $x = \Delta_s^{-2}$ the separation of the four areas is given by the curves

$$y = x \tag{18}$$

$$y = \frac{\sqrt{2 - x + 2\sqrt{1-2x}}}{\sqrt{x+4}} \tag{19}$$

$$y = \frac{\sqrt{2 - x - 2\sqrt{1-2x}}}{\sqrt{x+4}} \; . \tag{20}$$

The straight line determines the vanishing of the radicand of the inner square root, the curves of (19) and (20) the vanishing of the radicand of the outer square roots. From this discussion we expect two areas of coherent motion: area I for small values of λ_s, corresponding to the (quasi-) static case, and area IV for large values of λ_s, corresponding to the case of fast fluctuations.

The left column of Fig. 2 shows the time evolution of $\langle\rho_{11}\rangle$ for $\Delta_s = 0.5$, i.e. $\Delta_s^{-2} = 4$ and various values of λ_s. We see that for all values of λ_s the occupation probability shows an oscillatory behavior. For small λ_s-values the probability oscillates with two frequencies, for large λ_s only a single frequency survives. In Fig. 3 for $\Delta_s = 2$, i.e. $\Delta_s^{-2} = 0.25$ the series of λ_s-values shows the time dependence of occupation probabilities starting in region I, crossing II and III and ending in region IV. The left column starts with an occupation probability oscillating with two frequencies (region I). For medium values of λ_s the probability deacays exponentially, and for large λ_s-values it oscillates with a single frequency. A corresponding behavior is shown by the time evolution of $\langle\rho_{12}\rangle$, whose time evolution is determined also by the eigenvalues $R_{s9}...R_{s12}$.

3. Colored Noise: Optical Absorption

In the framework of linear response theory the optical absorption line shape of the dimer is given by

$$I(\omega) = Re \int_0^\infty e^{i\omega t} \langle \mu(t)\, \mu(0)\rangle \, dt \, . \tag{21}$$

Here $\mu(t)$ is the optical dipole operator and can be expressed by the creation and annihilation operators for a local excitation in the following way:

$$\mu(t) = \sum_n \mu_n \left(a_n^+(t) + a_n(t)\right) . \tag{22}$$

Taking into account the normalization of the line shape, we arrive at

$$I_{nz}(\omega) = \frac{1}{\pi} \sum_{m,n} Re \int_0^\infty e^{i\omega t}\, U_{mn}(t)\, dt \;=\; \frac{1}{\pi}\, Re \int_0^\infty e^{i\omega t}\, S(t)\, dt \;=\; \frac{1}{\pi}\, Re\, \tilde{S}(\omega) . \tag{23}$$

Here $U(t) = \overleftarrow{T} \exp(-\int_0^t dt' H(t'))$ describes the time evolution of the system and S(t) is defined below. To simplify the calculation we introduce the following variables

$$x \;=\; U_{11} + U_{12}\, , \quad y \;=\; U_{21} + U_{22} \tag{24}$$

$$\begin{aligned} S(t) &= \langle x + y \rangle \\ A(t) &= \langle \epsilon_1\, x\rangle + \langle \epsilon_2\, y\rangle \\ B(t) &= \langle \epsilon_2\, x\rangle + \langle \epsilon_1\, y\rangle \\ C(t) &= \langle \epsilon_1\, \epsilon_2\, x\rangle + \langle \epsilon_1\, \epsilon_2\, y\rangle \, . \end{aligned} \tag{25}$$

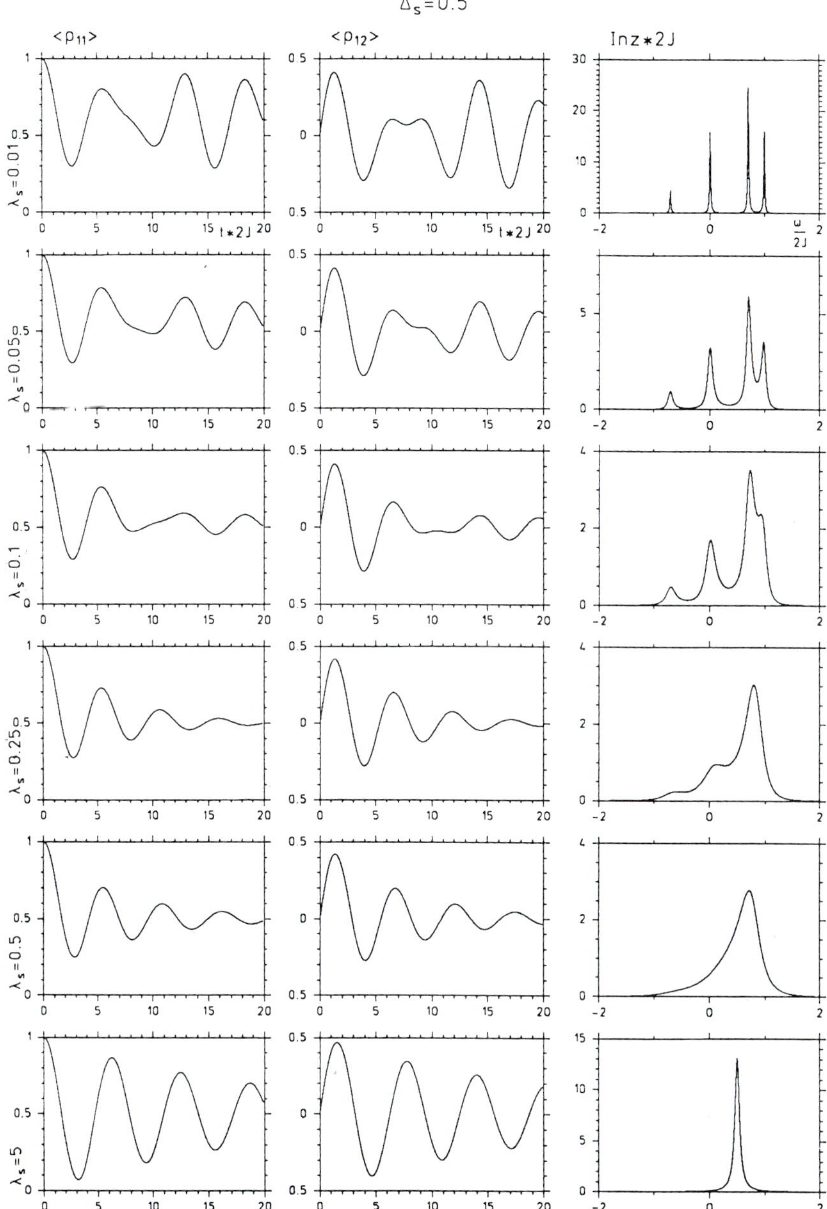

Figure 2: Excitation transport and optical absorption for $\Delta_s = 0.5$

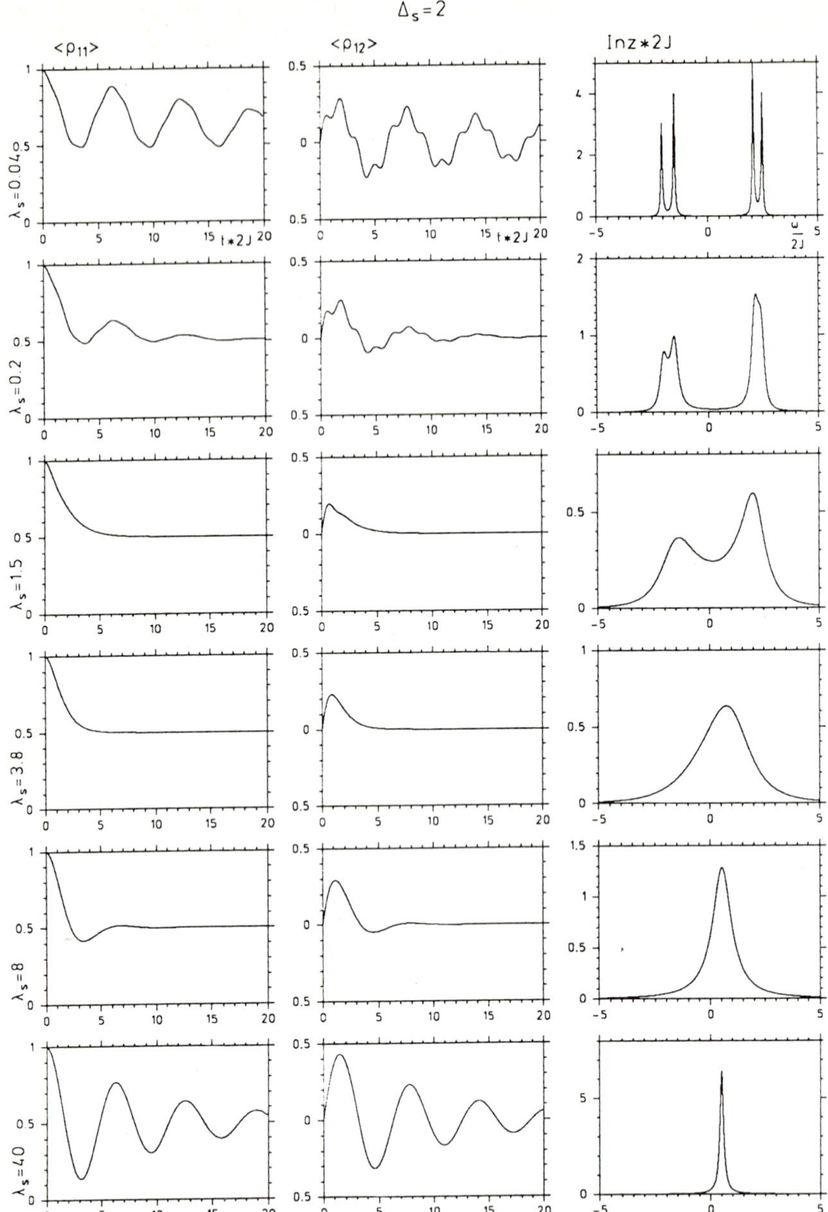

Figure 3: Excitation transport and optical absorption for $\Delta_s = 2$

The equations of motion for these quantities can be derived using again the theorem of Shapiro and Loginov:

$$\frac{d}{dt}\begin{pmatrix} S(t) \\ A(t) \\ B(t) \\ C(t) \end{pmatrix} = \begin{pmatrix} -iJ & -i & 0 & 0 \\ -i\Delta^2 & -\lambda & -iJ & 0 \\ 0 & -iJ & -\lambda & -i \\ 0 & 0 & -i\Delta^2 & -2\lambda - iJ \end{pmatrix} \begin{pmatrix} S(t) \\ A(t) \\ B(t) \\ C(t) \end{pmatrix} . \tag{26}$$

The line shape is obtained by Laplace–transforming the equations of motion and solving for $\tilde{S}(\omega_s)$ $(\omega_s = \omega/2J)$:

$$\tilde{S}(\omega_s)\,2J = 2\Bigg[- i(\omega_s - \frac{1}{2}) + \Delta_s^2\Big[- i\omega_s + \lambda_s + \frac{1}{4}\big[- i\omega_s + \lambda_s + \Delta_s^2[-i(\omega_s - \frac{1}{2}) + 2\lambda_s]^{-1}\big]^{-1}\Big]^{-1}\Bigg]^{-1} . \tag{27}$$

The optical line shape could also be calculated from the density matrix by again supplementing the Hilbert–space by the ground state of the dimer. As in the case of white noise, the eigenvalue equation for the density operator factorizes into a part describing the transport and a part describing the optical absorption. Therefore, also in the case of colored noise the two phenomena are mathematically independent.

The line shape for $\Delta_s = 0.5$ and various values of λ_s is shown in Fig. 4. For small values of λ_s we obtain a strongly structered line; the position of the peaks is determined by the random energy distribution $\epsilon \pm \Delta_s$, on account of the quasi-static dichotomic noise and their widths by the life-time of the energy state, i.e. by λ_s. With increasing values of λ_s the lines become broader and merge. For very large values of λ_s and finite values of Δ_s the lines show a narrowing, because the fluctuations are now so fast that the electronic degrees of freedom can no longer follow. The same line shapes are also shown in the right column of Fig. 2. In Fig. 3 the third column represents optical line shapes for $\Delta_s = 2$. For small values of λ_s we have a lineshape whose structure is determined by the quasi-static energy distributions. With increasing λ_s the peaks becomes broader and merge and for large values of the switching rate we have a dynamically narrowed line.

4. Concluding remarks

Fig. 2 shows that for small amplitudes of the energy fluctuations $\Delta_s = 0.5$ the excitation energy transport in the dimer occurs by damped oscillations for all values of λ_s and thus this transport can be denoted as coherent in the whole λ_s-range. For the larger amplitude in Fig. 3 we have two transitions between coherent and incoherent transport. The first corresponds to the transition between regions I and II in Fig. 1, the second to the transition into region IV. In the white noise case only the second transition is present.

In the right columns in Figs. 2 and 3 the optical lineshapes are pictured for the same values of the model parameter. For slow fluctuations we have the lineshape corresponding to the quasi-static distribution of energy levels, i.e an inhomogeneously broadened line. For fast fluctuations the line shows dynamic narrowing and a homogeneous width. Thus the model allows to describe a continuous transition between inhomogeneously and homogeneously broadened lines.

Comparing the transport and the optical lineshapes, we see an oscillatory behavior of the occupation probabilities when the optical line shape is narrow both in the inhomogeneous and homogeneous limits. When the full halfwidth of the line is equal to $4J$ (in the scaled units we are using this corresponds to a halfwidth of 2), the occupation probability decays without oscillations to the stationary value. Such a situation occurs in Fig. 3, but not in Fig. 2.

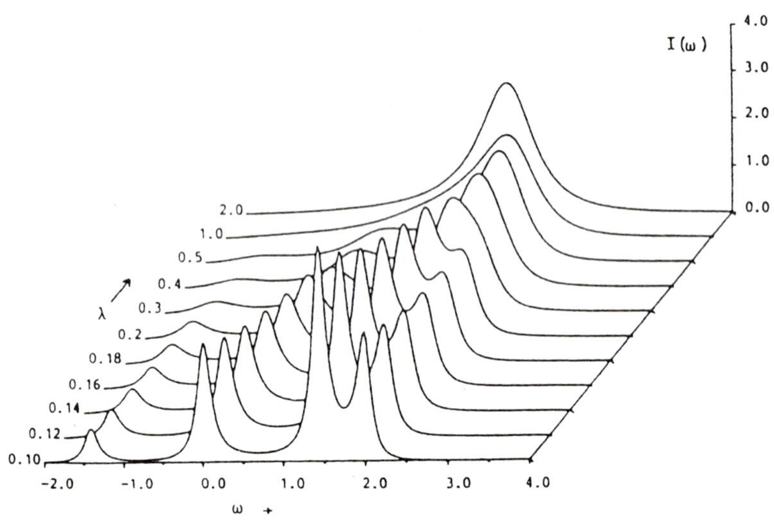

Figure 4: Optical line shapes for $J = \Delta = 1$ and $0.1 \leq \lambda \leq 2.0$

References

1. H. Haken, G. Strobl: in *The Triplet State*, ed. by A. B. Zahlan (Cambridge University Press 1967)
2. H. Haken, P. Reineker: Z. Physik **249**, 253 (1972)
3. H. Haken, G. Strobl: Z. Physik **262**, 135 (1972)

4. P. Reineker: in *Exciton Dynamic in Molecular Crystals and Aggregates*, Springer Tracts in Modern Physic, Vol 94, ed. by G. Höhler (Springer, Berlin 1982) p. 111
5. M.K. Grover, R. Silbey: J. Chem. Phys. **52**, 2099 (1969)
6. V.M. Kenkre: in *Exciton Dynamic in Molecular Crystals and Aggregates*, Springer Tracts in Modern Physic, Vol 94, ed. by G. Höhler (Springer, Berlin 1982) p. 1
7. Y. Inaba: J. Phys. Soc. Jpn. **52**, 3144 (1983)
8. K. Kassner, P. Reineker: Z. Physik B **59**, 357 (1985)
9. K. Kassner, P. Reineker: Z. Physik B **60**, 87 (1985)
10. K. Kassner: Z. Physik B **70**, 229 (1988)
11. V. Kraus, P. Reineker: Phys. Rev. A **43**, 4182 (1991)
12. B. Kaiser, A.M. Jayannavar, P. Reineker: J. Lumin. **43**, 73 (1989)
13. Ch. Warns: Diplomarbeit, Universität Ulm, 1992
14. B.W. Char, K.O. Geddes, G.H. Gonnet, M.B. Monagan, S.M. Watt: *MAPLE Reference Manual* (Waterloo Maple Software, Waterloo 1988)
15. V.E. Shapiro and V. Loginov: Physica **91** A, 563 (1978)

Part IV

Biology and Medicine

Coordination Dynamics of Human Brain and Behavior

J.A.S. Kelso

Program in Complex Systems and Brain Sciences,
Center for Complex Systems, Florida Atlantic University,
500 N.W. 20th Street, Boca Raton, FL 33431, USA

Abstract. New observations using a multisensor SQUID array are described showing that nonequilibrium phase transitions occur in both brain and behavior as a control parameter is systematically varied. Coherent brain and behavioral states are captured by the same kind of order parameter, linking very different material components in each case. In line with synergetics [1], such results indicate that the brain is a self-organizing, pattern forming system that can switch flexibly from one coherent state to another. In this respect, brain and behavior share common dynamics.

1 Introduction

How the nervous system orchestrates its many degrees of freedom in the service of behavioral functions is not understood. An increasingly voiced concern among neuroscientists is that although tremendous advances in knowledge of cellular and synaptic phenomena have occurred, insights into the organizational principles remain few. An often asked question is: How is it all coordinated? Following, indeed profoundly inspired by Haken's [1] synergetics, we pursue the possibility that principles of coordination may lie at the level of *patterns* and that a focus on pattern (especially the pattern dynamics) may provide the conceptual leap necessary to advance our understanding of how brain and behavioral events are coordinated. At the beginning of this century, the father of modern neurophysiology, Charles Scott Sherrington also emphasized the importance of pattern formation for understanding the brain. He described the brain as an enchanted loom where millions of flashing shuttles weave a dissolving pattern, always a meaningful pattern, but never an abiding one. It is only quite recently, however, that technology has provided us with tools of sufficient spatial and temporal resolution to observe dynamic patterns generated by brain cell ensembles. In a later section, we present some recent results using a multisensor SQUID (Superconducting QUantum Interference Device) array. SQUIDs allow access to the spatiotemporal patterning of neural activity generated by intracellular dendritic current flow in the brain. These signals are picked up directly, without having to stick electrodes in the brain itself. Because the skull and scalp are transparent to magnetic fields generated inside the brain and because the array is large enough to cover a substantial portion of human neocortex, this new research tool opens a (noninvasive) window into the brain's dynamic patterns and their relation to behavior.

But just as, if not more important than the technology itself, is the emergence of theoretical concepts of pattern formation and cooperative phenomena to deal

Springer Proceedings in Physics, Vol. 69
Evolution of Dynamical Structures in Complex Systems
Editors: R. Friedrich · A. Wunderlin

with large number of elements (such as neurons or neuron-like elements) along with mathematical and computational tools to model their behavior. As Haken [1] has repeatedly emphasized, in complex systems, regardless of level of observation, it is generally not possible, even if it were useful, to determine the detailed behavior of every degree of freedom. Indeed, often it is not even clear what the essential degrees of freedom are. The key step is to identify *collective variables* (or *order parameters*, cf. [1]) which capture the system's coordinative state (see sections 2 and 3 below). In matters concerning brain and behavior, this choice of collective variables must usually be based on empirical insights: collective variables define stable and reproducible relationships among the components of a behavioral or neural system and are, in general, function- or task-specific. Given a set of collective variables, the coordination dynamics may then be modelled as equations of motion of these collective variables (e.g. [2]). Our working hypothesis is that the linkage between events at a 'microscopic' level (e.g. of neurons and neuronal ensembles) and the 'macroscopic' behavioral level is by virtue of *shared dynamics*, not because any single level has ontological priority over another [3]. Thus the language of coordination dynamics may be seen to apply at multiple spatial and temporal scales in a level-independent fashion. In Section 4 below, we provide evidence suggesting an "order parameter isomorphism" between neural and behavioral events. With the aid of a multisensor SQUID array, we show that coherent states of both brain and behavior may be captured by the same kinds of order parameters or collective variables.

2 Strategic Approach to Finding Collective Variables

How do we find collective variables or order parameters for dynamic patterns on a chosen observational scale? Unlike much of contemporary brain research, but perfectly in line with synergetics [1], our focus is in and around phase transitions or bifurcations. Such critical points are significant because: (a) they allow a clear distinction between *patterns*, thus enabling us to identify the dimension on which patterns are defined; (b) predictions about the dynamics (e.g. stability, loss of stability) can be tested near critical points; and (c) instabilities allow the system to flexibly switch among multiple collective states. In other words, the dynamics provide generic mechanisms (e.g. phase transitions, intermittency) for entering and exiting coherent patterns. It is important to emphasize, however, that these "mechanisms" can be implemented neurophysiologically in very different ways (cf. [4]).

3 Elementary Coordination Dynamics

What does it mean for the nervous system and behavior to be subject to dynamics that govern its coordination activity? And what are the methodological and empirical consequences of this view? A major consequence is that *stability* is an essential property of a coordination pattern, directly observable as the ability to return to the pattern after a small perturbation, but most dramatically evident when stability is lost. Indications of loss of stability can be observed through various measures before pattern switching actually occurs.

To illustrate the key ideas and their implementation we describe a simple experimental model system which treats the problem of sensorimotor coordination as a pattern formation process ([5]). First we treat the behavioral level of description: later on we will use the same example to help elucidate the brain dynamics.

In the experiment, the subject's task was to flex the index finger in time to a computer generated auditory tone (or metronome) in either of two conditions: (1) on the beat, i.e. synchronize exactly with the tone; and (2) off the beat, i.e., produce a single manual response midway between two consecutive tones. For each condition, the rate of the auditory stimulus was increased every 10 cycles in .25 Hz steps from 1.0 Hz to 3.25 Hz. Instructions emphasized maintaining a 1:1 relation between the stimulus tone and the manual response. Forty consecutive trials were carried out for each condition, each trial lasting a little over a minute. The following phenomena were observed: (a) strong tendencies for phase- and frequency synchronization in both conditions; (b) a spontaneous transition from the syncopated, off-the-beat condition to the synchronized on-the-beat condition at a critical stimulus frequency; (c) desynchronization or loss of entrainment, characteristic features of which included occasional wandering in the phase relation between hand and metronome which increased with the movement frequency; and (d) hysteresis, under conditions when the direction of parameter change was reversed.

In this example, the relative phase, ϕ, between the rhythmically moving limb and the metronome, is a good collective variable characterizing the action-perception patterns. Determining the dynamics of collective variables, however, is nontrivial. The idea is to map empirically stable patterns at $\phi \simeq 0$ (on the beat) and $\phi \simeq \pi$ (off the beat) onto attractors of the collective variable dynamics. Obviously when several stationary patterns coexist, nonlinear models must be found. A general strategy ([2, 5, 6, 7]) is to assume sufficiently high order dynamics and expand the vector field of those dynamics to accommodate the patterns observed. *Figure 1* plots the vector field of the dynamics $\phi(x - \text{axis})$ versus $\dot{\phi}(y - \text{axis})$ for different parameter values using this minimality strategy of modeling only observed states:

$$\dot{\phi} = \delta\omega - a\sin\phi - 2b\sin 2\phi \tag{1}$$

where a and b are parameters that can be obtained from the experimental data and $\delta\omega$ expresses any frequency difference between participating components. The system defined by Eq. 1 ([5]) contains stationary patterns or fixed points of ϕ where the time derivative of $\dot{\phi}(\phi)$ crosses the $\phi - $axis. When the slope of $\dot{\phi}$ is negative the fixed points are stable and attracting; when the shape is positive the fixed points are unstable and repelling. Arrows in *Figure 1* indicate the direction of flow.

The model illustrates the following essential features:

- The concept of *collective variable* or *order parameter* ([1]), relative phase, ϕ, in this case corresponding to stable phase- and frequency-synchronized states. In *Figure 1a*, two such states exist, one close to in-phase and one close to anti-phase;

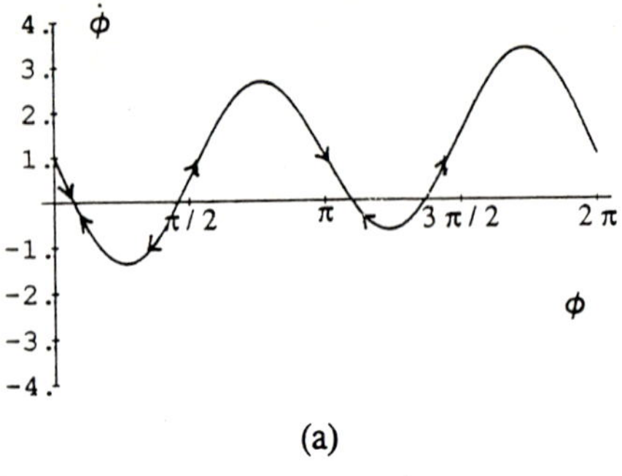

(a)

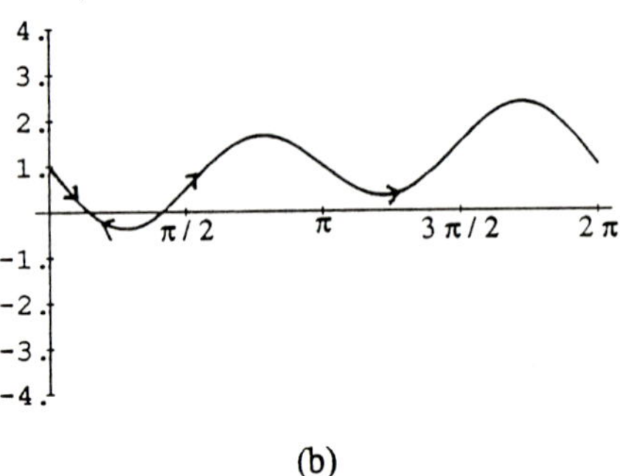

(b)

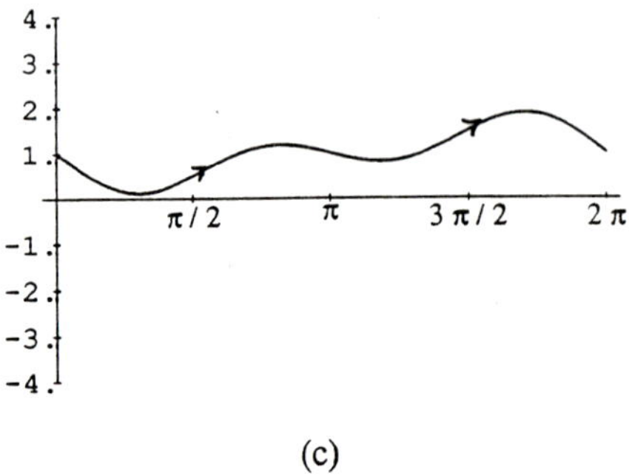

(c)

Figure 1: Plot of $\dot{\phi}$ vs ϕ in Eq. 1 for $a = .5$ Hz, $\delta\omega = 1.0$ Hz and $b = 1.0$ Hz (a), $b = 0.5$ Hz (b) and $b = 0.25$ Hz (c). See text for details.

- *Symmetry and Broken Symmetry.* When $\delta\omega$ is zero, the system (Eq. 1) is symmetric, that is to say, invariant under the operation $\phi \to -\phi$. In certain cases, for example, Kelso's ([8, 9]) bimanual coordination paradigm, left-right symmetry (similar limbs, same frequency) explains the position of the natural phases at $\phi = 0$ and $\phi = \pi$ [2]. But because metronome and limb are obviously not equivalent, $\phi \rightarrow -\phi$ symmetry can no longer be assumed. Specifically, because patterns with a systematic lead or lag were experimentally observed whose symmetry partners do not coexist at the same parameter values, the $\phi \rightarrow -\phi$ symmetry constraint cannot be applied to restrict the functional form of the dynamics. The consequences of this broken symmetry in the coordination dynamics are explored briefly below;

- The concept of *control parameter(s)*. As the ratio $\mid \frac{a}{4b} \mid$ corresponding to the experimental frequency is increased, the amount of attraction to the stable states decreases, until one (the less stable pattern near $\phi = \pi$, *Fig. 1b*) and ultimately both (*Fig. 1c*) cease to be stationary solutions. Control parameters thus move the system through its collective states;

- The concept of *bifurcation* or *phase transition*. Changes in coordination may be smooth or abrupt. Regardless, loss of stability is at the origin of coordinative change. Eq. 1 is not just a good description of action-perception patterns but contains additional properties that can be (and have been) systematically explored. Typically, these properties pertain to predicted behavior around nonequilibrium transition points and the effects of noise. Predicted features associated with growing instability are revealed in Eq. 1 by (1) *drift in the fixed points* (due, e.g. to intrinsic frequency differences between metronome and limb, $\delta\omega$) as the control parameter changes. For instance, the fixed point of ϕ near zero drifts along the x-axis toward $\pi/2$ and *vice-versa* (compare *Figs 1a* and *1b*); (2) slowly drifting fixed points signal upcoming *tangent* or *saddle node* bifurcations. Stable and unstable fixed points coalesce onto a saddle node leaving only one fixed point attractive state beyond (*Fig. 1b*) the critical point; (3) *critical slowing down*: as a coordinative state flattens out around the minimum (compare *Figs. 1b* and *1c*), it takes longer and longer for the system to return to its initially prepared state after it is pushed away by a small perturbation; and (4) *enhancement of fluctuations*: the effect of noise, which acts to continuously "kick" the system away from its attractive states, increases as the system approaches a critical point. Without reviewing the literature, all four of these predictions have been confirmed experimentally at the behavioral level in a large number of different experimental systems not only by ourselves (e.g. [9, 10, 11]) but also by others (e.g. [12, 13]).

- *Loss of entrainment*. When stationary solutions are no longer found, "running" solutions (ϕ ever increasing or decreasing) or, if the 2π periodicity convention is applied to ϕ, "wrapping" solutions occur (*Fig. 1c*). In this régime there is no longer any phase or frequency locking, a condition called loss of entrainment or desynchronization. Desynchronization may play a useful role in biological coordination, i.e. it does not always mean irregular behavior (see next). In Eq. 1 the exact point at which the detuning parameter, $\delta\omega$, causes running solutions and whether it appears at all depends on the other parameters a and b. This simply reflects the fact that both the frequency difference between components as well as the basic frequency of coordination are control parameters, affecting the onset of the running solution.

- *Intermittency.* Although the saddle nodes may vanish indicating loss of entrainment, the coordination system may stay near the previously stable, fixed point. Note that in *Fig. 1c* there is still *attraction* to certain phase relations even though the relative phase itself is unstable. Thus "wrapping" solutions can possess a fine structure, spending more time around relative phase values at which $\dot{\phi}$ is minimal (the remnant of the attractor). Such behavior is especially significant because it shows that although there is no strict mode locking, there is partial coordination interrupted occasionally, by phase wandering. This form of partial coordination was termed *relative coordination* by von Holst (1939) who attributed the phase attraction to a central control principle, the so-called magnet or M-effect.

Relative coordination has a theoretical interpretation in terms of intermittent dynamics ([5, 14]): In *Fig. 1* as the curve is lifted up from the x-axis a phantom fixed point appears. Motion hovers around this point most of the time but occasionally escapes away along the repelling direction. What happens after the escape depends on the attractor layout around the saddle node. In our case, due to the 2π periodicity of ϕ, the relative phase slips and is then reinjected. This phenomenon is called intermittency and represents one of the generic processes found in low dimensional systems near tangent bifurcations ([15]). The identification of relative coordination with intermittent behavior of periodic flows ([14]) is consistent with the emerging view that biological systems tend to live near boundaries separating regular and irregular behavior. Relative coordination *qua* intermittency allows for low energy, flexible switchings among metastable coordinative states.

In summary, our elementary coordination dynamics (Eq.1) contain (a) no coordination; (b) absolute coordination (when two or more components oscillate at the same frequency and maintain a fixed relation); and (c) relative coordination (the tendency toward phase attraction even when the component frequencies are not the same). All these spatiotemporal forms of organization have an explanation, namely, they are patterns that emerge in different parameter régimes of the identified coordination dynamics. Transitions between such régimes add richness to the behaviors possible. Such dynamics can express the coordination between (a) components of an organism ([2, 8, 9, 11]); (b) organisms themselves ([13]); and (c) organisms and their environment ([5, 12]). Attesting to its veracity is that all the effects predicted by the elementary coordination law (Eq. 1) have been found in experiments. We note that the collective dynamics specified by Eq. 1 can also be derived from the component level (e.g. [2, 5]) where the various coordination patterns arise due to nonlinear coupling among the elements involved.

4 Phase Transitions in the Brain

Universal features of self-organization, such as multistability, disorder-order and order-order transitions, intermittency and hysteresis can be found in perception, sensorimotor coordination, learning and development (see e.g. chapters in reference [16]). Thus, not only are physical, chemical and biological systems subject to dynamical laws, but psychology – the science of mind and behavior – is as well. Despite the fact that phase transition phenomena have been found also in artificial neural networks (e.g. [17]) no direct evidence, to our knowledge, has been reported for phase transitions or bifurcations in the human brain itself. This is surprising in light of the conclusion, some twenty years ago, that..."the *possibility* (italics ours) of waves, oscillation, macrostates emerging out of cooperative processes, sudden transitions, prepatterning etc. seem made to order to assist in the understanding of integrative processes of the nervous system that remain unexplained in contemporary neurophysiology" ([18]).

In recent work [19, 20] we have used dynamic brain imaging in order to observe rapidly changing, large scale patterns of neural activity in conjunction with the Kelso et al. ([5]) behavioral paradigm (cf. Section 3 above) to systematically probe stability and change in sensorimotor coordination. Specifically one can ask: 1) Do phase transitions, strongly indicative of cooperative and critical phenomena, exist in the brain and if so, what form do these transitions take? 2) Are predicted features of nonequilibrium phase transitions such as fixed point drift, enhancement of fluctuations and critical slowing down seen experimentally? 3) Can coherent spatiotemporal states of the brain be observed and can they be mapped onto behaviorally observed coordination modes? These questions are co-implicative: On the one hand, observables must be found that characterize coherent states of affairs in the brain. On the other, phase transitions offer a theoretically motivated experimental strategy for identifying coherent brain states and the observables that define them.

In our experiment, the 37-SQUID array was centered over left parieto-temporal cortex (see [19,20] for full description of apparatus, methods, analyses and results). As described earlier, the subject was asked initially to syncopate (off the beat) with an auditory stimulus, i.e., he was instructed to press a response key in between each stimulus. After 10 cycles or so, the stimulus was speeded up. However, the subject does not operate in this syncopated mode of coordination indefinitely. When the rate speeds up to a critical value, spontaneous switching to a synchronized, in phase mode of coordination occurs. Can this switching be seen in the brain itself?

In *Figure 2* we plot the relative phase of each sensor with respect to the stimulus overtime for each cycle at the stimulus frequency. The size of the squares corresponds to the amplitude of the Fourier component at that frequency. On the lower left, the relative phase of the averaged manual response with respect to the stimulus is plotted. In general, an abrupt change in relative phase occurs after about 30 cycles near the end of the third frequency plateau (around 1.75 Hz). This is a well-replicated bifurcation or phase transition in sensorimotor coordination discovered by ourselves some years ago [5].

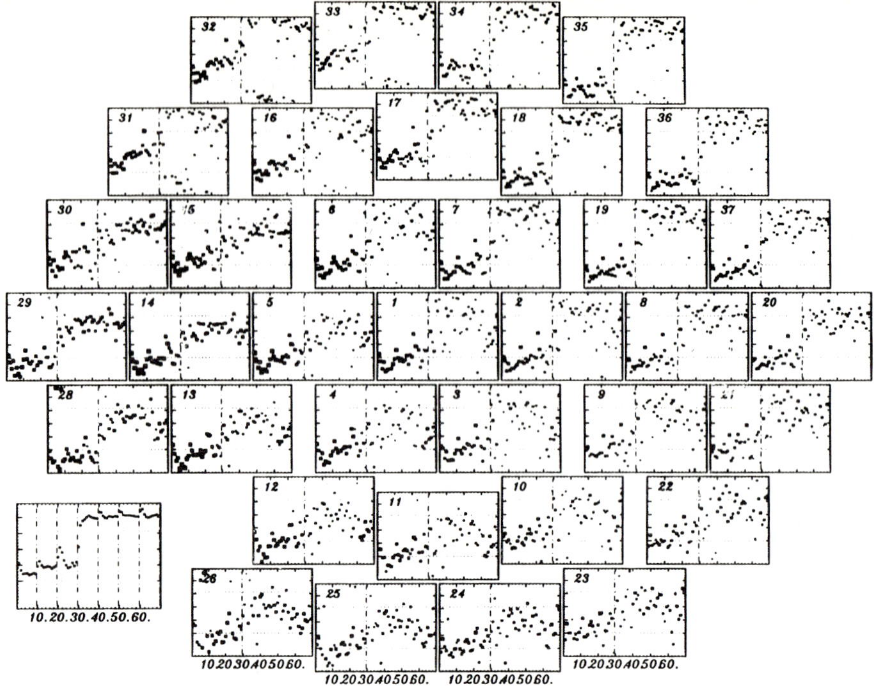

Figure 2: Relative phase evolving over time calculated at the stimulus frequency for each stimulus cycle (x-axis). Size of squares corresponds to the magnitude of the Fourier coefficient at the stimulus frequency. Lower left box is the produced sensorimotor behavior; other boxes are the generated brain signals for each SQUID sensor. Vertical lines indicate change in control parameter every 10 cycles. The space between horizontal dashed lines corresponds to a phase difference of 180 deg. See text for details.

But the most striking result is the relative phase data from the SQUID sensors themselves. Over much of the SQUID array a remarkable change in the phase relation occurs. In some of the posterior sensors the transition mirrors quite exactly the transition in behavior itself. Horizontal bars in *Figure 2* show that the magnitude of the transition, when it occurs in a given sensor, is always around π rad. Observed transitions are *spatial* as well as temporal: scanning left to right from sensor 30 to 37 the relative phase changes from linear (30, 15) to diffuse (6) to nonlinear (7, 19, 37). Note also that a significant increase in phase lag occurs at the first cycle of each frequency plateau and that the phase often drifts upward and fluctuates before the switch itself actually happens, typical signs of approaching instability.

Predicted features of nonequilibrium phase transitions such as *critical fluctuations* and *critical slowing down* ([1]) can be nicely seen by superimposing the relative phase time series of the behavior (open squares) and two representative

SQUID sensors (solid squares). As shown in *Figure 3*, the SQUID data, as expected, are somewhat noisier than the behavioral data. Nevertheless, it is easy to see that both brain and behavior are perturbed more by the *same* magnitude

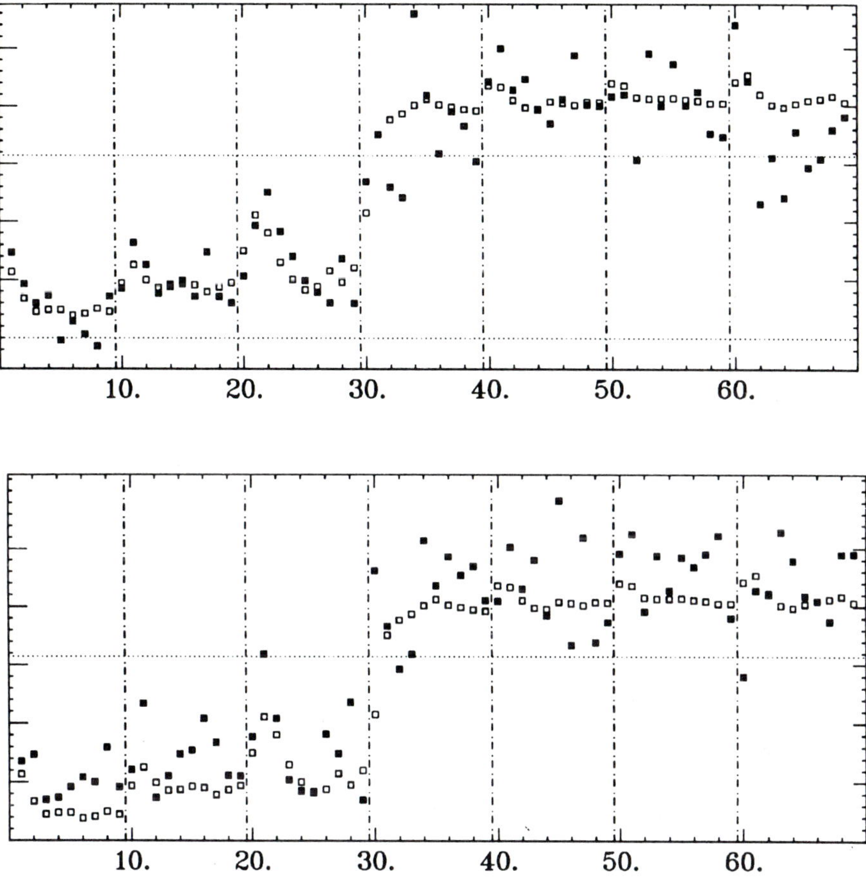

Figure 3: Superimposed relative phase (y-axis) calculated at the stimulus frequency for each cycle (x-axis) for the sensorimotor behavior over time (open squares) and sensors 28 (Top) and 20 (Bottom). Note the rather precise overlay of the two observables.

of perturbation (a parameter change of .25 Hz) as the critical point approaches. Moreover, critical slowing down is indicated by the fact that it takes longer to return to the pre-perturbation relative phase value as the transition draws near. Such loss of stability appears to be the main factor underlying qualitative change in both brain and behavioral patterns.

At this *mesoscopic* scale of description, i.e. at the level of each sensor (*Figs. 2* and *3*) relative phase proves to be an adequate order parameter. How then do we capture the *macroscopic* dynamics, i.e. the patterns of interrelations among all 37 SQUIDs? By decomposing the data from the entire sensor array into an equivalent set of spatial modes using the Karhunen-Loéve method ([21, 22]), it turns out that only a few modes (three to five) are sufficient to account for about 95% of the variance in the data ([19,20]). Much of the current literature is aimed at developing strategies to systematically model the spatial and temporal aspects of the brain dynamics (e.g. [23]). Following Friedrich, Fuchs & Haken ([22]) we aim to map these spatial modes onto dynamics using symmetry constraints and normal form theory. But it already seems quite clear that the essential macroscopic, large scale activity of the brain under the present experimental conditions can be captured by low dimensional, order parameter dynamics.

5 Conclusion

Some years ago, Haken ([24]) hypothesized that the brain is a self-organizing, pattern forming system that operates close to instability points, thereby allowing it to switch flexibly from one coherent state to another. In the last 12 years or so, my colleagues and I have tried to show that psychology and neuroscience are not outside the laws of pattern formation in nonequilibrium systems by adopting the strategies of synergetics and dynamical language. We have tried to develop, in Haken's words, a "flair" for identifying relevant dimensions in such complex systems. Now, through a certain kind of experiment which promotes a study of the evolution of pattern formation processes in time, and a new tool - the SQUID array - phase transitions in global patterns of cortical activity have been found. By showing that switching among relative phasing patterns takes the form of a critical instability our results confirm Haken's [24] thesis and suggest a new mechanism for the collective action of neurons in human cerebral cortex. A further interesting aspect of our results is that coherent states and state transitions in both brain and behavior are captured by the *same* collective variable, namely the spatiotemporal phase relations among (obviously very different) material components. We leave the fuller implications of this result for a more philosophical discourse. Nevertheless, by finding abstract, level-independent order parameters that characterize the behavior of brains and people, it seems we have taken a small step toward what it means to change one's mind.

Acknowledgements

The work reported here was supported by NIMH (Neurosciences Research Branch) Grant MH 42900, BRS Grant RR07258 and U.S. Office of Naval Research Contract N00014-88-J-119. I thank Armin Fuchs for discussion.

References

[1] H. Haken. *Synergetics, an introduction: Non-equilibrium phase transitions and self-organization in physics, chemistry and biology.* Springer, Berlin, 1983. (3rd edition).

[2] H. Haken, J. A. S. Kelso, and H. Bunz. A theoretical model of phase transitions in human hand movements. *Biological Cybernetics*, 51:347–356, 1985.

[3] J. A. S. Kelso and B. Tuller. A dynamical basis for action systems. In M. S. Gazzaniga, editor, *Handbook of Cognitive Neuroscience*, pages 321–356. Plenum, New York, 1984.

[4] G. Schöner and J. A. S. Kelso. Dynamic pattern generation in behavioral and neural systems. *Science*, 239:1513–1520, 1988a.

[5] J. A. S. Kelso, J. D. DelColle, and G. Schöner. Action-perception as a pattern formation process. In M. Jeannerod, editor, *Attention and Performance XIII*, pages 139–169. Erlbaum, Hillsdale, NJ, 1990.

[6] H. Haken. *Information and self-organization: A macroscopic approach to complex systems.* Springer, Berlin, 1988.

[7] G. Schöner, W. Y. Jiang, and J. A. S. Kelso. A synergetic theory of quadrupedal gaits and gait transitions. *Journal of Theoretical Biology*, 142(3):359–393, 1990.

[8] J. A. S. Kelso. On the oscillatory basis of movement. *Bulletin of the Psychonomic Society*, 18:63, 1981.

[9] J. A. S. Kelso. Phase transitions and critical behavior in human bimanual coordination. *American Journal of Physiology: Regulatory, Integrative and Comparative Physiology*, 15:R1000–R1004, 1984.

[10] J. A. S. Kelso, J. J. Buchanan, and S. A. Wallace. Order parameters for the neural organization of single, multijoint limb movement patterns. *Experimental Brain Research*, 85:432–444, 1991.

[11] J. A. S. Kelso and J. J. Jeka. Symmetry breaking dynamics of human multilimb coordination. *Journal of Experimental Psychology: Human Perception and Performance*, in press.

[12] R. H. Wimmers, P. J. Beek, and P. C. W. van Wieringen. Phase transitions in rhythmic tracking movement: A case of unilateral coupling. *Human Movement Science*, in press.

[13] R. C. Schmidt, C. Carello, and M. T. Turvey. Phase transitions and critical fluctuations in the visual coordination of rhythmic movements between people. *Journal of Experimental Psychology: Human Perception and Performance*, 16(2):227–247, 1990.

[14] J. A. S. Kelso and G. C. DeGuzman. An intermittency mechanism for coherent and flexible brain and behavioral function. In J. Requin and G. E. Stelmach, editors, *Tutorials in Motor Neuroscience*, pages 305–310. Kluwer, Dordrecht, 1991.

[15] P. Bergé, Y. Pomeau, and C. Vidal. *Order within chaos.* Hermann, Paris, 1984.

[16] H. Haken. Synergetics as a tool for the conceptualization and mathematization of cognition and behavior - How far can we go? In H. Haken and M. Stadler, editors, *Synergetics of cognition*, pages 2–31. Springer, Berlin, 1990.

[17] J. Shrager, T. Hogg, and B. Huberman. Observation of phase transitions in spreading activation networks. *Science*, 236:1092–1094, 1987.

[18] A. Katchalsky, V. Rowland, and R. Blumenthal. Dynamic patterns of brain cell assemblies. *Neurosciences Research Program Bulletin*, 12, 1974.

[19] J. A. S. Kelso, S. L. Bressler, S. Buchanan, G. C. DeGuzman, M. Ding, A. Fuchs, and T. Holroyd. Cooperative and critical phenomena in the human brain revealed by multiple SQUIDS. In D. Duke and W. Pritchard, editors, *Measuring Chaos in the Human Brain*, pages 97–112. World Scientific, New Jersey, 1991.

[20] J. A. S. Kelso, S. L. Bressler, S. Buchanan, G. C. DeGuzman, M. Ding, A. Fuchs, and T. Holroyd. Self-organizing Cooperative Dynamics of Human Brain and Behavior Revealed by Multisensor SQUID Array. submitted for publication.

[21] A. Fuchs, R. Friedrich, H. Haken, and D. Lehmann. In H. Haken, editor, *Computational Systems: Natural and Artificial*, pages 74–83. Springer, Berlin, 1987.

[22] R. Friedrich, A. Fuchs, and H. Haken. In H. Haken and H. P. Köpchen, editors, *Rhythms in Physiological Systems*. Springer, Berlin, in press.

[23] A. Destexhe and A. Babloyantz. *Neural Computation*, 3:145–154, 1991.

[24] H. Haken. Synopsis and introduction. In E. Basar, H. Flohr, H. Haken, and A.J. Mandell, editors, *Synergetics of the Brain*, pages 3–25. Springer, Berlin, 1983.

Brain Electric Fields and Brain Functional States

D. Lehmann

Department of Neurology, University Hospital,
CH-8091 Zürich, Switzerland

Abstract. General aspects of high level brain functions are discussed, stressing the indivisibility of emergent brain properties and the state-dependency of brain functions. Studies employing human brain electric field data need to use analysis techniques which describe system behavior on the level of the phenomenon under study. Some aspects of the concept-driven frequency transforms of EEG time series are useful. Phenomenology-oriented analysis includes space-oriented adaptive segmentation of EEG map series into sequences of microstates, yielding momentary global descriptors of brain functional state. Methods for the analysis of non-linear systems include the estimation of the local (single time series) or global (multiple time series) correlation dimension which can serve as state descriptor, yielding interesting new aspects of EEG data, including the increase of a local correlation dimension of EEG in schizophrenia.

1. Introduction

How does the brain work? There is a multitude of ways to record aspects of its functioning. It is a formidable task to integrate all the data into a common concept. This chapter offers general proposals, and discusses brain electric field data which might be useful for the construction of models of brain functioning. The references [1-39] refer to selected examples; they are not intended to present a comprehensive review.

2. A Complex Structural System and a Homogeneous Function

The brain has a very complex anatomical structure. Even though basic sub-units of cell ensembles ("columns") in different brain areas appear to be constructed according to a common rule (in the visual cortex, they differ somewhat), there are brain areas whose functions clearly differ and which are asymmetrically interconnected in the most intricate manner, and there are two largely symmetrically structured halves of the brain which, however, are partially asymmetric in function.

But the brain functions smoothly as a whole. It is known that there is a great amount of parallel information processing, but

Springer Proceedings in Physics, Vol. 69
Evolution of Dynamical Structures in Complex Systems
Editors: R. Friedrich · A. Wunderlin

the final product is a mentation or a behavior which typically appears to be coherent and homogeneous. Particularly striking is that, at the highest level, the internal aspect of the brain's activity is an apparently indivisible entity, consciousness, even though it might carry different contents in different modalities as conscious experiences. Moreover, if a local cortical lesion destroys a well-defined subproperty of the percept, the total conscious experience continues to be coherent, without awareness of the loss [4]. It is quite clear that a myriad of individual chemical and electric processes must work correctly in many brain areas for conscious experience and consciousness to emerge - but the end result appears as a whole. How does this come about?

3. Understanding an Emergent Property

Maybe this is an ill-posed question. Mario Bunge had pointed to the property of speed which is developed by an engine on a track, a property which is not present in any single mechanical or chemical element of the engine or the tracks, but which is present when a large, but still surveyable number of conditions is met by all involved elements: a satisfactory description of speed, using the molecules of the two systems involved, would have to include the description of all geometric interrelations between the components, but thereby would become very uneconomical, probably even practically impossible. Nevertheless, speed is no deep problem, and it is believed to be "understood". Understood, because the conditions and interrelations of higher-order entities of engine and track can be described successfully, and if the catalogue is fully met, speed is known to be present. However, the "understanding" is the fact that one knows that a one-to-one correlation exists between a particular description of the states of the components and their interrelations, and speed.

The issue of consciousness is different in some respects since it is the privileged, inner aspect of the system; but the speed example shows that attempts to describe ("understand") the emergence of a high-level property using very low-level components might be beyond practical reach, and that it would be wrong to conclude rashly that "understanding" is impossible because no reliable low-level description is available. I think that it further demonstrates that understanding something means that a complete and reliable set of rules describing the states of this something is known. Since there is always the possible desire for further regress to deeper levels, no final insight appears conceivable.

4. Control, Coordination and Organisation

Obviously, the multitude of structural sub-units in the brain are functionally coordinated in a superb fashion. It is unclear exactly how this is organized, or what the functional binding forces between different cell ensembles and different brain areas are. Relaxing the concept of central control (the proposed function of the reticular formation), the idea of self-organization between equal components has recently gained importance [see e.g. 19, 20].

It is also unclear in what way the final integration of the total activity is accomplished to establish priority for access of some information to the consciousness channel, and to produce coherent subjective percepts, decisions, and consciousness as such. Loose talk about "ensemble characteristics of activity", a tautology, merely says that things are what they are: the medium is the message. But maybe that is it.

If higher brain functions are to be assessed numerically, it appears desirable to use measures that consider the global character of these functions, in order to parallel the emergent features of the brain's activity on their appropriate level where the phenomenon is present.

5. The Brain Electric Field as Integrator

The electric field of the brain appears to be an appropriate candidate to reflect high-level properties of the system in an adequate manner. It is homogeneous, obviously generated by the very large mass of active elements, shows very quick changes on a time scale (down to milliseconds) that easily accommodates expected times of high-level brain processing. The brain's electric/ magnetic field, produced by all active neuronal elements, might even be the right a candidate for the brain's property of integration: if the medium is the message, the medium "whole brain electric/magnetic field" as an entity might well be the physical aspect of the mind.

As a sideline, I note that there has been a trickle of papers since the early 60's which imply that the brain electric field not only reflects information, but operates as modulating information on the brain [1, 15, 39].

6. Recording Electric Fields of the Brain

In practical terms, multichannel recordings of the human brain's electric field produce a unique window on the human mind: results of spectral analysis of the data are known to reflect with extreme sensitively all kinds of spontaneous and induced changes of the brain's functional state; the brain field data are continuously available, and they can be recorded with non-invasive techniques, producing minimal inconvenience to the subject.

The problem is less how to obtain the data, the challenge is to construct useful experimental conditions, and to work out ways in which to analyze and interpret the potentially available masses of data in an adequate fashion.

7. Periodicity of Brain Activity, and Global States

A dominant feature of brain electric activity over time is its striking periodicity. This periodicity is prominent as periodic oscillations of the polarity and global strength of the brain electric field. The oscillations are clearly evident in conventional EEG time series recordings where the frequency range of interest is between 0.5 and 40 Hz. It has long been known from single time series analysis that these oscillations are associated with grossly different functional states of the brain. Examples in the adult are the association of sleep with slower oscillations, wakefulness with faster oscillations, and of immature stages of development with slower oscillations, of mature stages with faster oscillations. The periodicities apparently reflect subtle differences of functional states [22]. As already touched upon above, many studies using spectral analysis of the EEG have shown correlations between various specific aspects of higher level information processing with specific characteristics of the brain electric signals [13, 16, 26, 35]. Obviously, the activity of the billions of individual neurons in the brain is organized in a way that gives rise to ruleful global, spatial and temporal electric features of the total sum of this activity as recorded from the intact scalp.

8. All Brain Information Processing is State Dependent

The periodicities define functionally different states. A prime feature of brain information processing is its dependency on functional state, unlike processing in machines. Identical information reaching the brain in different states will be treated in very different ways, and thus lead to very different behavioral results. The effect of this strategy-constraining and strategy-inviting consequence of state is enhanced by the state-dependency of memory access in living beings which includes an asymmetry of recall mechanisms; material stored in less highly organized states (e.g. sleep) cannot be read out during higher states (here: wakefulness), but vice versa. The consideration and identification of the momentary global functional state is thus a foremost task when comprehensive studies of information processing in humans are done.

9. Assessing Global Brain States

The classification of brain states using spectral EEG measures has been quite fruitful for the ordering of observed characteristics of brain information processing. However, in this approach there are two aspects which are not satisfying: the necessary time window of typically more than a second, and the transformation of the data into the frequency domain restrict the possible range of theories of functions. The time minimum requirement is basic even though momentary measures might be derived in the frequency domain (HILBERT transform),

An unattractive additional feature of the transformation and the derived measures is the very large mass of results which can be computed out of a data set (power, phase angles, coherences, covariances etc. for each frequency point). This leads to post-hoc clustering in order to shrink the result space to interpretable proportions. Given the mathematics-driven basic approach, later clustering of the results tends to be governed by formal aspects whose relations to brain mechanisms are very speculative.

The historical preference for frequency domain measures in EEG analysis has been questioned. An unattractive feature of the approach is the very large mass of results which can be computed out of a few time series: power, phase angles, coherence, covariance, etc for each frequency point, each finding asking for a definition of its putative functional role, and asking for some synthesizing hypothesis.

On the other hand, it is conceivable that the successful operationalization of the data by the transformation into the frequency domain might be irrelevant as to the nature of the data - just as a mountain range might be perfectly operationalized by a two-dimensional Fourier transform in terms of frequencies, amplitudes and phase angles, but these sets of perfect numerical descriptors reveal nothing about its nature.

10. Brain Field Phenomenology: Microstates

The issue of the time window has been addressed by us in work on functional microstates, where we consider the brain activity as a sequence of global functional states of which each reflects a particular step, and/or type, and/or content of information processing. No minimal time duration should be set for such states. Viewing the brain field data as a series of momentary maps of field distributions, it must be concluded that the global functional state has changed whenever the landscape of the brain map has changed, since different map landscapes must have been generated by different geometries of the active neuronal elements; activity of different neuronal ensembles can reasonably be supposed to be associated with different acts of information processing. The time resolution in this approach is maximal, i.e. each data point in time could be an individual state. We used such adaptive, space-oriented segmentation of spontaneous EEG maps series in a study on normal subjects; we found a mean duration of microstates as defined by their electric field landscape of 210 msec [28, 29; see also 31].

The temporal resolution required for the study of building blocks of higher processes such as awareness, conscious perception and voluntary decision making appears to be in the subsecond range. However, there is apparently not only a maximal duration of the basic microstates, there also is a necessary minimal time duration for a brain process to achieve consciousness. In several studies, the estimated times ranged around 200 to 500 msec [7, 29, 30, 32, 36]. Hence, searching for constituting elements which still carry the characteristic of consciousness within too short a time might not be successful; for instance, the early and brief "components" of brain processing of arriving information before about 150 msec after the stimulus apparently do not qualify for conscious experience. (As discussed above, hacking something complicated into very small pieces will eliminate its higher property).

Thus, "atoms of thought" might be identified by identifying the brain functional states using momentary field map landscapes, with leeway to recognize larger pieces. For this approach, a phenomenological operationalization of the measured brain fields and of the reported thoughts (mentations) or processing style or observed steps of processing has to be developed, then the relations between classes of field maps and mentation styles must be examined, and possible clustering must be investigated. In studies with B. Henggeler and A. Kofmel, we have obtained first results of correspondences between momentary brain field map classes and classes of mentations [25], but the complete system of brain microstates and corresponding mental states must still be worked out.

However, we hypothesize that the complete description of system behavior will hierarchically need first an identification of system state in terms of global periodicity of the electric field. This should be done by subjecting the curve of effective field value ("Global Field Power" [27] as a function of time to an adaptive segmentation procedure, for instance using some autoregressive approach. Then, within each time-extended segment, the space-oriented segmentation procedure will identify landscape-classified microstates, hereby attaining the desired, unrestricted time resolution, and giving proper attention to the space domain of the data.

11. Non-Linear Measures: Dimensional Complexity

Work on non-linear systems has made available a number of measures which assess properties of the data which are different from those measured by spectral analysis. Some of these approaches have in the last years begun to be applied to EEG data [e.g. 14]. In particular, the dimensional complexity (the correlation dimension [17, 18]) of single-channel and occasionally, of multichannel EEG time series has been investigated [2-4, 8, 9, 12, 34, 37].

It is interesting that the results reported thus far in the literature [12, 37; for a review: 2] have suggested some relation between predominant EEG frequency and EEG correlation dimension, even though the correlation dimension, of course, is not dependent from frequency: conditions associated with a predominant slow EEG activity such as slow wave sleep, 3/sec spike wave patterns, Creutzfeldt-Jakob disease show low correlation dimensions in the

EEG, and conditions with predominant faster activity (REM sleep, wakeful alpha EEG) show higher EEG correlation dimension. Reports on differences of correlation dimension between wakeful EEG during eyes closed and eyes open conditions, and during resting before and after a mental task [34] indicated that the correlation dimension might be a sensitive metric for the assessment of different functional states during wakefulness.

12. Correlation Dimension of Multichannel EEG Data: Local vs Global Measures

Appropriate treatments of the original EEG recordings of potential differences between two locations in the field produce time series which carry information about a single location in the field. Such treatments are the computation of voltages vs the average potential (average reference= spatial DC rejection), of the local gradients (the first derivative; a spatial high pass) and the current source density (the second derivative; a stronger spatial high pass filter). When the correlation dimension of these time series is estimated using Takens' method [38] to estimate the embedding dimension, a map could be constructed which displays the landscape of the local correlation dimensions. The question is whether there are relevant and consistent differences between different recording areas.

Following a suggestion by Eckman and Ruelle [11] about the treatment of simultaneous time series one might use the number of electrodes of a multichannel EEG recording as embedding dimension [2, 6, 8, 9]. This approach views the multichannel EEG as a sequence of momentary field maps, corresponding to the concept of a trajectory in K-dimensional state space (K=number of channels). It permits a quantitative, single value measure of complexity of the brain states' trajectory, the global correlation dimension that describes the ensemble characteristics of all recorded channels. In other words, this estimates the dimensional complexity of the series of momentary potential distribution maps into which the multichannel EEG recording can be transformed. If the analyzed epoch can be kept short enough to yield the time resolution which is adequate for the addressed problem, the resulting one-number descriptor for global brain state appears promising for comparisons between normal and disease conditions.

We have used these two approaches to numerically assess brain functional characteristics of global nature: the estimation of the local correlation dimension in a study of EEG in untreated acutely ill schizophrenic patients, and the estimation of the global correlation dimension in a study of EEG changes induced by a single dose of a cognition-enhancing medication.

12.1. Local Dimensional Complexity of the EEG of Schizophrenics

In a study with M. Koukkou (University Hospital of Psychiatry at Berne), J. Wackermann, I. Dvorak (Psychiatric Center at Prague) and B. Henggeler we analyzed 2 channels of EEG data from four drug-free groups of subjects: First episode (acute) never-treated schizophrenics, remitted schizophrenics after a first episode, neurotics, and controls.

The local dimensional complexity of left temporal-parietal and parietal-occipital EEG recordings was assessed by computing the correlation dimension of EEG data epochs of 20 second duration in each of six recording conditions: initial eyes closed, eyes closed after opening, and before and after two presentations of an auditory sentence. Two "bipolar" channels of EEG from electrodes over the temporal-parietal and parietal-occipital region of the left hemisphere were recorded (EEG derivations T3-P3 and P3-O1). The data thus are the estimates of the time-varying local voltage gradients of the brain's electric field in two regions. An overall bandpass of 0.5-30 Hz was used.

The subjects were 15 first episode acute schizophrenics before any medication, 12 other medication-free individuals clinically and socially remitted after a first schizophrenic episode, 17 medication-free, completely remitted neurotics, and 17 normal controls.

The local correlation dimension of each data epoch was computed using a recent program (COREX, by I.D. and J.W.) based on Grassberger and Procaccia's [17, 18] algorithm with some modifications [10]. In repeated computational runs, the correlation dimension was estimated for an embedding dimension. Each estimate was based on 5120 randomly chosen points on the trajectory that was constructed from the data in state space [38]. The repeated estimates were averaged separately for each embedding dimension. If the mean estimates did not exceed a preset range, the computation was stopped. Otherwise, the preset number of 20 runs was done. The final estimate of the correlation dimension was based on the "saturation curve" through the estimated correlation dimensions along the increasing embedding dimension. Ideally, the estimates first follow the diagonal axis and remain constant beyond some embedding dimension. In practice, especially if the data is noisy, the saturation is not perfect, the curve showing a slowly flattening increase. In the used version of the program COREX, the crossing point of the two quasi-linear segments of the saturation curve is determined; this value is used as final estimate of the correlation dimension. The parameters used were: 5120 data points per epoch (=20 sec at 256 s/sec), maximum embedding dimension=12, time delay tau=5, 20 computational runs using 8192 point pairs per run.

ANOVA's of the medians over subjects of the local correlation dimensions in either channel showed no significant differences between the six recording conditions across the four subject groups (Friedman ANOVA's, N=4, K=6, df=5, chi square =4.0 and =3.6, respectively for the two channels; both p>.5). We therefore computed, for each subject and each channel, the median value of the correlation dimension across all six recording conditions, and used these for further analysis.

The local correlation dimension of the temporal-parietal EEG recordings differed significantly between groups (ANOVA p<.001).

The first episode schizophrenics showed highest values (median=4.44), remitted cases (and the neurotics) intermediate values (median=4.18, 4.11, respectively) and normal controls lowest values (median=3.96). The differences were significantly between first episode schizophrenics and controls ($p<.004$) while neurotics differed from schizophrenics at $p<.002$.

In the parietal-occipital recordings, there was no overall significant difference between groups.

Differences of the local correlation dimension of the temporal-parietal vs the parietal-occipital EEG were significant between groups (ANOVA $p<.05$); first episode schizophrenics differed from controls ($p<.002$) and remitted patients ($p<.08$).

The higher dimensional complexity of functional brain mechanisms in acute schizophrenics than normals corresponds with the general clinical observation of loosened organization of thought processes in schizophrenia as e.g. the reality-uncontrolled hallucinations, and on the other hand, Bleuler's [5] "double bookkeeping" of events in the delusional world and the real world. The finding of increased complexity of schizophrenic brain functioning is also reminiscent of reports of certain superior mental abilities in the patients [e.g. 21, 33]. Since increased dimensional complexity in acute schizophrenics was found in one, not in the other of the two examined brain regions, we conclude that local differences between brain areas are possible.

12.2. Global Dimensional Complexity of Multichannel EEG: Drug Effects

In another study with J. Wackermann and I. Dvorak (Psychiatric Center at Prague) and C.M. Michel we examined changes of the global dimensional complexity of spontaneous EEG field map series after a single dose of a cognition-enhancing medication, NOOTROPIL (by UCB-Pharma).

In five normal young volunteers, four recordings of 16 simultaneous channels of EEG during resting conditions were obtained 1, 1.5, 2 and 2.5 hrs after intake of a single dose of Placebo or 2.9g or 4.8g or 9.6g of Piracetam (NOOTROPIL, UCB-Pharma), during each of four randomized sessions (double blind design). 40 second epochs were analyzed from each recording, drug level and subject (76 of the 80 records were available). The analysis was done on artifact-free epochs with an overall bandpass of 2-20 Hz, covering the traditionally relevant EEG frequencies.

The multichannel mode of the program COREX (by I. Dvorak and J. Wackermann) was used for the computation of the correlation dimension. The global correlation dimension is estimated directly in the state space which is defined by the number of recording channels. Using an original implementation, the program manages a large amount of computations in a reasonable time. The time to obtain the estimate of the correlation dimension in the multichannel mode for one of our data epochs (16 channels x 10000 values) was about one minute (on a 25 MHz 386 PC) using 50 repeated runs, with 8192 point pairs for each run (the final estimate of the global correlation dimension of a data epoch is the mean of the results of the 50 computational runs).

The results were combined for the two intermediate doses and averaged over repeated recordings. The final 5x3 result matrix thus consisted of mean correlation dimension values for each of three treatment levels (placebo, low dose, high dose) in each of five subjects. The dimensionality decreased from placebo (median=5.89) to low dose (median=5.72) to high dose (median=5.59), significant in a Friedman ANOVA (N=5, K=3, chi square=8.4, df=2, p<0.02), with significant differences between placebo vs high and low vs high dose.

Thus, the subtle change of brain global functional state which is induced by a single dose of Piracetam is reflected by the non-linear measure of global dimensional complexity of the multichannel EEG. The decrease of complexity by a reportedly cognition-enhancing drug might indicate a U-shaped relationship as the relation between arousal and performance accuracy, where increase of arousal from drowsiness to wakefulness improves performance, further increase to anxiety levels decreases performance. On the other hand, the results with EEG in schizophrenia reported above suggest that increased complexity of brain processes does not in all respects correlate with reality-close, improved performance.

13. Brain States: Descriptions and Functions

Higher brain functions doubtlessly involve extended areas of the brain to differing degrees. The highly coordinated, synergetic action of the large masses of individual neural elements at each moment in time results in emergent properties which cannot be described on the level of individual elements. Since all brain processing strategies are embedded into and constrained by particular global functional states, the state-dependency requires careful charting in observations, and attention in modelling [e.g. 22, 23]. The brain's electric field with its high time resolution, sensitivity to brain states, and availability of continuous, convenient recording is a rich source of data about high level information processing in the brain. The traditional data treatment in the frequency domain emphasizes a generally important aspect, the predominant periodicity, but is strongly concept-driven; also, for studies of "spontaneous" activity, the time resolution appears insufficient. Data-driven phenomenology such as field landscape assessment might usefully complement the frequency domain analysis, accentuating the spatial aspects and increasing the time resolution down to a range where physiological building blocks of mentation might be expected which still implement the full mentation aspect. Estimates of the dimensional complexity opened an independent route for global state assessments. Although in principle, the method might detect the number of constituting, active processes of the examined phenomenon, given the various methodological questions it appears more persuasive at present to read the correlation dimension values of brain field data as relative indicators [24, 34] of degree of system complexity. Such information can add valuable anchors to attempts in ranking, classifying and eventually, systematizing and modelling high level processing steps and styles. The reported observation of locally increased dimensional complexity of EEG in schizophrenia illuminates an aspect of this disease which has received little attention; it also reminds us that the global brain functional state must be conceived as a mosaic of local states, whose entirety eventually is a new entity in terms of function.

Acknowledgement. The reported research was partially supported by the Swiss National Science Foundation, the Sandoz Foundation, the EMDO Foundation, and the Hartmann-Müller Foundation.

References

1. Adey, W.R. (1989) Cell membranes, electromagnetic fields, and intercellular communication. In E. Basar and T.H. Bullock (eds): Brain Dynamics. Springer, Berlin. pp. 26-42.

2. Babloyantz, A. (1989) Estimation of correlation dimensions from single and multiple recordings - a critical review. In E. Basar and T.H. Bullock (eds): Brain Dynamics. Springer, Berlin, pp. 122-130.

3. Babloyantz, A. and Destexhe, A. (1986) Low dimensional chaos in an instance of epilepsy. Proc. Natl. Acad. Sci. USA 83:3513-3517.

4. Baumgartner, G. (1990) Where do visual signals become a perception? In J.C. Eccles and O. Creutzfeldt (eds): The Principles of Design and Operation of the Brain. Pontifica Academia Scientarum, Vatican. pp. 99-114.

5. Bleuler, E. (1916) Lehrbuch der Psychiatrie. Springer, Berlin.

6. Destexhe, A., Sepulchre, J.A. and Babloyantz, A. (1988) A comparative study of the experimental quantification of deterministic chaos. Phys. Lett. A 132: 101-106.

7. DiLollo (1980) Temporal integration in visual memory. J. exp. Psychol., Gen. 109: 75-97.

8. Dvorak, I. (1990) Takens versus multichannel reconstruction in EEG correlation exponent estimations. Phys. Lett. A 151: 225-233.

9. Dvorak, I. (1991) Time and space structure of the determinacy of brain electrical phenomena. In I. Dvorak and A.V. Holden (Eds.), Mathematical Approaches to Brain Functioning Diagnostics. Manchester University Press, Manchester. pp. 353-367.

10. Dvorak, I. and Klaschka, J. (1990) Modification of the Grassberger-Procaccia algorithm for estimating the correlation exponent of chaotic systems with high embedding dimension. Phys. Lett. A 145: 225-231.

11. Eckmann, J.P. and Ruelle, D. (1985) Ergodic theory of chaos and strange attractors. Rev. Mod. Phys. 57: 617-638.

12. Ehlers, C.L., Havstad, J.W., Garfinkel, A. and Kupfer, D.J. (1991) Nonlinear analysis of EEG sleep states. Neuropsychopharmacol. 5: 167-176.

13. Ehrlichman, H. and Wiener, M.S. (1980) EEG asymmetry during covert mental activity. Psychophysiology 17: 228-235.

14. Friedrich, R., Fuchs, A. and Haken, H. (1991) Modelling of spatio-temporal EEG patterns. In I. Dvorak and A.V. Holden (eds): Mathematical Approaches to Brain Functioning Diagnostics. Manchester University Press, Manchester. pp. 45-61.

15. Gavalas, R.J., Walter, D.O., Hamer, J. and Adey, W.R. (1970) Effect of low-level, low-frequency electric fields on EEG and behavior in macaca nemestrina. Brain Res. 18: 491-501.

16. Grass, P., Lehmann, D., Meier, B., Meier, C.A. and Pal, I. (1987) Sleep onset: factorization and correlations of spectral EEG parameters and mentation rating parameters. Sleep Res. 16: 231.

17. Grassberger, P. and Procaccia, I. (1983a) On the characterization of strange attractors. Phys. Rev. Lett. 50: 346-349.

18. Grassberger, P. and Procaccia, I. (1983b) Measuring the strangeness of strange attractors. Physica 9D: 183-208.

19. Haken, H. (1983) Advanced Synergetics. Springer, Berlin.

20. Haken, H. (1988) Information and Selforganization. Springer, Berlin.

21. Keefe, J. and Magaro, P. (1980) Creativity and schizophrenia: an equivalence of cognitive processes. J. Abnorm. Psychol. 89: 390-398.

22. Koukkou, M. and Lehmann, D. (1983) Dreaming: the functional state-shift hypothesis. A neuropsychophysiological model. Brit. J. Psychiatry 142: 221-231.

23. Koukkou, M. and Manske, W. (1986) Functional states of the brain and schizophrenic states of behavior. In C. Shagass, R.C. Josiassen and R.A. Roemer (eds): Brain Electrical Potentials and Psychopathology. Elsevier, Amsterdam. pp. 91-114.

24. Layne, S., Mayer-Kress, G. and Holzfuss, J. (1986) Problems associated with the dimensional analysis of electroencephalogramm data. In Mayer-Kress, G. (ed.): Dimensions and Entropies in Chaotic Systems. Springer, Berlin. pp. 246-256.

25. Lehmann, D. (1990) Brain electric microstates and cognition: the atoms of thought. In E.R. John (ed): Machinery of the Mind. Birkhäuser, Boston. pp. 209-244.

26. Lehmann, D. and Koukkou, M. (1973) Computer analysis of EEG wakefulness-sleep patterns during learning of novel and familiar sentences. Electroenceph. Clin. Neurophysiol. 37: 73-84.

27. Lehmann, D. and Skrandies, W. (1980) Reference-free identification of components of checkerboard-evoked multichannel potential fields. Electroenceph. Clin. Neurophysiol. 48: 609-621.

28. Lehmann, D., Brandeis, D., Ozaki, H. and Pal, I. (1987a) Human brain EEG fields: micro-states and their functional significance. In: H. Haken (ed): Computational Systems-Natural and Artificial. Springer, Heidelberg, pp. 65-72.

29. Lehmann, D., Ozaki, H. and Pal, I. (1987b) EEG alpha map series: brain micro-states by space-oriented adaptive segmentation. Electroenceph. Clin. Neurophysiol. 67: 271-288.

30. Libet, B. (1982) Brain stimulation in the study of neuronal functions for conscious experience. Hum. Neurobiol. 1: 235-242.

31. Merrin, E.L., Meek, P., Floyd, T.C. and Callaway, E. (1990) Topographic segmentation of waking EEG in medication-free schizophrenic patients. Int. J. Psychophysiol. 9: 231-236.

32. Michaels, C.F. and Turvey, M.T. (1979) Central sources of visual masking: indexing structures supporting seeing at a single, brief glance. Psychol. Res. 41: 1-61.

33. Place, E.J.S. and Gilmore, G.C. (1980) Perceptual organization in schizophrenia. J. Abnorm. Psychol. 89: 409-418.

34. Rapp, P.E., Bashore, T.R., Martinerie, J.M., Albano, A.M., Zimmerman, I.D. and Mees, A.I. (1989) Dynamics of brain electrical activity. Brain Topography 2: 99-118.

35. Ray, W.J. and Cole, H.W. (1985) EEG alpha activity reflects attentional demands, and beta activity reflects emotional and cognitive processes. Science 228: 750-752.

36. Reeves A. and Sperling, G. (1986) Attention gating in short-term visual memory. Psychol. Rev. 93: 180-206.

37. Röschke, J. and Aldenhoff, J. (1991) The dimensionality of human's electroencephalogram during sleep. Biol. Cybern. 64: 307-313.

38. Takens, F. (1981) Detecting strange attractors in turbulence. In D.A. Rand and L.S. Young (eds), Dynamical Systems and Turbulence. Lecture Notes in Mathematics 898. Springer, Berlin. pp. 366-381.

39. Wever, R.A. (1985) The electromagnetic environment and the circadian rhythms of human subjects. In M. Grandolfo, S.M. Michaelson and A. Rindi (eds): Biological Effects and Dosimetry of Static and Elf Electromagnetic Fields. Plenum, New York. pp. 477-523.

Synergetic Analysis of Human Electroencephalograms: Petit-Mal Epilepsy

R. Friedrich and C. Uhl

Institute of Theoretical Physics and Synergetics,
University of Stuttgart, Pfaffenwaldring 57/4,
W-7000 Stuttgart 80, Fed. Rep. of Germany

Abstract. The clinical diagnosis of epileptic seizures is usually based on an inspection of the electroencephalogram (EEG). The present paper is devoted to a description of a more refined analysis of EEG patterns of petit-mal epilepsy. The treatment is based on the synergetic approach to macroscopic patterns in complex systems which has been inaugurated by H. Haken. This approach aims at an understanding of both spatial as well as dynamical aspects of macroscopic EEG patterns.

1. Introduction

The possibility to measure the electric potentials generated by the human brain was discovered by the psychologist Hans Berger in 1929 [1]. Since then the scalp electroencephalogram has become a valuable clinical tool. It is helpful in the diagnosis of brain tumors, epilepsies, infectious diseases etc. [2]. On the other hand, since its origin is due to the collective behaviour of cortical neurons, it yields a direct signal of brain activity in form of spatio-temporal patterns.

The electroencephalogram is generated by the activity of the cerebral cortex which consists of about 10^{10} interconnected neurons. The surface of a single neuron is covered with about $10^3 - 10^5$ synapses, which transmit inputs form other neurons. Most neurons, about two-thirds, are vertically oriented pyramidal cells, with their apical dendrites oriented upwards and the somata and axons oriented downwards. The remaining neurons are stellate cells, which are arranged randomly. Furthermore, about 10^{11} glia cells are disposed between the neurons. Each cell of the cortex constitutes a dipole source located in a conducting medium. A change of intracellular potentials of the neurons caused by synaptic inputs from other neurons leads to the formation of extracellular potentials and to the generation of electrical currents. These currents may interpenetrate into the skull and in turn generate potential differences, which are measurable at the human scalp in the form of electroencephalograms (EEG). Under certain conditions the time signal of the potential difference between two electrodes exhibits pronounced oscillations, as for instance in the case of the so-called α-waves. α-waves are detectable from a resting person keeping his eyes closed. Furthermore, spatial mappings of the electric field distribution over the scalp show wave patterns with remarkable spatial coherence [3].

Springer Proceedings in Physics, Vol. 69
Evolution of Dynamical Structures in Complex Systems
Editors: R. Friedrich · A. Wunderlin

Although the mechanisms leading to a polarization of a single neural cell are well-understood, the processes eventually resulting in cooperative spatial and temporal patterns remain unexplained. It was H. Haken [4] who has drawn attention to the fact that the emergence of spatio-temporal patterns can be viewed as a process of selforganization in close analogy to the emergence of ordered patterns in other synergetic systems, e.g. in the laser. This suggestion has important consequences for the investigation of brain waves. Since selforganization occurs due to the emergence of order parameters, which enslave the remaining degrees of freedom [5], [6], a macroscopic approach to the analysis of complex systems, as outlined by H. Haken in his book "Information and Self-organization" [7], should be applicable to the investigation of spatio-temporal EEG patterns.

In recent years research interest has been devoted predominantly to the properties of the EEG signal obtained by measuring the potential difference between two locations on the scalp. The time series have been analysed by the standard methods of dynamic systems theory, like the evaluation of correlation dimensions or Lyapunov exponents [8], [9]. However, the suggestions of H. Haken on the selforganization of brain waves implicitly points out the possibility to perform a more detailed analysis. This analysis aims, according to the macroscopic treatment of complex systems [7], at an identification of order parameters and the determination of their dynamics. The approach has so far been applied to α-waves as well as to brain waves arising during petit-mal epileptic seizures [10] – [14]. It takes into account the significant spatial patterns of the brain waves and aims at an understanding of both spatial and temporal aspects of the EEG.

The present article is devoted to the synergetic analysis of spatio-temporal EEG patterns arising in petit-mal epilepsies. As emphasized by A. Babloyantz et al. [8] the corresponding time series do possess a fractal dimension between two and three. Furthermore, the patterns are spatially coherent [3], [10], [12], [13]. Therefore, one has to deal with spatially coherent patterns undergoing a complex evolution in time. It is known from various synergetic systems, especially from pattern formation in hydrodynamics (see e.g. [15]), that such kind of behaviour usually arises due to mode interactions.

The paper is organized as follows. As a starting point we shall briefly discuss the spontaneous emergence of low-dimensional chaotic behaviour in complex systems in order to emphasize the method of analyzing these patterns. This approach is applied to two data sets of EEG patterns of petit-mal epilepsy derived from two different persons. We shall identify order parameters, which turn out to be amplitudes of certain spatial modes, which allow a reconstruction of the total pattern. Then the dynamics of the patterns is examined and is shown to belong to the class of Shil'nikov-attractors. We shall specify a dynamical system, which yields an accurate reconstruction of the spike-wave behaviour of the order parameters as well as, in connection with the spatial modes, a reconstruction of the total patterns. Then we shall pursue, based on a criterion first considered by Shil'nikov [16], the question whether the dynamics is chaotic or time periodic. The main results are summarized in section 5.

2. Low dimensional chaos in spatially extended systems

The main characteristics of the EEG patterns during epileptic seizures as well as in the case of α-waves (see [11], [12]) is the existence of spatially coherent patterns undergoing a complex time evolution. Frequently, the emergence of low dimensional chaotic behaviour in spatially extended systems is due to "mode interaction". Then the spatio-temporal patterns can be treated in accordance with the "slaving principle of synergetics" [5] – [7]. We also refer to the article by A. Wunderlin in this volume [17]. In this case the state vector $\mathbf{q}(\mathbf{r},t)$ of the system, which obeys a nonlinear evolution equation of the form

$$\dot{\mathbf{q}}(\mathbf{r},t) = \mathbf{N}[\mathbf{q}(\mathbf{r},t),\nabla] \tag{1}$$

turns out to be a superposition of unstable modes, $\mathbf{v}_i^u(\mathbf{r})$, and stable modes $\mathbf{v}_i^s(\mathbf{r})$:

$$\mathbf{q}(\mathbf{r},t) = \sum_i u_i(t)\mathbf{v}_i^u(\mathbf{r}) + \sum_i s_i[u_j(t)]\mathbf{v}_i^s(\mathbf{r}) \quad . \tag{2}$$

As a consequence of selforganization the amplitudes of the stable modes $s_i[u_j(t)]$ are enslaved by the "order parameters" $u_i(t)$, the amplitudes of the unstable modes: There exists a relationship of the form

$$s_i = s_i[u_j(t)] \tag{3}$$

and the order parameters obey a closed set of evolution equations:

$$\dot{u}_i(t) = p_i[u_j(t)] \quad . \tag{4}$$

If one or two modes become unstable, a time independent or a time periodic behaviour arises. In the case of three or more order parameters chaotic behaviour can be expected to occur, if the dynamics is nonvariational. The total pattern exhibits spatial coherence, since it is a superposition of few spatial modes. It should be stressed that the complex temporal behaviour emerging due to mode interaction is a cooperative behaviour of the whole system: An isolated part of the system or a single subsystem will usually not undergo chaotic behaviour on its own.

From the study of a variety of nonequilibrium systems it has become evident that the chaotic behaviour generated by mode interactions is frequently related with the existence of homoclinic or heteroclinic closed loops in the phase space of the order parameters, which exist for a definite value of the control parameters. These loops are composed of trajectories connecting unstable fixed points. That means that the pattern corresponding to the fixed point is destabilized with respect to a different spatial configuration. If no other stable spatial configuration can be achieved by the system, a continuous evolution of the pattern takes place, which may be rather complex.

An important example of such a homoclinic loop, which is of considerable interest with respect to the dynamics of petit-mal epilepsy, as we shall see in

Fig. 1: The Γ_0^+ and the Γ_0^- homoclinic orbits involved in Shil'nikov attractors.

the following, has been investigated by Shil'nikov [16]. He considers a three-dimensional dynamical system, which possesses a homoclinic orbit biasymptotically approaching an unstable focus (see fig. 1). The flow in the immediate vicinity of the saddle focus is linear and is given by

$$\begin{aligned} x(t) &= x(0)\sin(\omega t + \varphi_0)e^{\rho t} \quad, \\ y(t) &= y(0)\cos(\omega t + \varphi_0)e^{\rho t} \quad, \\ z(t) &= z(0)e^{\lambda t} \quad, \end{aligned} \tag{5}$$

where x, y denote the coordinates in the two dimensional manifold and z the coordinate in the one dimensional manifold of the saddle focus. Due to Shil'nikov there exists a countable infinity of hyperbolic unstable limit cycles, if the following condition ("Shil'nikov's condition") is fulfilled:

$$|\rho|/|\lambda| < 1 \quad . \tag{6}$$

Due to the countable infinity of unstable limit cycles the behaviour of the phase point is irregular: It approaches a certain limit cycle along its stable manifold and departs along the unstable manifold approaching a different limit cycle. This may result in an erratic wandering of the phase point from limit cycle to limit cycle. If Shil'nikov's condition is not fulfilled, $|\rho|/|\lambda| > 1$, the homoclinic orbit is deformed into a stable limit cycle if the dynamical system is disturbed in such a way that the unstable and stable manifolds of the saddle focus intersect transversally.

We mention that there are two different types of homoclinic orbits, denoted as Γ_0^+, Γ_0^-. In case of the Γ_0^+- orbit the two dimensional manifold of the saddle focus is a stable manifold ($\rho < 0, \lambda > 0$), for the Γ_0^-- orbit it is an unstable manifold ($\rho > 0, \lambda < 0$) (see fig. 1). For more extended discussions see e.g. [18].

3. Analysing Spatio-Temporal Patterns

The understanding of the emergence of chaotic behaviour in spatially extended systems due to mode interaction can be exploited with respect to a macroscopic analysis of EEG patterns in the following way. The fact that usually only few order parameters $u_i(t)$ are highly excited, whereas the contributions of stable modes are of higher order, allows one to determine the unstable modes from an experimental time series using the method of Karhunen-Loève, which is frequently also denoted as principal component analysis (see e.g. [7]). In the case of EEG analysis the patterns are described by a vector $\mathbf{V}(t)$, whose components consist of the potentials $V_i(t)$ measured at the i-th electrode. This vector is decomposed into a set of modes $\mathbf{v}^\alpha$ according to

$$\mathbf{V}(t) = \sum_\alpha \eta_\alpha(t)\mathbf{v}^\alpha \quad , \tag{7}$$

where the spatial modes are determined by the extremum condition

$$< \left(\mathbf{V}(t) - \sum_\alpha \eta_\alpha(t)\mathbf{v}^\alpha\right)^2 >= \frac{1}{T}\int_0^T dt\{\mathbf{V}(t) - \sum_\alpha \eta_\alpha(t)\mathbf{v}^\alpha\}^2 = min! \quad . \tag{8}$$

The modes, which extremalize this expression, are the eigenvectors of the correlation matrix

$$\begin{aligned} C_{ij} &= < V_i(t)V_j(t) > \quad , \\ \sum_j C_{ij}v_j^\alpha &= \lambda^\alpha v_i^\alpha \quad . \end{aligned} \tag{9}$$

The eigenvalues λ^α are proportional to the time average,

$$\lambda^\alpha =< \eta_\alpha(t)^2 > \quad , \tag{10}$$

if the modes $\mathbf{v}^\alpha$ are properly normalized. The eigenvalue λ^α is therefore a measure of the mean intensity through which the mode $\mathbf{v}^\alpha$ contributes to the total pattern. If there is a gap in the spectrum of λ^α indicating a separation of the modes into dominant modes and modes, which are negligible, we may identify the dominant modes as order parameters.

The dynamics of the patterns can subsequently be represented in the space spanned by the amplitudes of the dominant modes. A determination of the order parameter equation completes the macroscopic analysis of the experimentally derived spatio-temporal patterns. In the case of dynamics generated by mode interactions the structure of the order parameter equations can be established with the help of symmetry arguments (for the case of α-waves this has been done in [11], [13]) as well as well-known classifications of phase portraits of order parameter equations for instabilities with low codimension.

4. Analysis of Human EEG's: Epileptic Seizures

Epileptic seizures are usually divided into two classes. In the case of *partial seizures* the epileptic activity is localized at one or several epileptogenic foci. *Generalized seizures* are characterized by global seizure activity involving both cerebral hemispheres. Petit-mal epilepsy belongs to the class of generalized seizures. It is usually related with an absence which lasts a few seconds. It shows

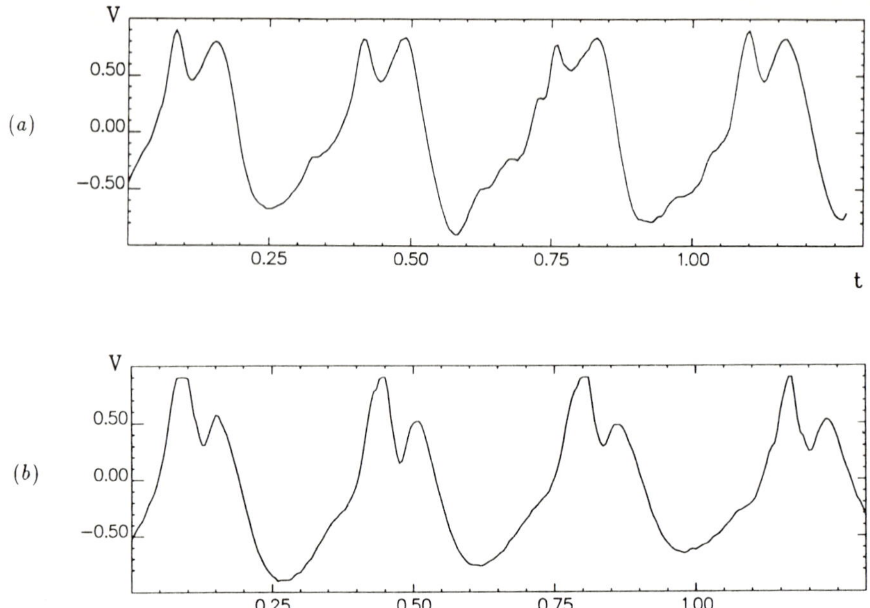

Fig. 2: Spike-Wave behaviour of the electric potential during petit-mal epilepsy: a) data set A, b) data set B.

up in EEG signals by a pronounced spike-wave behaviour (see fig. 2). In this section we shall present in parallel an analysis of two data sets of EEG during petit-mal epileptic seizures derived from different persons. The data sets have been provided to us by Prof. D. Lehmann. The maps were derived at locations based on the standard international 10-20 system [19]. Figure 2 exhibits the time series generated by the potential difference between two electrodes. The data show the pronounced spike-wave behaviour which in clinical application is the diagnosis of petit-mal epilepsy. Usually there are three spike-wave cycles per second.

Figures 3a, 3b demonstrate the evolution of the patterns during one cycle. (A movie [20] yields a direct impression of the characteristics of the spatio-temporal evolution). The patterns appear to be spatially coherent. In case of the data set A the patterns consist of two regions of opposite polarity. During the spike segment small oscillations of the patterns around a mean orientation are observed. During the time segment of the relaxation wave the pattern undergoes a reversal of polarity. Subsequently, a second polarity change drives the system back to a state roughly similar to the initial one and the next cycle is initiated. In case of the data set B the evolution of the patterns is quite different although the time series of fig. 2 exhibits a similar spike-wave characteristic.

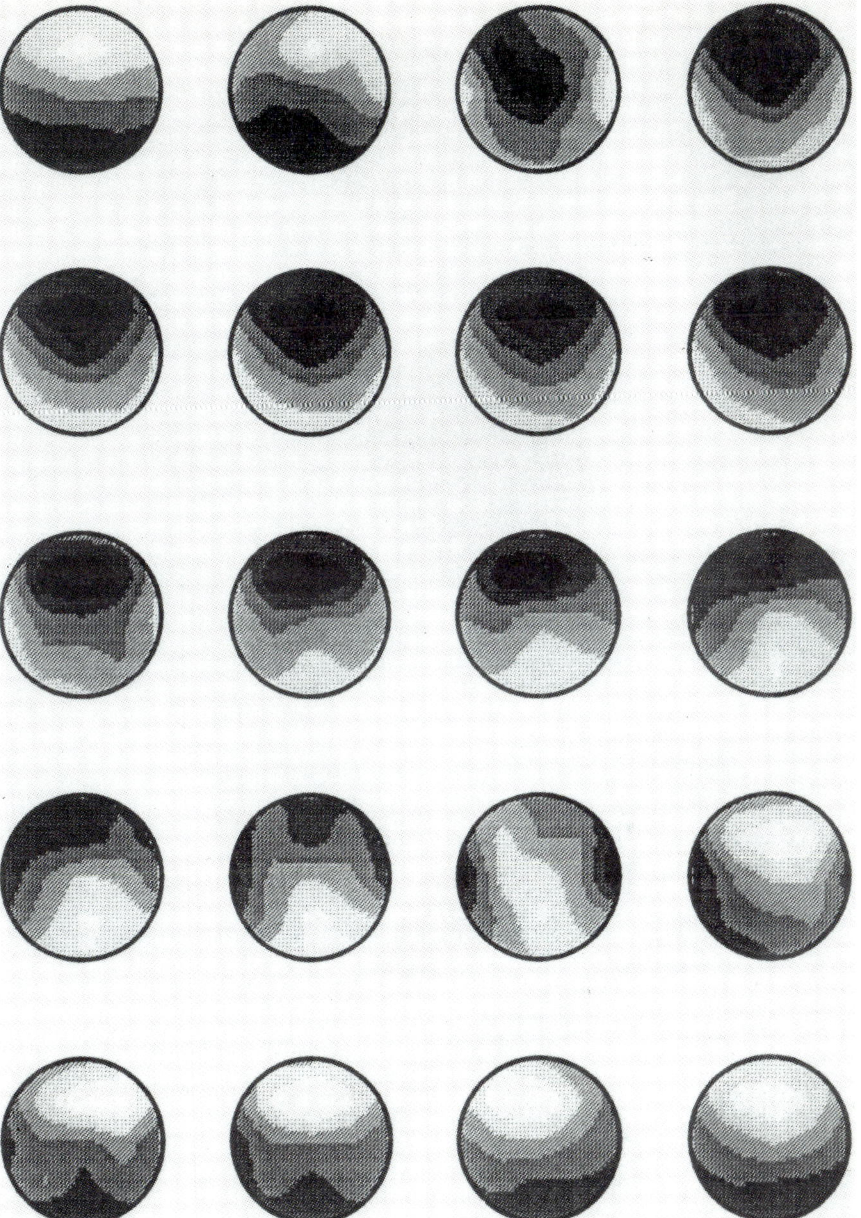

Fig. 3a: Temporal evolution of the electric field patterns during one spike-wave cycle (data set A).

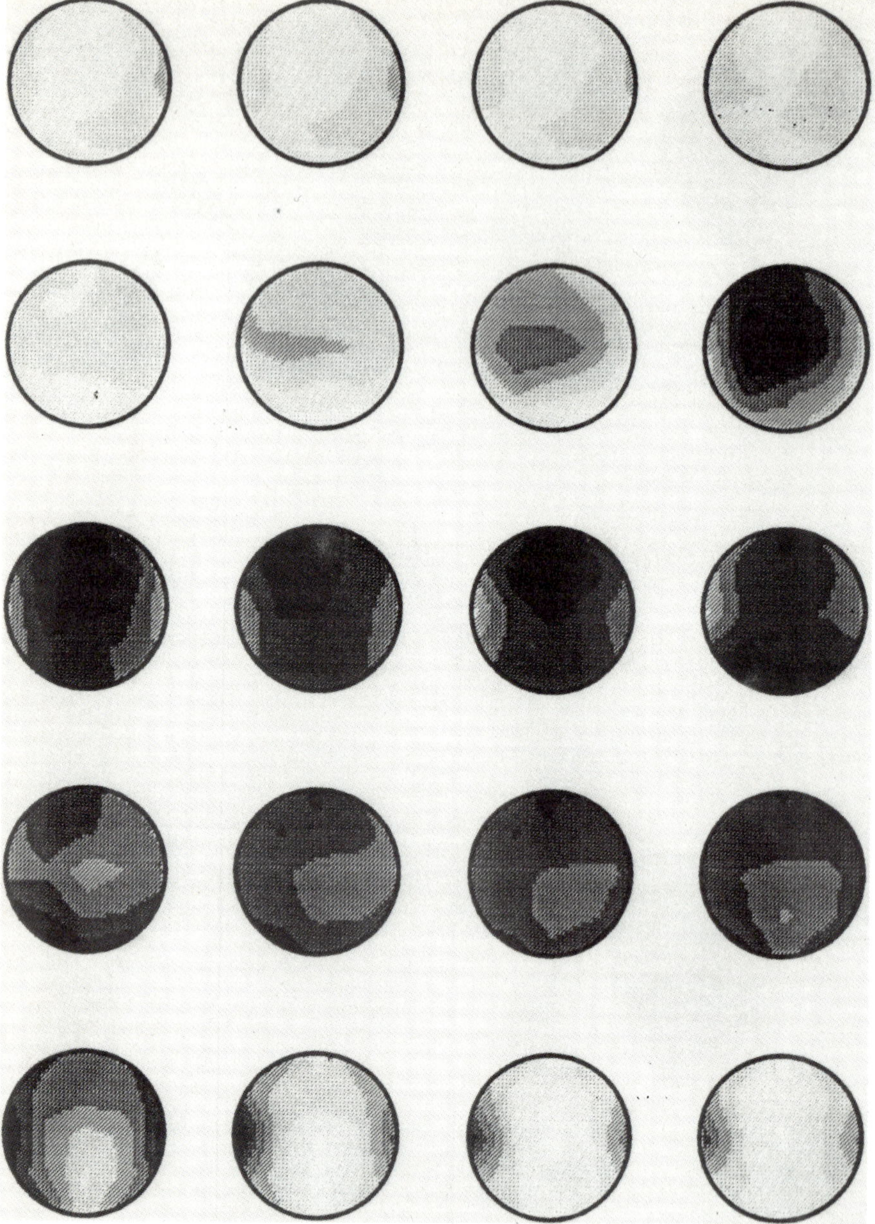

Fig. 3b: Temporal evolution of the electric field patterns during one spike-wave cycle (data set B).

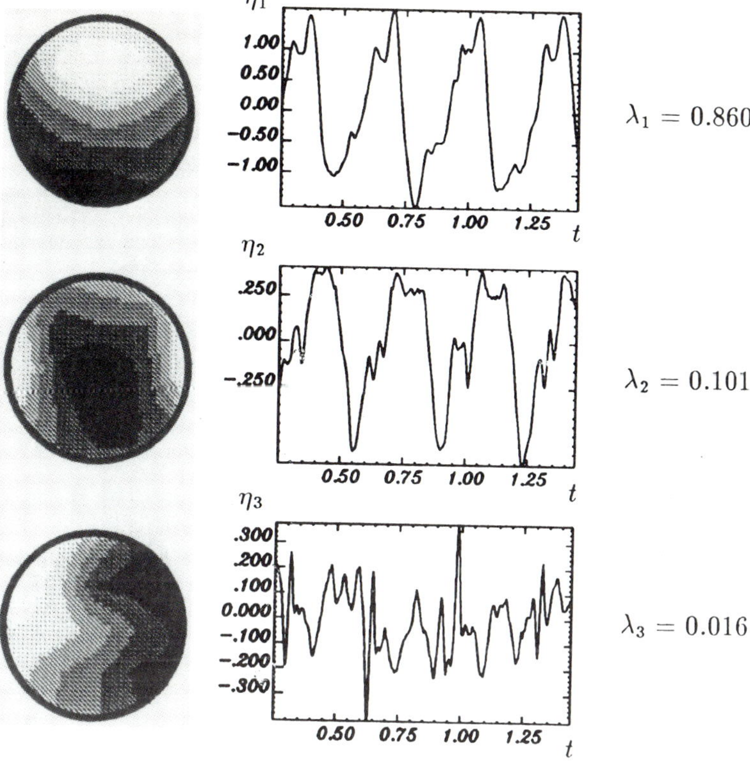

Fig. 4: The modes of the Karhunen-Loève decomposition and the temporal evolution of their amplitudes (data set A).

A previous analysis of the data set A [12] has been based on a Karhunen-Loève expansion of the patterns in order to identify order parameters. It has turned out that the first three modes cover about 97 per cent of the total potential $\mathbf{V}(t)$. The contributions of higher modes are not only negligible in the time mean, as follows from the Karhunen-Loève method, but also at each time instant. This may be shown by a reconstruction of the patterns from the first three modes (see fig. 4). As a main result it has become evident that the amplitudes of the modes are loosely characterized by a relationship of the form

$$\dot{\eta}_1(t) \approx const.\eta_3(t) \quad . \tag{11}$$

Such a relationship for the data set A is shown in fig. 5. This relationship seems to be crucial for the dynamics of the patterns. Therefore, we have decided to incorporate it directly by determining a more suitable set of modes already from the beginning. To this end we determine the spatial modes (which we shall denote by $\mathbf{w}^{\alpha}$) in the following two steps:

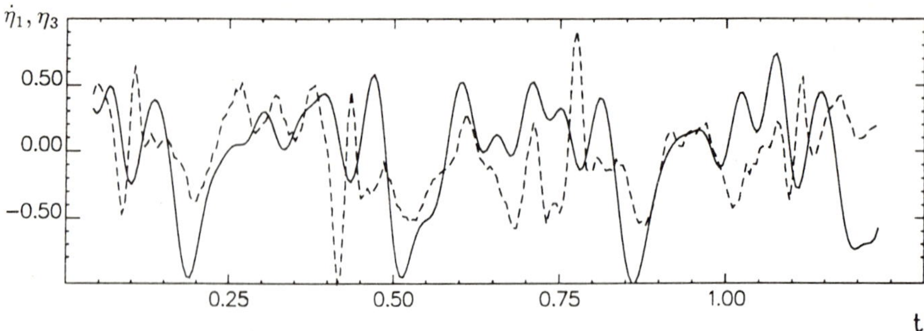

Fig. 5: The time derivative of the amplitude of the first Karhunen-Loève mode $\dot{\eta}_1(t)$ (solid line) and the amplitude of the third Karhunen-Loève mode η_3 (dashed line) of data set A.

Step 1: The most dominant mode $\mathbf{w}^1 = \mathbf{v}^1$ is determined by the method of Karhunen-Loève.
Step 2: The higher modes $\mathbf{w}^\alpha$ are determined in such a way, that their amplitudes $\xi_\alpha(t)$ approximate the relationship

$$\xi_{\alpha+1}(t) = \frac{d^\alpha}{dt^\alpha}\eta_1(t), \qquad \alpha \geq 1 \quad . \tag{12}$$

This is achieved by the extremum condition

$$< \left(\mathbf{V}(t) \cdot \mathbf{w}^{\alpha+1\dagger} - \frac{d^\alpha}{dt^\alpha}\eta_1(t)\right)^2 >= min! \quad . \tag{13}$$

Variation of this expression with respect to the adjoint eigenvectors allows us to express the amplitudes $\xi_\alpha(t)$ and the adjoint modes $\mathbf{w}^{\alpha\dagger}$ by the amplitudes η_α and the modes $\mathbf{v}^\alpha$ of the Karhunen-Loève method:

$$\xi_{\alpha+1}(t) = \sum_{\beta=2}^{N} \eta_\beta(t) < \eta_\beta(t)\frac{d^\alpha}{dt^\alpha}\eta_1(t) > / < \eta_\beta(t)^2 > \quad , \tag{14}$$

$$\mathbf{w}^{\alpha+1\dagger} = \sum_{\beta=2}^{N} \mathbf{v}^\beta < \eta_\beta(t)\frac{d^\alpha}{dt^\alpha}\eta_1(t) > / < \eta_\beta(t)^2 > \quad . \tag{15}$$

It turns out that in both cases the first three modes determined in this way yield an accurate representation of the spatio-temporal dynamics and simultaneously fulfill the conditions (12). Figures 6a, 6b demonstrate these relationships, which turn out to be crucial for an understanding of the spatio-temporal dynamics resulting from the mode analysis. The spatio-temporal patterns are determined with sufficient accuracy by the superposition

$$\mathbf{V}(t) = \eta_1(t)\mathbf{v}^1(\mathbf{r}) + \dot{\eta}_1(t)\mathbf{w}^2(\mathbf{r}) + \ddot{\eta}_1(t)\mathbf{w}^3(\mathbf{r}) \quad . \tag{16}$$

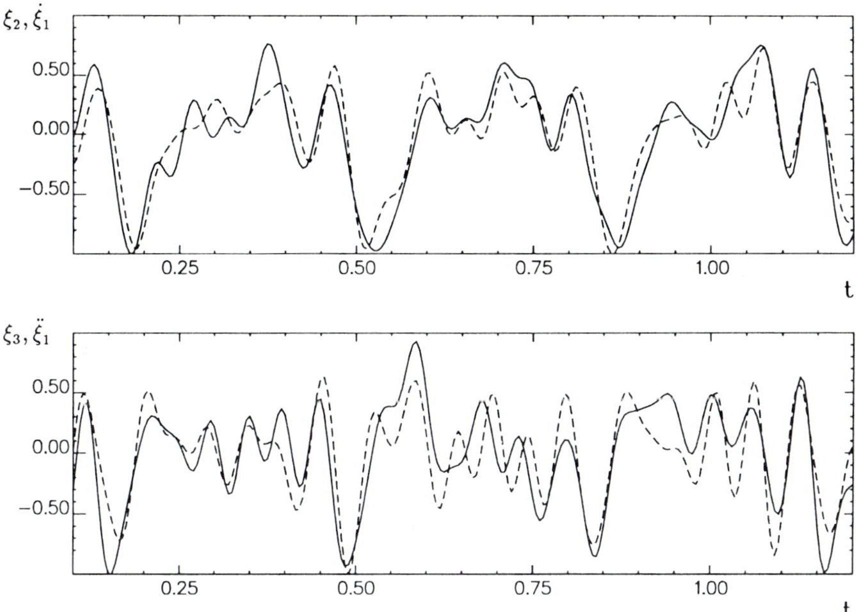

Fig. 6a: The amplitudes ξ_2 and ξ_3 (solid lines) compared to $\dot{\xi}_1$, $\ddot{\xi}_1$ (dashed lines) of data set A.

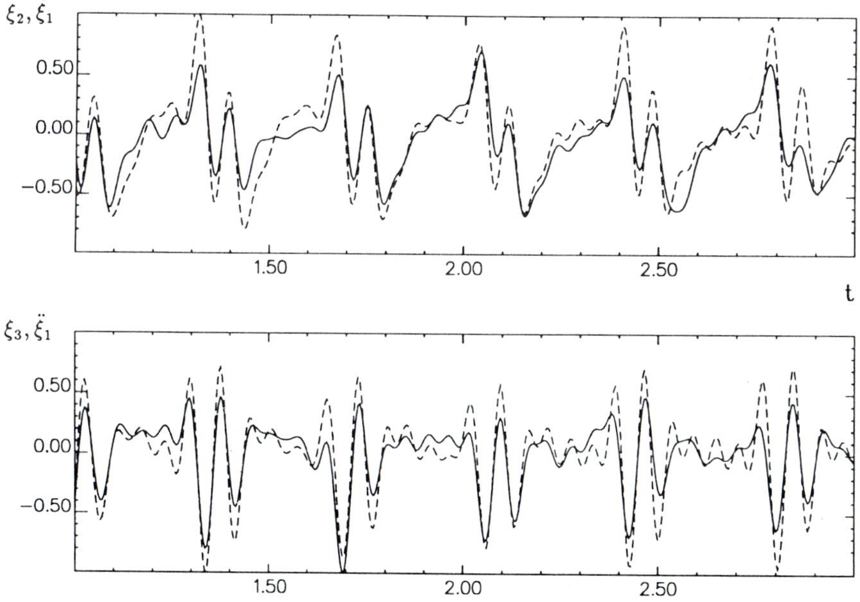

Fig. 6b: The amplitudes ξ_2 and ξ_3 (solid lines) compared to $\dot{\xi}_1$, $\ddot{\xi}_1$ (dashed lines) of data set B.

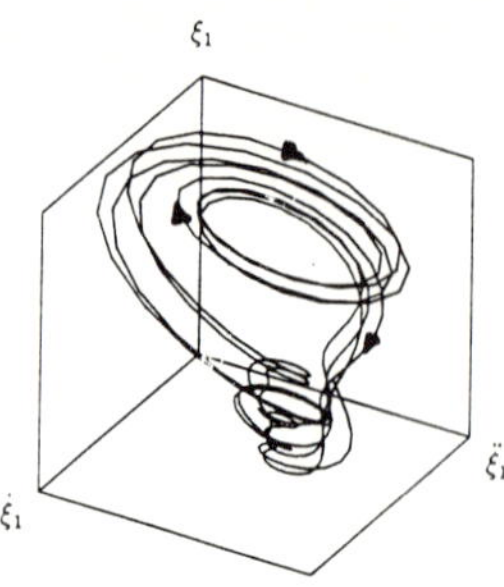

Fig. 7: Representations of the phase space spanned by the coordinates $\xi_1, \dot{\xi}_1, \ddot{\xi}_1$ (data set B).

A reconstruction of the dynamics of the amplitude $\eta_1(t)$ will complete our analysis.

4.1 Reconstruction of the Dynamics

Since it is possible to represent the patterns according to (16), it is straightforward to reconstruct the attractor in a space spanned by the coordinates η_1, $\dot{\eta}_1$, $\ddot{\eta}_1$. Figures 7, 8a yield the reconstruction for the two data sets. It is evident that in both cases the geometry of the dynamics corresponds to the class of dynamical systems investigated by Shil'nikov [16]. However, whereas in case of the data set A (fig. 8a) the two dimensional manifold of the fixed point is an unstable manifold, this manifold is a stable manifold in case of the data set B (fig. 7). This is a remarkable result since it demonstrates that the synergetic analysis yields a finer distinction of petit-mal epilepsies.

Due to the relationship (12) we may postulate that the dynamics is generated by a dynamical system of the form

$$\begin{aligned} \frac{d}{dt}\xi_1(t) &= \xi_2 + h_1(\xi_1,\xi_2,\xi_3) \quad , \\ \frac{d}{dt}\xi_2(t) &= \xi_3 + h_2(\xi_1,\xi_2,\xi_3) \quad , \\ \frac{d}{dt}\xi_3(t) &= h_3(\xi_1,\xi_2,\xi_3) \quad . \end{aligned} \tag{17}$$

Neglecting the functions $h_i(\xi_1,\xi_2,\xi_3)$ we obtain a linear dynamical system which arises for a certain codimension-III instability exactly at the tricritical point (see [21]). Close to such an instability a transformation to normal form yields an order parameter equation for the amplitude ξ_1 only:

$$\begin{aligned} &\frac{d^3}{dt^3}\xi_1(t) + \mu_2\frac{d^2}{dt^2}\xi_1(t) + \mu_1\frac{d}{dt}\xi_1(t) + \mu_0\xi_1(t) = c+ \\ &\quad +\ddot{\xi}_1(t)F_1[\xi_1(t), \dot{\xi}_1(t)^2 - 2\xi_1(t)\ddot{\xi}_1(t)] + \\ &\quad +\dot{\xi}_1(t)F_2[\xi_1(t), \dot{\xi}_1(t)^2 - 2\xi_1(t)\ddot{\xi}_1(t)] + F_3[\xi_1(t), \dot{\xi}_1(t)^2 - 2\xi_1(t)\ddot{\xi}_1(t)] \quad . \end{aligned} \tag{18}$$

Here, μ_0, μ_1, μ_2 denote the deviations from the tricritical point and the nonlinearities $F_i[\xi_1(t), \dot{\xi}_1(t)^2 - 2\xi_1(t)\ddot{\xi}_1(t)]$ are polynomials in their arguments [21].

A systematic derivation of the dynamics should involve a determination of the parameters μ_0, μ_1, μ_2, c as well as the nonlinearities $F_i[\xi_1, \dot{\xi}_1^2 - 2\xi_1\ddot{\xi}_1]$ directly from the experimental time series. A choice of the order parameter equation in the form

$$\frac{d^3}{dt^3}\xi_1(t) + \mu_2\frac{d^2}{dt^2}\xi_1(t) + \mu_1\frac{d}{dt}\xi_1(t) + \mu_0\xi_1(t) = -\beta\xi_1(t)^3 \quad , \tag{19}$$
$$\mu_2 = 1.428, \qquad \mu_1 = 4, \qquad \mu_0 = -6.1, \qquad \beta = 4$$

already yields a phase portrait which approximates the one obtained from the data set A quite accurately, as can be seen in fig. 8b. The specification of this order parameter equation for the amplitude $\xi_1 = \eta_1$, together with the representation (16) of the spatial patterns, completes the macroscopic analysis of the EEG patterns for the data set A.

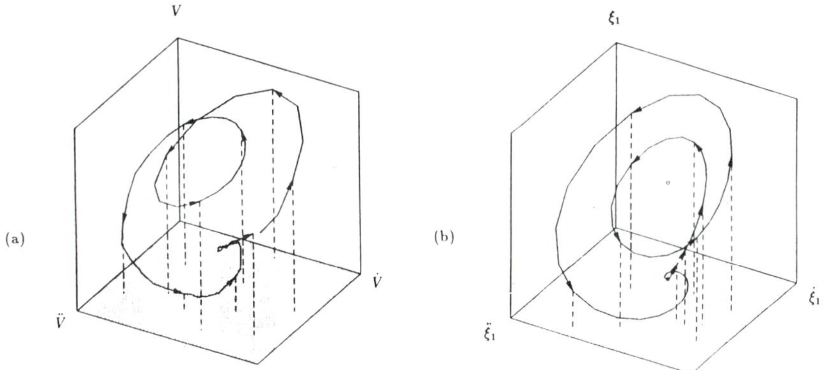

Fig. 8: (a) Experimentally derived phase portrait (data set A), (b) reconstruction of the dynamics from the order parameter equation (19)

4.2 Shil'nikov's Condition

The determination of correlation dimensions of epileptic time signals [8] yields a fractal value between two and three indicating a chaotic temporal evolution. However, there has been some scepticism about the determination of fractal dimensions from a time series, which, as in the present case of epileptic seizure, is not stationary during a sufficiently long period [22]. Therefore, it would be desirable to decide whether the underlying dynamics is chaotic or not by other means. In the present case this may be decided by considering whether the Shil'nikov-condition (6) is fulfilled.

In the immediate vicinity of the saddle focus the dynamics is approximated by a linear differential equation of the form

$$\frac{d^3}{dt^3}\xi_1(t) + \mu_2\frac{d^2}{dt^2}\xi_1(t) + \mu_1\frac{d}{dt}\xi_1(t) + \mu_0\xi_1(t) = c \quad . \tag{20}$$

The coefficients can be determined by a least square fit from the following extremum condition

$$\frac{1}{t_1 - t_0}\int_{t_0}^{t_1} dt\{\frac{d^3}{dt^3}\eta_1(t) + \mu_2\frac{d^2}{dt^2}\eta_1(t) + \mu_1\frac{d}{dt}\eta_1(t) + \mu_0\eta_1(t) - c\}^2 = min \quad , \tag{21}$$

where $\eta_1(t), \dot{\eta}_1(t), ..$ are determined from the experimental time series. $[t_0, t_1]$ is the interval, where the assumption of linearity of the flow in the vicinity of the saddle focus holds. Variation with respect to the coefficients c, μ_i yields an inhomogeneous linear algebraic set of equations, which can be solved numerically. Figure 9 shows a comparison of the fitted solutions with the experimental time series. From the coefficients μ_i, c we may evaluate the linear growth rates λ, ρ by a straightforward calculation and can subsequently investigate Shil'nikov's condition.

An estimation of the ratio $|\rho|/|\lambda|$ can be calculated for several spike-wave complexes. Results for the data set B are contained in table 1. We mention that the frequency of the spiral motion in the stable manifold is rather constant, whereas the growth rates do fluctuate due to the instationarity of the signal. However, the Shil'nikov-condition for the existence of strange dynamics is fulfilled. We should like to mention, that, whereas a possible nonstationarity of the signal, which is at least significant at the beginning and end of a petit-mal seizure epoche, is likely to show up in the metric properties like fractal dimension etc., the criteria of Shil'nikov for the existence of chaos is a robust indication for the existence of chaotic behaviour.

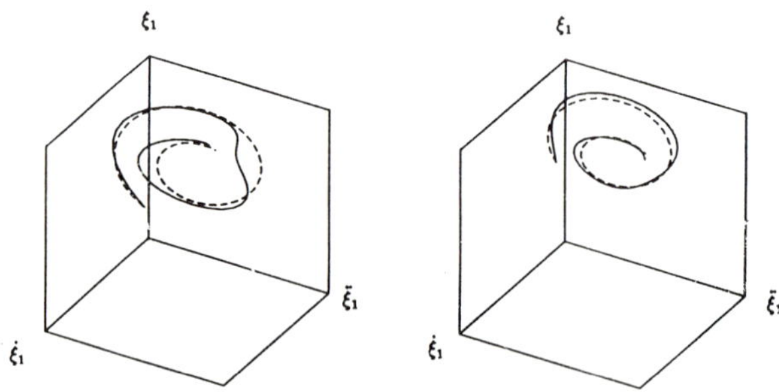

Fig. 9: Approximation of the experimental trajectory (solid line) obtained by solving the linearized order parameter equation (20) (dashed line) in the vicinity of the saddle focus (data set B).

t_0	t_1	ρ	λ	ω	$\delta = \rho/\lambda$
0.30	0.42	0.072	−0.339	1.28	0.212
0.63	0.75	−0.008	−0.175	0.93	−0.046
0.97	1.07	−0.095	0.164	0.91	0.594
1.33	1.42	−0.079	0.382	0.93	0.207
1.68	1.78	−0.080	0.168	0.91	0.476
2.04	2.15	−0.165	0.433	1.00	0.381
2.42	2.54	−0.092	0.313	1.09	0.294
2.78	2.92	−0.105	0.254	0.95	0.413
3.17	3.27	−0.045	0.039	0.86	1.154
3.55	3.77	−0.060	0.069	0.88	0.870

Tab. 1: Results of the least square fit in the vicinity of the saddle focus for several time intervals $[t_0, t_1]$ of data set B.

5. Summary

Based on the analysis of two data sets derived from different persons we have established the following properties of brain waves during petit-mal epilepsies:
a) The patterns can be represented by three spatial modes. The amplitudes of these modes are related through

$$\xi_2 = \dot{\xi}_1, \qquad \xi_3 = \ddot{\xi}_1 \quad . \tag{22}$$

b) The dynamics can be reconstructed in a phase space spanned by the amplitudes of the three modes.
c) The dynamics belong to the class of Shil'nikov-attractors. Data set A shows a strange attractor related with a Γ_0^-- homoclinic orbit. Data set B is related with a Γ_0^+-homoclinic orbit.
d) For the data set A we have given a dynamical system, which accurately reproduces the experimentally derived phase portrait.
e) For the data set B we have confirmed Shil'nikov's condition and have shown the existence of chaos.

Summarizing we would like to emphasize that the methods of macroscopic synergetics developed by H. Haken can be successfully applied to the analysis of EEG patterns.

Acknowledgement. We wish to thank L. Borland, K. Scheller, and A. Greiner for helpful suggestions during the preparation of the manuscript.

References

[1] H. Berger, 'Über das Elektrenkephalogramm des Menschen', Arch. Psychiatr. Nervenkr. **87**, 527-570 (1929)

[2] J. Kugler, *Elektroenzephalographie in Klinik und Praxis* (Georg Thieme-Verlag, Stuttgart, 1981)

[3] D. Lehmann, 'Multichannel Topography of Human Alpha EEG Fields', Electroenceph. clin. Neurophysiol. **31**, 439-449 (1971); 'Human Scalp EEG Fields: Evoked, Alpha, Sleep, and Spike-Wave Patterns', in: *Synchronization of EEG Activity in Epilepsies*, eds. H. Petsche and M.A.B. Brazier (Springer-Verlag, Berlin, 1972)

[4] H. Haken, in *Synergetics of the Brain*, eds. E. Basar, H. Flohr, H. Haken, and A.J. Mandell (Springer-Verlag, Berlin, 1983)

[5] H. Haken, *Synergetics. An Introduction* (Springer-Verlag, Berlin, 1983)

[6] H. Haken, *Advanced Synergetics* (Springer-Verlag, Berlin, 1987)

[7] H. Haken, *Information and Selforganization* (Springer-Verlag, Berlin, 1988)

[8] A. Babloyantz, 'Strange Attractors in the Dynamics of Brain Activity', in: *Complex Systems - Operational Approaches*, ed. H. Haken (Springer-Verlag, Berlin, 1985)

A. Babloyantz, C. Nicolis, and M. Salazar, 'Evidence of chaotic dynamics of brain activity during the sleep cycle', Phys. Lett. **A 111**, 152 (1985)

A. Babloyantz and A. Destexhe, 'Strange Attractors in the Human Cortex', in: *Temporal Disorder in Human Oscillatory Systems*, eds. L. Rensing, U. an der Heiden, M.C. Mackey (Springer-Verlag, Berlin, 1986)

A. Babloyantz and A. Destexhe, 'Low-dimensional chaos in an instance of epilepsy', Proc. Natl. Acad. Sci. **83**, 3513-3517 (1986)

[9] G. Mayer-Kress and J. Holzfuss, 'Analysis of the Human Electroencephalogram with Methods from Nonlinear Dynamics', in: *Temporal Disorder in Human Oscillatory Systems*, eds. L. Rensing, U. an der Heiden, M.C. Mackey (Springer-Verlag, Berlin, 1986)

S.P. Layne, G. Mayer-Kress, and J. Holzfuss, 'Problems Associated with Dimensional Analysis of Electroencephalogram Data', in: *Dimensions and Entropies in Chaotic Systems*, ed. by G. Mayer-Kress, (Springer-Verlag, Berlin, 1986)

[10] A. Fuchs, R. Friedrich, H. Haken, and D. Lehmann, 'Spatio-Temporal Analysis of Multichannel α-EEG Map Series', in *Computational Systems - Natural and Artificial*, ed. by H. Haken (Springer-Verlag, Berlin, 1987)

[11] R. Friedrich, A. Fuchs, and H. Haken, 'Synergetic Analysis of Spatio-Temporal EEG Patterns', in *Nonlinear Wave Processes in Excitable Media*, eds. A.V. Holden, M. Markus, and H.G. Othmer (Plenum, New York, 1991)

[12] R. Friedrich, A. Fuchs, and H. Haken, 'Modelling of Spatio-Temporal EEG Patterns', in *Mathematical approaches to brain functioning diagnostics*, eds. I. Dvorak, A. V. Holden (Manchester University Press, 1991)

[13] R. Friedrich, A. Fuchs, and H. Haken, 'Spatio-Temporal EEG Patterns', in *Synergetics of Rhythms in Biological Systems*, eds. H. Haken and H.P. Köpchen, (Springer-Verlag, Berlin, 1992)

[14] C. Uhl, Diplom thesis, Stuttgart 1991

[15] R. Friedrich, M. Bestehorn, H. Haken, 'Pattern Formation in Convective Instabilities', Int. J. Mod. Phys. B Vol. 4, No. 3 365-400 (1990)

[16] L.P. Shil'nikov, 'A case of the existence of a countable number of periodic motions', Sov. Math. Dokl. **6**, 163-166 (1965); Math. USSR Sbornik **10**, 91 (1970)

[17] A. Wunderlin, 'On the principles of synergetics', this volume

[18] P. Glendinning, C. Sparrow, 'Local and Global Behavior near Homoclinic Orbits', J. Stat. Phys. **35**, 645-696 (1984)

[19] E.J. Speckmann, *Experimentelle Epilepsieforschung* (Wissenschaftliche Buchgesellschaft, Darmstadt, 1986)

[20] Movie *Spatio-Temporal EEG Patterns*, by R. Friedrich, A. Fuchs, and H. Haken

[21] C. Elphick, E. Tirapegui, M.E. Brachet, P. Coullet, and G. Iooss, 'A simple global characterization for normal forms of sinular vector fields', Physica D **29**, 95

[22] J.P. Eckmann, D. Ruelle, 'Fundamental Limitations for Estimating Dimensions and Lyapunov Exponents in Dynamical Systems', preprint IHES/P/90/11

Synergetics, Resonance Phenomena and Brain Internal Codes

E. Başar, C. Başar-Eroglu, T. Demiralp, and M. Schürmann

Institute of Physiology, Medical University Lübeck,
Ratzeburger Allee 160, W-2400 Lübeck, Fed. Rep. of Germany

Abstract

In this paper, resonance phenomena of the brain are discussed in the context of synergetics. Additionally, a possible role of resonance phenomena in brain's internal coding is tentatively described.

1. Introduction

The **formation of temporal and spatial patterns of neural activities** is one of the problems treated by the discipline called **"synergetics"** /1,2/. Spontaneous electrical activity of the brain also gives rise to such patterns. Such records are called **electroencephalograms (EEG).** This activity can be measured by using either intracranial macroelectrodes (50-100 μm) implanted in nervous masses or by using scalp electrodes. The term **"evoked potential" (EP)** is used for responses elicited by means of an external stimulus. In this paper, EP frequency components are tentatively interpreted as possible elements of brain's internal coding.

1.1. Relation between EEG and evoked potentials

There are several interpretations concerning the genesis of EPs and their relation to the spontaneous EEG (for an overview, see e.g. /3/).

According to our working hypothesis, the brain is a dynamic system having the property of continuously and spontaneously changing state. "At any instant, each state variable has a rate of change which depends on the current state of the system. Therefore if we stimulate or excite such a dynamic system (brain), it is to be expected that for every state a different **excitability** or different **response susceptibility** should exist. Accordingly, we use single records of brain responses together with the brain electrical activity prior to stimulation" /4/. At present, these "single sweep records" would be called records of **"induced rhythmicities"** /5/. "Our approach to the understanding of neural populations is derived from a combined analysis of EEG and EPs. This combined analysis provided us with results which could be interpreted by using analogies from laser physics, magnetism theory, concepts of statistical physics and dynamic systems theory and on cooperative phenomena in populations of neurons /6,7,8,9/. Accordingly, the new discipline 'synergetics', which established links between dynamic system theory and statistical physics, provided us with useful integrative concepts and methods for the understanding of evoked potentials and the brain waves"/4/.

Springer Proceedings in Physics, Vol. 69
Evolution of Dynamical Structures in Complex Systems
Editors: R. Friedrich · A. Wunderlin

1.2 Evoked potentials: resonance phenomena in the brain - transitions to coherent states - reduction in entropy

According to the working hypothesis outlined above, EPs can be considered as manifestations of **resonance properties** of the EEG. Furthermore, we stated that general resonance phenomena do reflect most dominating transfer functions of the brain. We defined the resonance phenomenon as the responsiveness of brain structures where the EP contains frequency components already present in spontaneous EEG activity; responses are time-locked, magnified in amplitude and frequency-stabilized. In other words: "resonance" and/or "selectivity" is defined as the ability of brain networks to facilitate (or activate) electrical transmission within determined frequency bands, when a stimulation is applied to the brain /6/. This working hypothesis was based on our own measurements of the cat and human brain, and took into consideration experimental findings of the groups of Spekreijse and Van der Tweel /10/ and of Lopes da Silva and Storm van Leeuwen /11,12/.

In terms of synergetics, **coherent states** in the electrical activity of the brain can be obtained by the application of an external stimulus. This is an analogy both to a simple harmonic oscillator and to the classical examples of synergetics in the field of **laser physics** /1,2/: the transition from an incoherent to a coherent state is also a feature of the laser at the emission threshold. In parallel to this analogy, the external stimulus giving rise to the EP is regarded as the **control parameter** or **order parameter.**

1.3 Evoked potentials and single cell recordings

At present, resonance phenomena of the brain merit increasing consideration. This is due to several recent neurophysiological investigations both at the cellular level and in EEG and magnetoencephalographic (MEG) measurements.

Some examples may demonstrate the essential progress made at the cellular level. The concept of Llinás /13/, who demonstrated in single cells 6 and 10 Hz intrinsic resonances, opened a new window to our knowledge at the level of field potentials and single neuronal recordings. Studies by Gray and Singer /14,15/ showed 40 Hz coherent oscillations in the visual cortex to light stimulation. Their measurements - confirmed and extended by Eckhorn et al /16/ - provided another important step forward to indicate the relation between EEG and single unit recording. These advances have been accompanied by a number of measurements at the EEG and MEG level. In some of these studies, the approach mentioned above (1.2) has been applied: For example, Narici et al /17/ have used the concept of resonance behaviour for magnetic activity of the brain in 6 and 10 Hz frequency ranges.

Induced rhythmicities in magnetoencephalographic recordings have also been investigated by Saermark et al /18/ and Mikkelsen et al /19/. Similar approaches have been used in EEG-EP studies /20,21/.

In the past few years, links between synergetics and electrical activity of the brain have also been established in the context of nonlinear and chaotic dynamics. This topic, however, is beyond the scope of this paper. The reader is referred to the literature (for a collection of papers, see e.g. /22/.

4. Resonance phenomena and brain coding

Now let us consider three different brain structures or networks, A,B, and C, which have excitabilities in the four frequency ranges of 4 Hz, 10 Hz, 20 Hz, and 40 Hz, which I will call theta, alpha, beta and gamma. Let us assume that structure A can be excited and can excite structure B, which in turn can excite structure C. When structure A excites structure B, there will be a transmission of signals in one of these frequency channels from structure A to structure B; and structure B, if not already excited in a different frequency, will then excite structure C in the same frequency range (see Fig. 1).

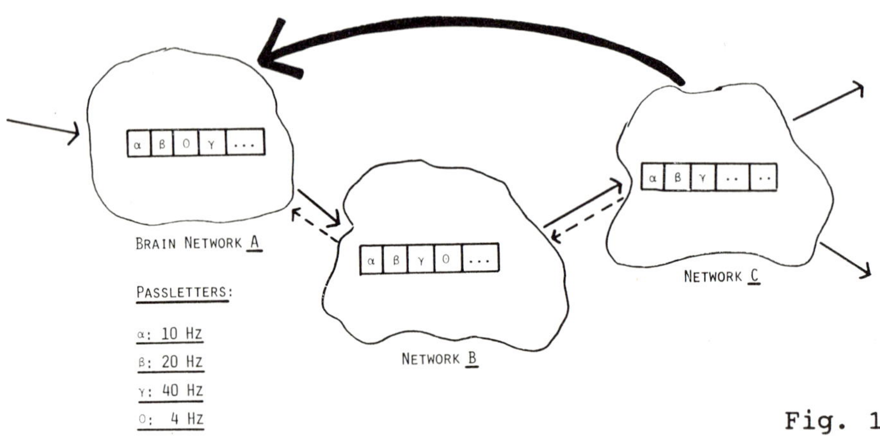

Fig. 1

Let us now give another name to the specific resonant modes, calling them "passletters", because we have seen that a brain structure into which activity in the 10 Hz frequency range is entering will, by an excitability rule, resonate in that range. In this case the passletter would be the excitability at 10 Hz. A second passletter is at 20 Hz and the third passletter is at 40 Hz. The term "password" or "pass charateristic" has already been used by Barlow /23/, who tried to develop a theory of passwords and pass characteristics in neural networks. Our theory is somewhat different from his but we use his expressions.

According to empirical results, theta, alpha, beta, and gamma resonances do occur in the brain. Therefore we can suppose that **the passletters represent resonant modes that help to optimize signal transfer between various brain structures.** This idea of a way in which brains may work was suggested by the excitability rules that I derived from experiments.

Now we will go further and speculate that the passwords alpha, beta, gamma, and theta represent resonant modes that are **almost invariant.** However, passletters may perhaps combine during the transfer of messages through various brain networks to create a great number of less elementary or complex combination which we may call passwords. For example, there can be only alpha transmission in all three of the structures A,B, and C, or there can be a combination of alpha and beta (passwords α,β). A still more complicated password would be alpha-beta-gamma (α,β,γ) and so on (see Fig. 2).

COMBINATION OF PASSLETTERS DURING TRANSFER OF MESSAGES THROUGH VARIOUS BRAIN NETWORKS MAY CREATE A GREAT NUMBER OF ELEMENTARY (AT LOWER LEVEL) OR COMPLEX (HIGHER LEVEL) PASSWORDS:

α

α β

α β γ

β θ γ

.

.

α β θ γ δ ... ETC.

IN TURN, THIS FACT MAY CREATE AN ENSEMBLE OF PASSWORDS FOR HETEREGENOUS MESSAGES (OR FOR COMPLEX SENSORY/COGNITIVE INPUTS) BY USING REVERBERATION OF PASSWORDS, POSSIBLY VIA AUTOEXCITATION.

Fig. 2

The information capacity of a structure can be important, especially when it is busy, for if the information capacity of structure B is limited, or if structure B is busy at the time, the transmission in one or another of these special channels will not be optimal /24/. This fact in turn may lead to the creation of an ensemble of passwords for complex sensory-cognitive inputs and other heterogeneous messages, by using reverberations of passwords, possibly via auto-excitation. How might such auto-excitation occur? One possibility is that structure A could excite structure B again, as in limbic system networks, for example. In such a case, **it is possible that as a consequence of a preliminary input signal, the ensemble of neural structures A, B, and C could start to reverberate.** Such reverberation would be possible only if all the structures had the ability to resonate in the same frequency channels.

Finally, one can speculate further about why it would be advantageous for a brain to transmit by means of resonance phenomena. We have thought of one possibility. The simplest invariant transfer functions are probably represented in the brain by resonant networks, but this **fixed hardware may create a much richer array of useful software.** That is, it could be an economical principle of the brain to achieve good internal communication first by using a small number of similarly structured network as channels for a small number of frequencies, and then to combine the elementary passletters into a large number of compound passwords that could transmit a wide variety of complex patterns.

The building of such patterns is, as stated above, probably due to cooperative phenomena of activity in large neural populations. We therefore tentatively assume that the use of synergetics concept is of basic importance for understanding the EEG phenomena related to brain function.

Acknowledgements

The research presented in this study was financially supported by DFG grant Nr. Ba 831/5-1 and Volkswagen-Stiftung grant Nr. I/67678.

References

/1/ Haken H. Synergetics: An introduction. Berlin Heidelberg New York: Springer, 1977

/2/ Haken H. Synopsis and introduction. In: Başar E, Flohr H, Haken H, Mandell AJ, eds. Synergetics of the brain. Berlin Heidelberg New York: Springer, 1983: 3-27

/3/ Ruchkin DS. Measurement of event-related potentials: signal extraction. In: Picton TW, ed. Human event-related potentials (EEG handbook, revised series, vol. 3). Amsterdam: Elsevier, 1988: 7-43

/4/ Başar E. Synergetics of neuronal populations. A survey on experiments. In: Başar, E, Flohr H, Haken H, Mandell AJ. eds. Synergetics of the brain. Berlin Heidelberg New York: Springer, 1983: 183-200

/5/ Başar E. Brain natural frequencies are causal factors for resonances and induced rhythms. In: Başar E, Bullock TH, eds. Induced rhythms in the brain. Boston: Birkhäuser (in press)

/6/ Başar E. EEG-Brain dynamics. Relation between EEG and brain evoked potentials. Amsterdam: Elsevier, 1980

/7/ Başar E, Gönder A, Ungan P. Important relation between EEG and brain evoked potentials. I. Resonance phenomena in subdural structures of the cat brain. Biol Cybernetics 1976; 25: 27-40

/8/ Başar E, Gönder A, Ungan P. Important relation between EEG and brain evoked potentials. II. A systems analysis of electrical signals from the human brain. Biol Cybernetics 1976; 25: 41-48

/9/ Başar E, Demir N, Gönder A, Ungan P. Combined Dynamics of EEG and evoked potentials. I. Studies of simultaneously recorded EEG-EPograms in the auditory pathway, reticular formation and hippocampus of the cat brain during the waking stage. Biol Cybernetics 1979; 34: 1-19

/10/ Spekreijse H, van der Tweel LH. System analysis of linear and nonlinear processes in electrophysiology of the visual system. Proc R Neth Acad Sci C 1972; 75: 77-105

/11/ Lopes da Silva FH, van Rotterdam A, Storm van Leeuwen W, Tielen AM. Dynamic characteristics of visual evoked potentials in the dog. I. Cortical and subcortical potentials evoked by sine wave modulated light. Electroencephalogr Clin Neurophysiol 1970; 29: 246-259

/12/ Lopes da Silva FH, van Rotterdam A, Storm van Leeuwen W, Tielen AM. Dynamic characteristics of visual evoked potentials in the dog. II. Beta frequency selectivity in evoked potentials and background activity. Electroencephalogr Clin Neurophysiol 1970; 29: 260-268

/13/ Llinás RR. The intrinsic electrophysiological properties of mammalian neurons: insights into central nervous system function. Science 1989; 242: 1654-1664

/14/ Gray CM, Singer W. Stimulus-specific neuronal oscillations in the cat visual cortex: a cortical function unit. Soc Neurosci Abstr 1987; 404: 3

/15/ Gray CM, Singer W. Stimulus-specific neuronal oscillations in orientation columns of cat visual cortex. Proc Natl Acad Sci USA 1989; 86: 1698-1702

/16/ Eckhorn R, Bauer R, Jordan W, Brosch M, Kruse W, Munk M, Reitboeck HJ. Coherent oscillations: A mechanism of feature linking in the visual cortex? Biol Cybern 1988; 60: 121-130

/17/ Narici L, Pizzella V, Romani GL, Torrioli G, Traversa R, Rossini PM. Evoked alpha and mu rhythm in humans: A neuro-magnetic study. Brain Res 1990; 520: 222-231

/18/ Saermark K, Mikkelsen KB, Başar E. Magnetoencephalographic Evidence for induced rhythms. In: Başar E, Bullock TH, eds. Induced Rhythms in the Brain. Boston: Birkhäuser (in press)

/19/ Mikkelsen KB, Saermark K, Lebech J, Bak C, Başar E. Selective averaging in auditory magnetic field experiments. Proceedings of the VIIth International Congress of Biomagnetism, New York, August 1989

/20/ Dettmar P, Volke HJ. Time-varying spectral analysis of single evoked brain potentials. In: Klix F, Näätänen R, Zimmer K, eds. Psychophysiological approaches to human information processing. Amsterdam: Elsevier, 1985: 225-233

/21/ Klauck U, Heinrich H, Dickhaus H. Classification of EP's with respect to EEG activity. Proceedings of the North Sea Conference for Biomedical Engineering 1990 (in press)

/22/ Başar E. Chaos in Brain Function. Berlin Heidelberg New York: Springer, 1990

/23/ Barlow HB. Possible principles underlying the transformations of sensory messages. In: Rosenblith WA, ed. Sensory communication. Cambridge: MIT Press, 1961: 217-234

/24/ Başar E. EEG-Dynamics and evoked potentials in sensory and cognitive processing by the brain. In: Başar E, ed. Dynamics of sensory and cognitive processing by the brain. Berlin Heidelberg New York: Springer, 1988: 30-55

Synergetics, Self-Simplification, and the Ability to Undo

O.G. Meijer[1] *and R. Bongaardt*[1;2]

[1]Department for the Theory and History of Movement Sciences, Vrije Universiteit Room A 614, Van der Boechorststraat 9, NL-1081 BT Amsterdam, The Netherlands

[2]Unit for Theoretical Psychology, Vrije Universiteit Room A 144, De Boelelaan 1111, NL 1081 HV Amsterdam, The Netherlands

Abstract

It is argued that Bernstein created a coherent science of biological movement. He failed, however, to completely escape mechanicism. Haken's synergetics is interpreted as the science which can solve that problem: order arises stochastically as self-simplification, i.e., the macroscopic reduction of microscopic degrees of freedom. Synergetics has been successfully applied to biological movement. In the present paper, two further steps in that process are outlined. Since biological self-simplification is not only behavioural but also structural, it is firstly argued that differential perturbation resistance will be met upon probing structure. The most perturbation resistant layers are the ones that embody self-simplification. Two examples are given from the neurosciences. Secondly, the exact determination of the control landscape is often difficult, the more so since transitions look different at different levels. But the creating of transitions requires surplus-energy, and energy can be measured. A line of thinking is sketched, in which Parkinsonism is understood as the inability to create transitions, or, in other words, as lacking the ability to undo. At the end, one is left with more questions than answers. Maybe, however, the questions have become more interesting.

1. Bernstein and the Coherence of Movement Science

To date, there exist computer programmes which embody the ability to play top-level chess. But while infants all over the world display considerable skill in building their block towers, computer-controlled robots to do the same are still incredibly difficult to design. Apparently, in biological movement there is 'something' that still escapes the scientific community at large.

Nikolai Aleksandrovich Bernstein (1896-1966) took his starting point in the analysis of everyday movements. In doing so, he succeeded in formulating an intrinsically *coherent set of problems* which, when taken together, can serve to define 'movement science' [1-3]. Ironically, however, much of Bernstein's readership has treated his problems separately [4], thereby missing the coherence of his understanding. Bernstein's stressing the impossibility of one-to-one relationships in his *physiology of coordination* has served to inspire one group of movement scientists, i.e., 'action system theorists'. On the other hand, his emphasizing the role of models of the future in what he called the *physiology of activity*, is in agreement with the

Springer Proceedings in Physics, Vol. 69
Evolution of Dynamical Structures in Complex Systems
Editors: R. Friedrich · A. Wunderlin

central tenets of their adversaries, i.e, 'motor systems theorists'. Whereas the role of stochasticity in both, has received little attention from either.

In the present section the development of Bernstein's science of biological movement will be described, the most important point being the impossibility of linear coordination and control. In section 2 it will be argued that Haken's synergetics can be understood as the science of 'self-simplification', essential to the further development of movement science. In section 3 the synergetics of physical self-organization is briefly mentioned and the importance of biological structure is emphasized for any synergetic understanding of biological self-organization. In section 4 two examples of structural self-simplification are taken from the neurosciences. Finally, in section 5 it is shown that synergetics may lead to a new understanding of Parkinsonism.

1.1 Nonlinearities in the Physiology of Coordination

Bernstein studied hammering movements in an industrial setting. He discovered that the (topologically) 'same' movement was never executed in the (metrically) 'same' way (Fig.1). If one assumes that the central impulse is the same for every (topologically) same movement, the old idea of one-to-one relationships between the central impulse and the metrics of the ensuing movement, Bernstein argued, must be wrong. Since we are dealing with *changing elastic properties* of the locomotor apparatus, Bernstein concluded that this is necessarily so: in biological movement, we have to do with *nonlinear relationships.*

In 1940, Bernstein stated:

> the change in muscle tension leads to a movement, and the movement causes ... further changes in muscle tensions. ... Mathematical analysis of such a relationship reveals that it ... *does not allow for an unequivocal dependence between force and movement,* that is to say, that one and the same sequence of forces can, when repeated ... , lead to different movements [1, p.22; our translation from the German edition, italics in the original].

And he was right.

It is impossible to linearly coordinate and control changing elastic properties. The present authors want to stress that Bernstein definitively made this point. Hence, the nonlinearity of coordination and control should be a starting point for all further research on biological movement.

The changing elasticity of the locomotor apparatus is not the only reason for nonlinearities. Bernstein argued that the central impulse also has to squeeze itself between the actual changes in external forces, and the 'desired' changes in movement. Taken together, these internal and external nonlinearities present themselves as *context-conditioned variability,* a phenomenon which has as its consequence that the metric effect of the central impulse can never be known in advance. Continuously,

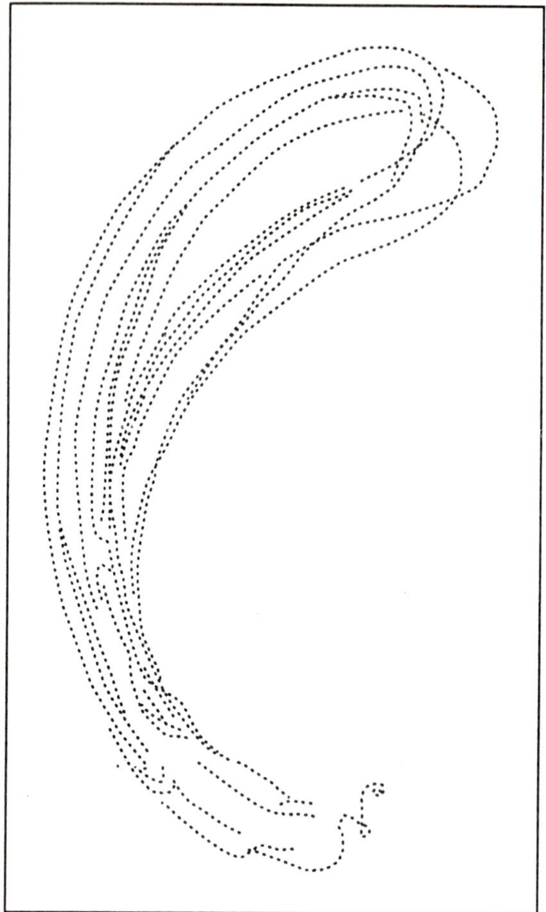

Fig. 1. Cyclogram of repeated movements [after 1, p.90]

thus, the organism needs sensory information so as to correct for any differences between the actual and the desired situation. Accordingly, Bernstein considered the closed *ring* as the basic unit of coordination.

Even complete sensory information, however, is not enough. Context conditioned variability confronts the moving organism with an abundance of *degrees of freedom*. In order to cope with them, the organism must have at its disposal an abundance of degrees of freedom [5,6]. Bernstein never offered a quantitative analysis of the degrees of freedom problem. Nor could he. There exists no definite limit X on the number of degrees of freedom a system can flexibly coordinate or control. But *if the organism were able to simplify its problem, then, it would just simplify its problem.* We argue, maybe somewhat surprisingly, that this is the logical structure of Bernstein's solution to the degrees of freedom problem.

Whenever there are too many degrees of freedom, one has to reduce them, i.e., by relating them in such a way that coordination emerges. This is how Bernstein conceived the possibility of (learning to) control. He coined the necessary relationships *functional matrices* or *synergies*. These synergies embody the higher-order, essential, variables and ignore the large number of lower-order, non-essential variables, i.e., those that are not crucial to the functional execution of the actual movement.

1.2 Models of the Future in the Physiology of Activity

Slowly, Bernstein came to understand that even the use of synergies in his 'physiology of coordination' was insufficient. It allowed for an understanding of what animals do and also, possibly, how they do it but not, emphatically not, why they do it. And in some elusive way, the how and the why are intimately related. In starting to build a theory of the why—the 'physiology of activity'—Bernstein realized that a reevaluation of the underlying mechanisms was due. Having been a pioneer of cybernetics in the 1930s, in the 1960s Bernstein came to doubt its relevance to biology:

> It is difficult to say whether or not the "honeymoon" between these two sciences is over, and with it their common quest for and use of analogies and other similarities; but *problems that suggest an opposite line of development* have been increasingly coming to the fore in recent scientific literature: is there, after all, a ... difference in principle between living and non-living systems, and if there is, where does the "watershed" forming the boundary between them lie [2, p. 542; our italics]?

Cybernetic control, the later Bernstein would argue, allows for the construction of homeostatic devices, but does not generate, in principle, the kind of 'negentropic' behaviour that living organisms display all the time. Cybernetics emphasizes reactivity, but behaving organisms are not, or not only, reactive. Behaving organisms are active. Their behaviour is negentropic by the very fact that they actively solve the problems they continue to encounter:

> The motor activity of organisms is of enormous biological significance—it is practically the only way in which the organism not only interacts with the surrounding environment, but also actively operates on this environment ... . If movements are classified from the point of view of their biological significance ... , it is clear that on the first level of significance we have acts which solve one or another particular *motor problem* which the organism encounters [2, pp. 343-344; italics in the original].

Bernstein's notion of 'motor problem' implied a difference between actual and desired relationships. Apparently, Bernstein argued, the organism creates a model of its actual past-present. This model is then evaluated against a stochastic *model of the future*. To Bernstein it was obvious that such a model of the future firstly must be based upon the model of the past-present, there being no other conceivable source; secondly that it must also be different, since extrapolation to the future entails uncertainty. A problem is a problem if, and only if, one does not know its solution.

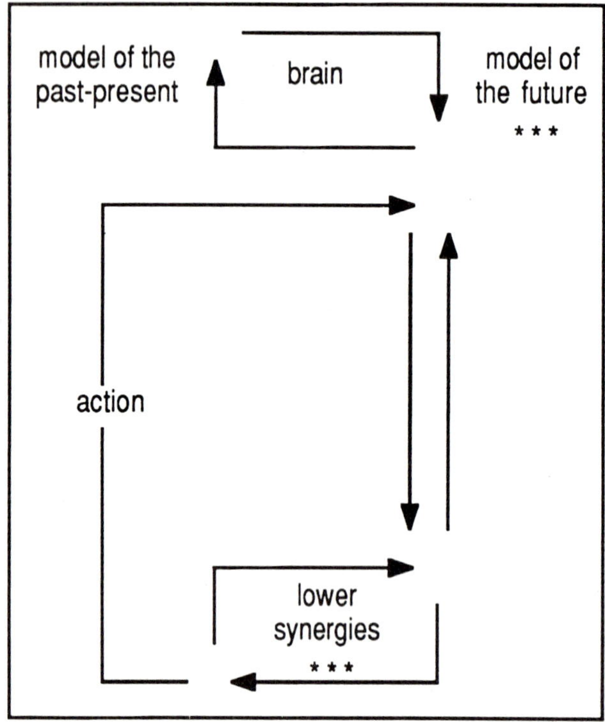

Fig. 2. Stochasticity (***) in Bernstein's physiology

This, then, is how Bernstein conceived biological 'activity', or the ability of living organisms to 'anticipate'. The brain functions in two different ways. It constructs models of the past-present as well as stochastic extrapolations of these (models of the future). Any difference between the two not only constitutes a problem but also entails a probability distribution of its solution. This probability distribution will lead to the construction of a motor 'engram' (program).

If the problem is new, the organism will try to reduce as many degrees of freedom as possible: the stochasticity of its behaviour remains contained in the choice of a specific model of the future (physiology of activity). However, as soon as the organism is able to construct the relevant synergies (physiology of coordination), it may start to stochastically play with the non-essential variables. It then varies the metrics while the topology is kept fixed. The original probability distribution (which solution to choose) is now replaced by another one (how to optimize the solution). In Bernstein's theory, exploratory behaviour of living organisms, able to cope with the unpredictable, is contained in the stochastic properties of both the models of the future that control their behaviour, and the richness of the degrees of freedom they use to coordinate that behaviour (Fig. 2).

2. 'Order' as Self-Simplification

In hindsight, one can easily recognize traces of the past in Bernstein's theoretical framework. His initial creation of 'cybernetics' for the physiology of coordination, and his later use of concepts such as 'engram' in the physiology of activity, remind one exactly of the kind of mechanicism he wanted to transcend.

In the late eighteenth century, 'biology' arose as the science of the non-mechanical. Ironically, however, mainstream twentieth century biology embraces a kind of mechanistic physicalism which the physicists themselves appear to have long abandoned [7]. Although Bernstein was aware of this problem, the phraseology of his solutions suggests that he did not go far enough.

2.1 Self-Simplification Replaces Maxwell's Demon

The historically problematic relationship between biology and the second law of thermodynamics, can be used to analyse the problem of mechanicism.

It is commonly accepted that the increase of entropy in isolated systems implies an increase of *disorder*. The increase of *order* in the evolution and development of biological systems, then, suggests an ability to escape the second law [8]. The fact that biological systems are open systems, fails to offer sufficient explanation for this amazing ability, that is to say, as long as it remains unclear *how* the openness of biological systems allows for the creation, and maintenance, of order.

The relationship between biology and the second law has been the subject of a long series of confusing debates. The present authors want to argue that this confusion arose because 'order' was implicitly seen as high-dimensionality imposed by superior intelligence. In this sense, a pack of cards is 'ordered' if an outside intelligence imposes extrinsic dimensionality to the pack by, for instance, recognizing and using the symbolic meanings of the numbers.

Imagine an isolated vessel with water where the overall temperature is more or less the same, all over the vessel. If now it is split into two compartments by a semi-permeable membrane, with Maxwell's demon standing at the pores, allowing fast molecules to go to one side, slow ones to the other, then the two halves of the vessel will come to have distinctly different temperatures. Obviously, such a compartition of temperature is not what we observe in the physical world; or in biology. In terms of 'order', however, biology abounds with semi-permeable membranes on the two sides of which distinct differences are to be found—osmotic differences, differences in chemical concentration, electrical differences.

In physics, it has been shown that Maxwell's demon must spend more *free energy* to recognize and distribute the molecules than the amount that would come to be available from the ensuing temperature difference. The demon has to bring in this free energy all by itself, a process whereby its own entropy will increase drastically. In fact, the demon's entropy will increase by a higher amount than the decrease of the entropy of the water. Maxwell's demon, therefore, does not falsify the hard-core meaning of the second law, i.e., that the entropy increases in isolated systems.

But the use of the word 'demon' also suggests some analogy to the intelligence that is able to order a pack of cards, as if the demon has intelligent *knowledge*, allowing it to effortlessly 'read' the speed of the molecules. The demon, we argue, was implicitly seen as an intelligent controller with more knowledge than the entities under its control. It may have been precisely this implicit understanding of the nature of order that forced twentieth century biology to be mechanicistic—relying as it did on stupidly passive matter, supervised by intelligent control devices.

Time and time again crystallization was taken to offer a clear exception. In crystals, one can see the emergence of order without any supervision. On the other hand, once crystals have formed they can be understood from a *static* point of view, whereas such freezing out of *process* is rarely seen in biology [9]. Chemical auto-catalysis, therefore, is biologically more interesting, illustrating as it does the emergence of non-supervised order in on-going processes. Indeed, at the end of the sixties and the early seventies, several demonstrations of auto-catalytic processes took the biological community by surprise [10, 11].

The Belousov-Zhabotinsky reaction and Prigogine's Brusselator revealed how, under certain conditions, order could emerge, and be maintained, in non-supervised cycles of chemical reactions. And multicellular morphogenesis as well as differentiation in the normally unicellular slimemold *Dictyostelium discoideum*, was shown to be dependent upon the concentration gradient of the relatively simple substance cAMP (cyclic adenosine monophosphate). Analysing the origin of life, Eigen [12] conceived auto-catalytic cycles of proteins and nucleic acids in a 'primordial soup'. Such cycles are extremely vulnerable since their constituents can be washed away or degraded: unless, that is, some structural coherence emerges such as a membrane [13], and a 'code' develops [14] which couples to the nucleic acids the ability to reproduce the proteins whenever required.

'Order' in these cases does not come from the outside but from the internal relationships, an idea which was precipitated by von Bertalanffy's general systems theory [15], where 'system' had been defined as a set of elements *together with their relationships*. Apparently, there are systems which, under the appropriate conditions, create, and maintain, order *by themselves*. Theoretical biologists were quick to realize that the concept of 'order' itself had changed. What kind of 'knowledge' is, for example, DNA supposed to contain? At a *microscopic* level, it is certainly 'just' another molecule, as such not different from the proteins. At a *macroscopic* level, however, it is a control structure. But not by knowing more than the rest of the cell; it is a control structure by knowing less. DNA embodies the primary sequence of proteins but is entirely oblivious to, for instance, the intricacies of cell-metabolism, the composition of the membranes, or even its own three-dimensional architecture. When compared to the astounding complexity of the living cell, DNA is of a low-dimensional nature. It embodies an essential loss of detail. Inasmuch as the microscopy of the cell allows for DNA-controlled structural 'order', it does so by allowing for a tremendous reduction of degrees of freedom. In more general terms, living order emerges from, and can be defined as, *self-simplification*, that is to say: the *non-supervised macroscopic reduction of microscopic degrees of freedom.*

In 1973, Pattee argued that this is necessarily so [16]. Living species are problem solvers within a multidimensional search space. The more multi-dimensional this search space is, the longer it takes to reach an optimal solution. Just as in Bernstein's degrees of freedom problem, there is no quantitatively definable maximum of the number of dimensions that can be conquered. But any search in a highly multidimensional space tends to lead to the stalemate of sub-optimal solutions. If now, the multidimensional microscopy were to self-simplify into a low-dimensional macroscopy, the search could be continued within that macroscopic self-simplification. Stalemate will now be avoided and the system gains a flexible stability which it did not have before. That is to say, as soon as the living system has developed such a self-simplification, it will abide by it.

In the above sense, the origin of life is a paradigm of self-simplification. DNA embodies, so to speak, a high-level low-dimensional 'eigenmodel' of the living cell. It is clear, at least for DNA, that such an eigenmodel emerges stochastically: DNA contributes to the ability to search by means of its stochastic instability (mutations), subject to functional evaluation (selection). Analogously, in any Bernstein-inspired physiology of movement, one has to look for stochastically arising self-simplification, i.e., the lower synergies or the models of the future.

2.2 Synergetics Comes to Revisit the Second Law

In a way, similar claims had already been made in Romantic *Naturphilosophie* [17] but at the time no mathematics existed to analyse such phenomena. Later, Poincaré laid the foundations of the mathematics in question [18], but the then dominant 'scientific materialists' had returned to mechanicism. And at the turn of the century [19], physics became so occupied with modelling the atom, and biology with the mechanisms of heredity, that neither paid much attention to Poincaré's initiative. In the Soviet Union, however, where after the Revolution matter had become the source of all intelligence, a (deterministic) mathematics of 'dynamical systems' was further developed, and flourished [20]. It is as if the rest of the world needed similarly shattering experiences before *the nature of stability and its transitions* was recognized as the most pressing problem of all [21].

In 1971, Glansdorff and Prigogine [22] pointed to the analogies between different manifestations of the emergence of stability. In 1972 [23], Thom presented a general theory of 'catastrophes' which could be applied to the development of biological form. And from 1969 onwards, Hermann Haken [24-26] undertook to show that a *stochastic* theory of dynamical systems could actually be applied to natural systems. And could, when elaborated, turn the metaphors of the past into a systematic revealing of the formal analogies of 'synergetics'.

In synergetics the nature of order is studied, as are the mechanisms of its emergence, and the transitions between its different forms.

In 1969, Haken was engaged in the analysis of laser light. Imagine a vessel with excitable atoms [5]. There are two mirrors at the end faces of the vessel. When energy is pumped into the system, the excitation of the electrons leads to the emitting of light waves which, through the mirrors, remain inside the vessel for a relatively long time. When the pumping rate is low, the light waves are unordered. At some higher pumping rate, however, the system becomes unstable. Waves will start to force the emission of light with the same phase. Several such phases will stochastically compete until the system self-simplifies into a single ordered state. A low-dimensional macroscopic *order parameter* now not only emerges from the microscopy, but also 'enslaves' the microscopic elements. In other words, self-simplification not only serves as a simplified macroscopic description of the complicated microscopy, but also constrains the behaviour of that microscopy. The behaviour of self-simplified systems is different from that of systems which have not self-simplified.

Up to a point, synergetic order parameters are *perturbation resistant*, i.e., they respond to perturbations with a short-lived disturbance, to then return to their original value. That is to say that, within limits, the order maintains itself. There are, however, external changes which will induce the system to make a 'transition' from its original state to a new one. But again, the new state is a self-simplification which will, within limits, be perturbation-resistant.

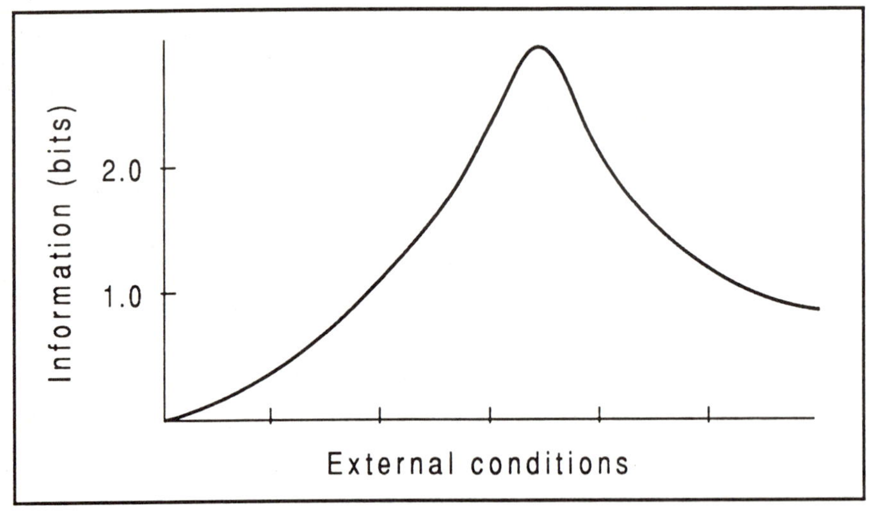

Fig. 3. From classical equilibrium to the ordered state [5]

Classical treatments of the second law of thermodynamics focused on the microscopy of systems. Classical equilibrium is that macrostate which can be realized by the highest number of microstates. Microscopic information, therefore, is at its maximum at equilibrium. On the other hand, synergetic order can be understood as macroscopic information [27]. In classical equilibrium, there is only one macrostate; hence, at equilibrium macroscopic information equals zero. If a relevant parameter is then increased in such a way that a transition occurs, the resulting instability implies a rather dramatic increase in macroscopic information (Fig. 3), which finally is reduced when the system settles into one of, say, the two ordered macrostates that are possible. Classically, one would say that the entropy increases during this *reduction of the information*. The synergetic notion of macroscopic 'information', however, could also be taken to imply that the entropy is high if the number of possible macrostates is high, an interpretation which would lead to the opposite conclusion, i.e., that the entropy increases upon any *increase of the information*. Indeed, in 1990, Haken and Wunderlin stated:

> If we interpret the [macroscopic] information as entropy, the rather striking conclusion offers itself that a ... transition from an unordered to an ordered state implies an *increase* of the entropy within the system. It is usually assumed that the emergence of order implies a *decrease* of entropy. At least for the entropy contribution of the order parameter, this assumption appears to be wrong ... [5, p. 16; italics in the original].

Of course, earlier attempts have been made to explain the natural emergence of order [28, 29]. Order has been 'explained' as resulting from chance fluctuations; as deriving from highly improbable initial conditions; or as the effect of the expansion of the universe, physical forces creating coherence before entropy has had the time to strike. The problem with such theories is their inability to explain how the emergence of order can be a *reproducible* phenomenon. The above suggests that the crucial point resides in accepting the ontological reality of macroscopic self-descriptions, i.e., self-simplifications. The emergence of macroscopic order parameters implies that the microscopy starts to behave differently. The system no longer settles in classical equilibrium. It reduces its own degrees of freedom and may develop a form of order which, within limits, maintains itself.

Now that a general theory has been formulated to understand the stochastic emergence of macroscopic order, we may even decide to completely free ourselves from the confusion, and refer entropy back to its original realm: that of classical thermodynamic equilibrium [30]. It may be a wee bit vacuous to conclude that natural systems settle in those states which are able to, and do, reproduce themselves, i.e., the classical equilibrium state, or, under the appropriate conditions, specific forms of macroscopic order (self-simplification). But as soon as the ontological reality of macroscopic self-simplification is accepted, the gap between physics and biology appears to have never existed. 'Order' is no longer seen as high-dimensionality imposed by superior intelligence. Hence, mechanicism is no longer needed, and Bernstein's watershed, indeed, is no watershed-in-principle—it is one in degree.

3. Quantifying the Qualitative

According to dynamical systems theory, the state x of any deterministic differentiable system can change over time ($\dot{x}$) in three different ways. If it changes as a function of itself only, the system is *autonomous* [31]. If it changes as a function of external events only, it can be said to be *heteronomous* which appears in the equations as a complete dependence upon time (t). All other relevant systems can be said to be *mixed.*

3.1 Synergetics and the Physics of Self-Organization

On-going natural processes require the availability of energy. In Haken's synergetics [26], the available energy is part of the relevant external conditions which are captured by one or a few *control parameters,* say: ε.

This notion of 'control' is unusual because the to-be-controlled is not specified by it! Although a synergetic control parameter is time-dependent ($\varepsilon = \varepsilon(t)$), it is still meaningful to say that the system specifies its own changes over time, given a specific value of the control parameter:

$$\dot{x} = f(\varepsilon, x),$$

where ε appears as a constant. In other words, given the appropriate conditions, the elements of certain natural systems will cooperate in such a way that the behaviour of their macroscopy, captured in one or a few order parameters, can be shown to be autonomous. Such natural systems are said to be 'self-organizing'.

Why now would it be meaningful to treat *mixed* systems as if they were autonomous, and reserve the term 'self-organization' for systems which, on the face of it, do not self-organize? This is because the system self-simplifies into one or a few *macroscopic* levels at which the system's behaviour is *oblivious to considerable ranges of change* in the values of the control parameters.

The most conspicuous difference between the macroscopy of self-organizing systems and the deterministic microscopy of Laplace's dream is, therefore, not what these systems macroscopically do, but exactly what they don't do, what they don't respond to, what they are oblivious to.

Self-organizing systems gain 'freedom' not by doing more than ordinary systems, but by doing less, i.e., by *constraining their own degrees of freedom* [16]. It is our contention that this paradox is a hallmark of living systems. Goethe's "In der Beschränkung zeigt sich der Meister" or the Wu-Wei (Do-Nothing) of Taoism, do not refer to inactivity but to active restraint. Even here, we argue, the watershed between physics and biology is one of degree only.

In synergetics, the control parameters of self-organizing systems shape a 'potential landscape' in which the order of the system can be visualized as a rolling ball. This ball will 'search' for minima in the landscape and can, therefore, be trapped in a local minimum (Fig. 4).

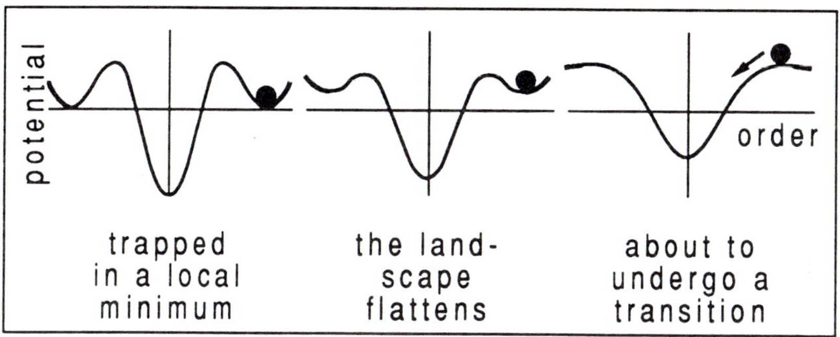

Fig. 4. Changes in a potential landscape [5]

If a 'trapped' system were able to reach a deeper minimum, it would be able to lower its energy. But for self-organizing systems a rather amazing paradox presents itself: *in order to lower its energy, a system may have to first reach a higher-energy state before it can settle into the lower one.* Synergetic systems need surplus-energy in order to take the 'hurdles' which separate them from neighbouring minima.

Several models have been formulated to understand how systems can free themselves from the deadlock of a local minimum, Pattee's evolutionary argument for self-simplification being a case in point. In all models, surplus-energy is needed to overcome the hurdle. This energy may be used to, for instance, create a higher-level self-simplification, or to 'shake' the system, or to 'melt' the landscape as in annealing [32]. In Haken's synergetics, any self-organizing system will undergo a *transition* if it finds itself within a critical range of values of the control parameter(s) (Fig. 4). The system starts to fluctuate ('critical fluctuations'), and the landscape flattens in such a way that the system returns more slowly to its original state upon perturbation ('critical slowing down'). The critical fluctuations, together with the flattening of the landscape, then, allow the system to reach a new ordered state.

Recent analysis [5] has shown that, again, the given value of the control parameter does not pre-specify the behavioural outcome. Neither could any deterministic differential equation. Rather, it is the system itself which *stochastically* explores its search space. If s^* denotes a stochastic function, then the state of synergetic systems changes over time as:

$$\dot{x} = g(\varepsilon, x, s^*).$$

3.2 Structure-Function Duality in Living Organisms

The popularity of the laser example, consistent with the above equation, can easily mislead one into believing that synergetics is a form of applied physics. But the laser is a homogeneous system which biological and social systems are not. Indeed, Haken and Wunderlin [5, p. 9] stressed that no "one-to-one correspondence between a specific system such as the laser and any other one such as a neuronal network" is to be expected.

In the laser, all self-simplification is behavioural: given the values of the control parameters, the system develops one or a few collective behaviours which after a transition disappear without a trace. On the other hand, much of biological self-simplification is *structural*, i.e., the potential to produce certain behaviours solidifies into the organization of structure. It is by using the organization of structure that the biological system can preserve parts of its behavioural history and anticipate parts of its behavioural future. The organization of structure, thus, embodies a still higher time scale than that of behavioural order parameters. Conceptually, we argue, structural self-simplifications have to be distinguished from order parameters.

The structure of DNA, for example, is as such not a component of the macroscopic behaviour of the living cell. DNA allows for the emergence of specific order parameters but, in our opinion, it would not be fruitful to regard DNA as being an order parameter. Its time scale is larger than that of any cellular behaviour. From the viewpoint of behaviour, DNA is not even a self-simplification; it is of a higher-dimensional nature than any set of on-going cellular behaviours. It is only when structure is analysed that DNA can be shown to be a paradigm of self-simplification.

In the laser, the microscopy and the macroscopy are mutually enslaved. On the other hand, structural self-simplifications, such as DNA, neural networks, or governments, have the ability to stochastically change relatively 'on their own'. In their role as self-simplifications, they sometimes change independently of the underlying microscopy—of which, of course, they are also parts! In biological systems there is, thus, much more *free play* between macro and micro than in the case of the laser. It is this free play which allows the biological system to, in a sense, escape the here and now, and to stochastically develop a 'model of the future'. Recently, Robert Rosen [7] has argued that this is a defining characteristic of life.

All biological systems are self-organizing systems but not all self-organizing systems are biological systems. For biology, the gain of theorizing in terms of ordinary physical self-organizing systems is considerable, but still too small. The stochastic emergence of structural self-simplification has to come in before physical behaviour can turn into biological *function*. Although we find it extremely hard to define 'structure' or 'function' we want to at least argue that function cannot appear before (stochastically changeable) structure appears, and that structure should be regarded as the solidification of potential function.

The biological appearance of 'structure-function' duality depends upon the presence of material heterogeneity. This heterogeneity, then, leads to a specifically biological kind of *partitioning* which allows the organism to parse its behaviour into manageable units. In the living cell, for instance, protein synthesis is something different from, and is done somewhere else than, mutation repair. Herbert Simon's 'watch-maker argument' [33] reveals the advantages of such partitioning: while certain parts are kept stable, other parts can be creatively changed by evolution or development. Again, in order to promote life, nature has to avoid the stalemate of multidimensional search spaces.

The partitionings of biological systems are non-rigid. We have to do with a 'fluid' partitioning where neither structural nor functional boundaries can be analysed from a static point of view. A change in the organization of structure may imply a change in function, whereas any functional change can influence the organization of structure. The relationships between structure and function may even turn out to be different under different circumstances. To the researcher of biological structure-function duality, therefore, the 'fluidity' of partitioning may have rather surprising consequences: the localization of structural self-simplifications may turn out to be different in a different functional context.

In the study of biological self-organization, the organization of structure must be related to the stability of behaviour such that stable behaviour can be recognized as function, and the structure in question as an actual self-simplification of the living organism. Layer upon layer of structure-function are to be probed under different circumstances. To such a methodology, we argue, somewhat analogous to Bernstein's physiology of activity, the organism presents itself as a *Swiss roll* (a sweet roll with concentric rings of different consistency).

In any given functional context, some of the layers will turn out to be more perturbation-resistant than other ones. The hard layers, then, are the ones that reveal structural self-simplification, their hardness being revealed by the stability of function. But under different conditions it may turn out to be the case that layers which were 'hard' before, are now the 'soft' ones, and vice versa. Studying biological systems under always the same conditions, therefore, will fail to reveal how the structure-function of such systems is organized.

4. Circles of Analysis in Motor Control

Until well into the twentieth century, motor control was understood as detailed instruction 'from above' [34, 35]. Bernstein's analysis, however, proves such an approach to be wrong: any theory of motor coordination and control has to take the nonlinearities into account.

The history of understanding rhythmic biological movements in terms of externally modulated ('forced') oscillators, goes back to at least the 19th century [36]. In 1928, Van der Pol and Van der Mark modelled the heartbeat as a relaxation oscillation [37], and in 1931, Adrian and Buytendijk presented evidence that respiration is co-controlled by a relatively autonomous oscillator in the *medulla oblongata* (the lower brain stem) [38]. It was in the same period that Bernstein started to develop his physiology of coordination.

Around 1950, Bernstein ran into trouble with the Pavlovians and was prohibited from continuing experimental research. He then started his theoretical work on the physiology of activity. Later, Bernstein was rehabilitated, and in the 1960s a special movement laboratory was created within the Moscow Institute for Information Transmission. A rather amazing period of Bernstein-inspired productivity followed.

It was in the late 1970s [39] that similar activity started in the West. Since then, it has been shown that a physics of self-organization constitutes an essential starting point for movement science [40]: it takes the nonlinearities into account and shows, in its own way, *what* it is that moving animals are doing. It has been established that such a physics can actually be applied to biological movement [41, 42]. New facts have been successfully predicted such as, for instance, an increase in the symmetry of human hemiplegic gait upon increasing speed, with hysteresis upon subsequently slowing down [43].

Synergetics provides the formal analogies between different manifestations of the emergence of order and its transitions. While not unrelated to the above approaches, it finds itself at a higher level of abstraction—focusing, for instance, on changes in the behaviour of oscillators over time, rather than satisfying itself with the exact description of any particular oscillator. Cooperation between Hermann Haken's group and that of Scott Kelso has introduced this higher-level approach to movement science [44]. In studying the transition from out-of-phase to in-phase finger movements, critical fluctuations and critical slowing down could be demonstrated.

In 1988, Schöner and Kelso presented a general methodology for the study of the macroscopy of biological movement and, in terms of the present paper, the underlying structure-function duality [45]. The most important point in that operational approach was the pinpointing of transitions. At the time, neither a specific synergetic understanding of the biology of structure-function duality could be reached, nor a precise determination of the control landscape. The present authors are not able to solve these urgent problems. It will be tried, however, to somewhat further the arguments.

4.1 Fel'dman's Mass-Spring Hypothesis

In a first circle of analysis, movements can be filmed to then interpret the changes over time in terms of *body-related space*. In such an understanding, movements can be categorized as either 'discrete', i.e., going somewhere, or 'rhythmic', i.e., repeating something.

In the 1960s, Anatoly Fel'dman undertook to analyse discrete movements in terms of Bernstein's physiology of coordination [46]. In a rather primitive experimental setting, experimental subjects were asked to hold their arm as if carrying, say, a tray. They were asked to remain upright, and not to focus their attention on the arm but to retain the same 'feeling' about it. When a load was added, the arm would lower a bit; when then the load was reduced, the arm went up a bit; when, finally, the load was removed, the arm returned to its original position.

Apparently, the system is able to retain the position of the arm, notwithstanding temporary disturbances by external perturbations. In that respect, Fel'dman argued, the arm behaves as a damped mass-spring which is able to 'find' its equilibrium point no matter, within certain limits, what happens on the way. The central impulse just has to control the equilibrium point while the system itself is able to coordinate its trajectory. In Bernsteinian terms, the equilibrium point actualizes the essential variable while the precise trajectory of reaching it, resides in the non-essential variables. Moreover, if the appropriate amount of 'negative damping' (energy) is available, the mass-spring can produce non-linear, self-organizing oscillations [47, 48]. Fel'dman's *mass-spring* hypothesis, therefore, allows for the understanding of discrete as well as rhythmic movements.

The validity of Fel'dman's mass-spring hypothesis has been confirmed in, for instance, a series of animal experiments, performed by Bizzi and his co-workers [49, 50]. One burning question, however, remained: How can the central impulse actually 'set' equilibrium points? In terms of direct muscle innervation, this problem appears to be insurmountable: it would simply not work to pre-specify a certain level of activity for alpha neurons (neurons to the muscle fibres) or even gamma neurons (neurons to the muscle spindles) *since the recruitment of specific sets of muscle fibres has to change over time.* Muscle fibres may, for example, pass a mechanical axis during the on-going movement so that their function is reversed!

The second circle of analysis, i.e., the structure-function implementation of the mass-spring and its related oscillators, has been made accessible only recently, by experiments from, again, the Bizzi group [51, 52]. Upon microstimulation of the ventral roots (alpha neurons) at the appropriate level in the decerebrated frog, a leg will move. If now, the leg is given another initial position, and the experiment is repeated, it will move again but to *a different place.* Apparently, the ventral roots are structurally (in terms of muscle fibres) but not functionally (in terms of body-related space) perturbation resistant (Fig. 5). Upon microstimulation of the more central gray matter (the premotor area), however, the situation is reverse (Fig. 5)!

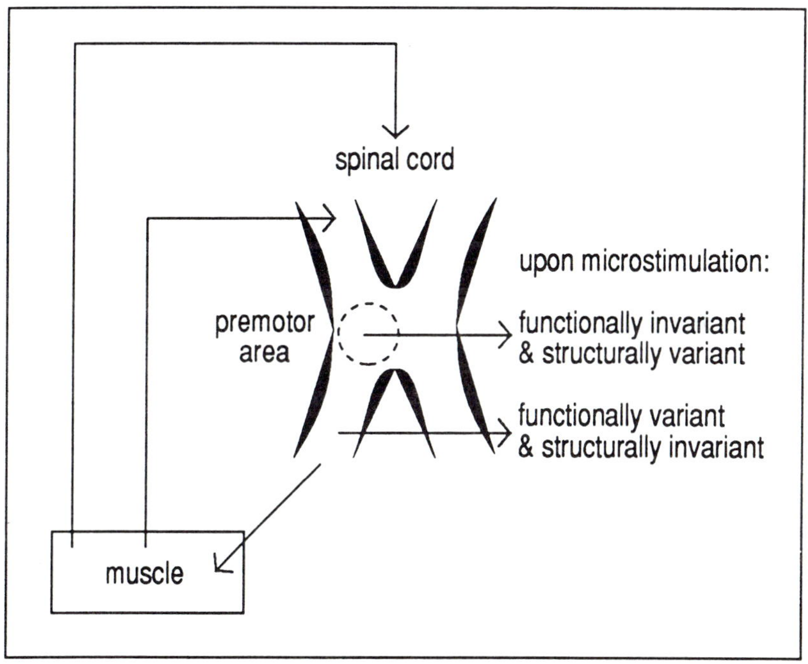

Fig. 5. Swiss roll structure-function in the spinal cord

If the premotor area is stimulated under different initial positions of the leg, *the same body-related place* will be reached notwithstanding the fact that for each initial position a different sequence of alpha-gamma activations will be needed. Hence, *part of the central gray matter of the spinal cord is functionally and not structurally specific*. In terms of our 'differential perturbation resistance' argument, it is this part which embodies the relevant self-simplification while the interneurons and the ventral roots take care of the non-essential degrees of freedom. In the functional context of these experiments, the premotor area represents one of the hard layers of the Swiss roll.

Finally, thus, the mass-spring hypothesis has been connected to spinal cord structure-function duality.

Not, of course, without new problems arising. How is it, for instance, that this functional specificity translates into the recruitment of alpha and gamma neurons? Maybe a principled reformulation of computer models of ventral root function [53] is due, or maybe it will turn out that the premotor area contributes first and foremost to the changes of the (sensory) dorsal root [54], which then 'attracts' the limb. Is varying the initial position of the leg sufficient to determine the control landscape? Probably not! It may very well turn out to be the case that a different functional context will lead to different results. In the experiments so far, the movements of the leg are unrestricted; it is quite conceivable, to be established upon further experimentation, that when an external force is applied to the leg, the map will be found at a different location. Would there then be stochasticity in the location of the map, and if so, would the organism be able to control the amount of stochasticity? The Bernsteinian answer to these last two questions would probably be: Yes!

Whatever the outcome of the continuing search, however, remarkable progress has already been made in creating the possibility of a specific synergetic understanding of structure-function duality in the analysis of biological movement; in terms of body-related space, that is. In the decerebrated frog, there is a map in the premotor area of the spinal cord, the structural organization of which allows the organism to coordinate the invariant reaching of a specific point in body-related space by the unrestricted leg. Apparently, the premotor area embodies a structural self-simplification, and there exists a level of biological functionality which has to be understood as reaching a specific point in body-related space.

4.2 A Strange Hierarchy

Pieces of the spinal cord can be taken from an animal, to be kept alive in a petri dish under appropriate chemical conditions. It has been known for decades that these isolated pieces exhibit rhythmic electrical activity which, in a rather ungraspable sense, is said to be 'similar' to the rhythmic activity of the intact spinal cord during actual (whole animal) rhythmic movements. The relevant networks have been termed CPGs ('central pattern generators') [55], which the majority of researchers conceive as being responsible for the control of rhythmic movement.

For an appropriate third circle of analysis in motor coordination and control, one has to accept that an isolated piece of spinal cord is able to produce rhythmic activity. But contrary to the belief of many [56], there can be no little boxes exactly specifying what the limbs have to do. Movement behaviour is non-linear and much too flexible for the rigidity of anatomically defined 'music boxes'. On the other hand, a walking animal is, indeed, a walking animal, and 'something', of a relatively high level, should remain the same in order for walking to be walking. Movement behaviour is reproducible and much too topologically invariant for a nervous system which does not structurally self-simplify.

Imagine a sitting cockroach who is eating something. The cockroach is oblivious to almost everything in the surround—our senses are the watch-dogs of our tranquillity. But watch-dogs bark whenever danger appears and the cockroach is equipped with vibrating sense organs, the *cerci*, which are specifically tuned to such air vibrations as usually coincide with the approach of a cockroach-eater. Upon such vibrations, the threshold of cerci-related nerve cells is exceeded, and a 'message' is created which is passed on to a central 'command' cell (or system of cells). This often-used term 'command' appears to refer to superior intelligence but the command cell is actually just a central threshold device, which sends an inhibitory signal to the CPGs of all on-going activities and simultaneously stimulates a CPG for flying.

The nature of such stimulation has recently been revealed in elegant research by, among others, Libersat et al. [57]: the rhythm of the stimulation is as irrelevant as that of somebody's hand when that person starts playing with a yo-yo by just throwing it down. In other words, the stimulation is a non-specifying 'kick'. This kick will arrive at the CPG, and activity ensues to bend the wings simultaneously. Whether or not the signal for wing activity is 'written' on Bizzi's 'map' in the premotor area, is a matter that remains to be decided. Anyhow, based upon Bernstein's analysis of non-linearities, we argue that the signal for wing activity cannot exactly pre-specify what will happen. The wings, therefore, are also 'kicked' whereby the first phase of a behavioural transition comes to an end (Fig. 6).

The rhythm of the wings, which will stretch again due to their elasticity, is then passed *upwards* to reach the CPG. It is our understanding that the CPG and the wings *negotiate* about the rhythm to be. After two or three wing beats, this rhythm becomes mutually entrained and a rhythmical stimulation is passed *upwards* to reach the command cells. The major point of the Libersat et al. study is that the command system does not pre-specify the rhythm to-be, but kicks the CPG to pick up the ensuing rhythm after a couple of wing beats.

Research by, again, the Libersat group has shown that the command system then passes the rhythm on to the sense organs which in their turn become entrained [58]. If the cockroach were to retain its original sensitivity to air vibrations, it would not even be able to fly—since the flying itself already produces such air vibrations.

The fact that the command cells pass on their rhythm to the sensors shows that *switching to a new activity re-establishes the threshold and rhythm of the sense organs.* This ends the second phase of the transition (Fig. 6). The cockroach is now flying. It will pick up information that allows for low-dimensional modulation of its flight: more to the right or left, more upwards or downwards, faster or slower. The active CPG embodies exactly this flexibility: the rhythm can be modulated to adapt the flight to whatever relevant events occur. The flexibility of CPGs [59] allows for functional adaptation, given a specific behavioural state. This state will continue

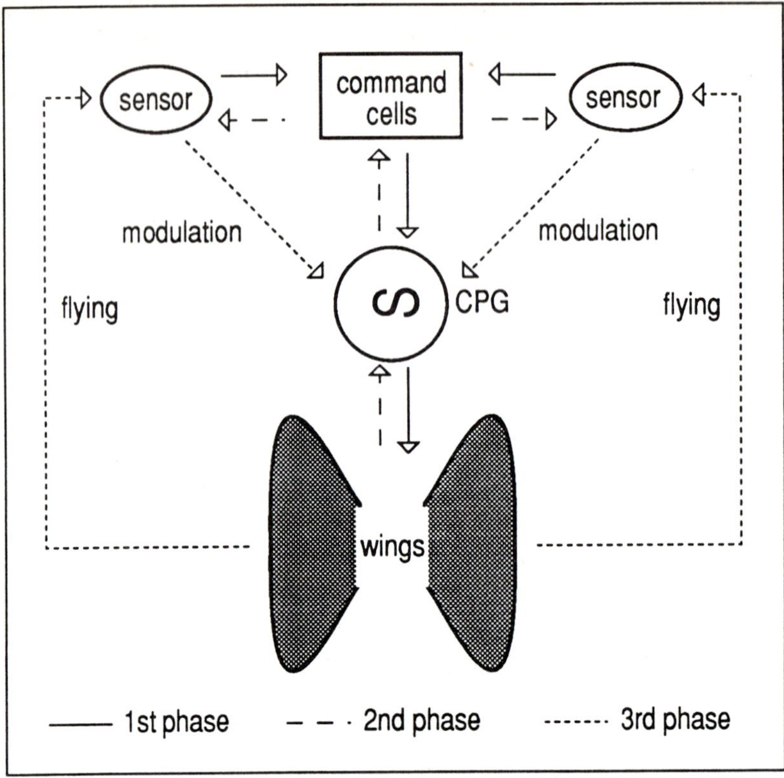

Fig. 6. Swiss roll structure in starting to fly

until internal or external events will tell the cockroach that it is time to do something else.

The literature on CPGs is overwhelmingly rich and complicated. We refrain from analysing the extreme difficulty of finding an appropriate low-dimensional description of the relevant chemistry [60] or of the dynamics of the the overall-functioning of the relevant networks [61, 62]. In fact, we have hardly discussed CPGs at all. But in the sensors and the command system we have found the structural self-simplification that appears to control them.

In the fourth circle of analysis, it is, again, possible to film the movements. Indeed, we have to do with a strange hierarchy: we are back to where we were, but a change in principle has taken place. We no longer observe displacements in body-related space, but 'functional' activities in the language of ethology, such as walking, swimming, or flying. 'Function' which concerned place in body-related space in our interpretation of the ventral root, now becomes overall behavioural activity. In order to avoid confusion, we will coin it 'function$_2$', to be distinguished from the earlier 'function$_1$'.

Function$_2$ can be triggered from anywhere in the outer circle, but the senses and the command system appear to be the most perturbation resistant, and therefore to embody the relevant structural self-simplification. *From the perspective of function$_2$, the CPG embodies the non-essential variables.* From the perspective of function$_1$, however, the CPG has to embody (changeable) sequences such as, in human gait, left leg / right leg / left leg, etc. It may turn out to be the case that the level of the CPG embodies a specific self-simplification, i.e., in terms of body-related sequences. We are aware of the fact that this may raise more questions than answers, but the new questions are certainly more interesting than the old idea of a music box. So far, we have been unable to infer from the literature what exactly it is that always remains the same during flying or, for instance, walking. But we have been able to show, at least for the flying cockroach, that these overall behavioural states are organized by a single low-dimensional structure-function duality.

5. New Hope for Patients?

Earlier, we pointed out that Bernstein wondered if there is a "difference in principle between living and non-living systems, and if there is, where does the "watershed" forming the boundary between them lie?" [2, p. 542]. Synergetics has shown that the difference is mainly a difference in degree: pumping energy into natural systems can lead to the emergence of a macroscopy which is able to stochastically choose between different forms of order. These principles are applicable to biological systems which thus, paradoxically, gain degrees of freedom by reducing them.

On the other hand, we have tried to understand the watershed: biological organisms have a stochastic structure-function duality in a Swiss roll type of partitioning. In order to relate to their environment, biological systems thus, paradoxically, make separations. Of course, this does not make them non-physical in principle, but it distinguishes them from ordinary self-organizing physical systems.

To the student of movement, it becomes of paramount importance to rethink what lies beyond the watershed. We, for ourselves, have long thought that the hallmark of animal movement resides in the ability to (re-)produce serial order. Obviously, physical systems can display serial order but we felt that animals are more able to do so. For a long time, people have thought that DNA contains 'programs' for development, and, in often similar terms, movement scientists have claimed that movement is controlled by a 'motor program'. Even Bernstein's notion of 'engram' is reminiscent of this idea of a structure which contains the serial order to-be-produced. Synergetics, in its emphasis on self-organization, is strongly opposed to such ideas. Presently, we are of the opinion that the problem of serial order may be solved by invoking a third paradox.

DNA's role in starting the synthesis of a particular protein has been relatively well studied. Some metabolic change in the cell leads to the deactivation of a repressor gene. In its turn, this deactivation leads to activity of the operator gene, and at the end of the cascade a protein is synthesized. The normal situation, therefore, is that genes are suppressed. Amazingly, chemical inhibition has received much less theoretical and empirical attention than chemical catalysis. It is, however, the ability of the genome *not* to start the synthesis of a particular protein which renders the system so uncannily controllable.

In the nineteenth century, Ivan Sechenov argued that the function of the brain is largely to inhibit actions that otherwise would naturally occur. Although the details proved more intricate than Sechenov could foresee, it has, indeed, been shown that inhibition plays a crucial role in the nervous system. This fact reveals an important point. The cockroach who starts to fly, first actively inhibits other on-going activities. Before doing something, the cockroach must *undo* what was going on. In order to be able to do, living organisms must be able to 'undo'.

Moving animals allow themselves to be guided by a low-energy perceptual field [6]. The free energy of this field is insufficient to do the work of moving, but animals store their own free energy. We suggest that this free energy is not only spent in moving but also in undoing the on-going activities in order to take the hurdle of a transition. As to human movement, it is our conviction that this may reveal itself in the frequency of the heart beat, oxygen consumption, the production of anaerobic metabolites, or in some psychophysiological measures, blinking the eyes being a possible candidate for monitoring transitions. Careful application of the methods of exercise physiology, or psychophysiology, therefore, may help to exactly determine the nature of the control landscape.

5.1 Parkinsonism as the Inability to Undo

Parkinson's disease results from a degeneration of the *substantia nigra,* an area in the *mesencephalon* (upper brain stem) which is located not far from where visual and auditory information enters the central nervous system, and from where the cerebellum, after having digested vestibular information, passes its modulatory signals on to the descending tracts. The substantia nigra sends signals to the circuitry of the basal ganglia, located in the brain, not far from where olfactory information enters the central nervous system. The circuitry of the basal ganglia has been claimed to be responsible for the major sequencing of movement behaviour (do this first, and then that).

Symptoms similar to those of Parkinson's disease, appear under a large variety of circumstances. The set of symptoms as such, is known as 'Parkinsonism'. When a patient with Parkinsonism sits, it is difficult to rise. When the patient stands upright, it is difficult to start moving. When the patient walks, it is difficult to stop. The characteristic gait of persons with Parkinsonism is quick, with small steps, whereby the feet remain close to the ground. Persons with Parkinsonism who write, write too small. Patients who tell a story, don't have the appropriate facial expression. Patients who want to pick something up, start to tremble. While in bed, persons with Parkinsonism have great difficulty in turning over.

To date, no satisfying hypothesis has been formulated to low-dimensionally capture all these symptoms. Over the past decades, research has focused on the importance of *dopamine,* the neurotransmitter of which Parkinson's patients produce too little. Indeed, when dopamine or related substances are administered, the symptoms are alleviated. Until, that is, chemotherapy is no longer effective which often occurs after about ten years. The implanting of dopamine producing fetal cells appears to meet with at least the same limitation. The original optimism after the discovery of dopamine's importance [63] is, therefore, being replaced by the wish to start new lines of research. Maybe, then, a synergetic understanding of the nature of Parkinsonism could offer new hope to the patients.

In terms of synergetics, a striking phenomenon in Parkinsonism is the patients' inability to take the hurdles of transitions. The majority of the symptoms entail that the patient is unable to undo, unable to stop sitting, to stop walking, or to stop trembling. A possible second dimension can be found in the apparent lack of energy: the patient walks without vigour, and writes too small. Certainly, not all the symptoms of Parkinsonism can easily be brought under this tentative low-dimensional description. In particular, the problems with axial musculature (such as when turning over in bed) appear to represent another, so far not-elucidated dimension. But for the time being, understanding Parkinsonism as the low-energetic inability to undo, already creates new possibilities for intervention.

In 1984, Forssberg et al. [64] presented evidence that patients benefit when visual rhythms are offered to them. White sheets of paper were put on the floor and while patients were walking over these, their gait improved (vertical displacements increased). In 1988, Eni [65] observed that (vertical) gait parameters improved while patients listened to J.S. Bach's 'Sleepers Awake', an effect that remained present for some time after stopping the music. In 1989, Frischer [66] established that the rhythm of a metronome could reduce the tremor. Frischer speculated that patients with Parkinsonism are unable to internally generate the appropriate sequences.

Fig. 7. Pushing the patient over the hurdle of a transition

It appears that patients with Parkinsonism may be 'trapped' by sensory rhythms or sudden transitions in sensory information. How otherwise, we argue, could one explain the amazing inability of some patients to go through an open door, or the fact that a sudden visual event in the corridor behind it can 'pull' them through the opening? Offering rhythms and/or sensory transitions to patients with Parkinsonism may help them in their everyday actions. It is possible, we argue, that sensory rhythms acquire such dominance that a rather flat landscape is created for as long as the rhythm is going on. The hurdle of going through an open door for the visually controlled patient, will not present itself to the patient who walks on a musical rhythm. Or when the patient 'freezes', it is possible that a sudden visual transition just offers the appropriate push over the hurdle of a transition which would otherwise be insurmountable (Fig. 7).

5.2 Weaver Mice and the Problem of Parsing

Presently, experiments are being carried out by Robert Wagenaar and Richard van Emmerik at the Vrije Universiteit in Amsterdam, where the movements of patients with Parkinsonism are filmed, and displacements are analysed in order to pinpoint transitions. It may be possible not to limit such an analysis to the level of displacements in body-related space.

If our understanding of the nature of transitions is correct, and if, indeed, transitions occur at all levels, then one thing that all transitions will have in common, is an increase in the use of free energy. We speculate that such an increase will reveal itself to physiological monitoring. This would allow one to know the transitions even before one has exactly established what the order and control parameters are. Moreover, it has the advantage of intuitive elegance. If it could be shown, for example, that Parkinson's patients need surplus-energy to pass through an open door, the nature of their problems would be much easier to understand.

The group of researchers working with John Fentress at Halifax, Nova Scotia, has a long history of studying sequences in animal behaviour [67, 68]. Weaver mice suffer, among other defects, from dopamine deficiency. Their behaviour is presently being studied at Fentress's lab by Valery Bolivar [69]. Weaver mice start grooming as do all other mice: they lick their paws and then make a slow C-shaped movement around the cheeks, at irregular moments interrupted by very fast oscillatory movements around the tip of the nose. That is how far they come. Weaver mice have not been seen to start washing their ears. In normal grooming behaviour there is a general sequence from front to back. The Fentress group discovered that this sequence is 'parsed' into manageable units within which the sequence may change. A weaver mouse may go from the cheeks to the nose and then back to the cheeks again, but they do not reach their ears. In terms of the present paper, there are minor as well as major hurdles to take. It may be the major hurdles which are so difficult.

Patients with Parkinsonism sometimes report to benefit from changing their parsing. One may, for instance, ask the patient to rise, walk through the door, and then go to the toilet *as if this whole sequence were one action*. Possibly, in such a case, the patient's model of the future lacks the major hurdles which are otherwise so difficult to take. Informal research, however, has also suggested that the opposite can work. One may, for instance, ask the patient to first bend forwards, then move a leg, then bend forward again, then move the other leg, then orient towards the door, etc., *as if this whole sequence was an almost infinite chain of subactions*. Possibly, in such a case, the patient's model of the future is crammed with minor hurdles, none of which is as high as they normally are. Again, we argue, monitoring the spending of free energy could help to reveal the nature of the landscape created.

What, then, are the differences between this global and speculative understanding-in-principle, and the traditional idea of a 'motor program'? If programs are supposed to linearly pre-specify every detail of the movement, then the concept of 'program' is of no relevance to biological movement. It has been suggested that self-organizing synergies exist in the periphery, to be linearly switched on or off by the controlling program [70]. Fentress's research has revealed that such an approach is wrong also. Still, sequencing is a real biological phenomenon. The most appropriate metaphor we have come across so far, is that of the synergetic potential landscape. This landscape can be created and changed but, whatever its form, it has to be travelled through. We express the hope that our thoughts on 'synergetics, self-simplification, and the ability to undo' may be helpful to at least somewhat further the understanding of such travelling.

Acknowledgements

This paper was written in honour of Hermann Haken on the occasion of his 65th birthday. The authors shamelessly used the intellectual fruits of, among others, the following students and ex-students: Paul Derksen, Jolande Jurrius, Clemy van Koningsbergen, Claudine Lamoth, Frank Posthumus, Chris Sybrandy, and Wybren Zijlstra. We gratefully acknowledge highly stimulating discussions with Hermann Haken, and with, among others, Valery Bolivar, Avis Cohen, Richard van Emmerik, Anatoly Fel'dman, John Fentress, Wim van der Grind, Hans Lakke, Claire Michaels, Howard Pattee, Hans van Rappard, Robert Rosen, Beatrix Vereijken, Robert Wagenaar, Eric Wolters, and Arne Wunderlin. Helpful comments to earlier versions of this paper were given by, among others, Fons Blankendaal, Claire Michaels, Lieke Peper, Hans van Rappard, and Robert Wagenaar.

The writing of this paper took place during discussions between the authors. That was where it all went wrong.

References

1. L. Pickenhain, G. Schnabel (eds.), *Bewegungsphysiologie von N.A. Bernstein* (Johann Ambrosius Barth, Leipzig 1988)
2. H.T.A. Whiting (ed.), *Human Motor Actions—Bernstein Reassessed* (North-Holland, Amsterdam 1984)
3. L. Pickenhain: In *Complex Movement Behaviour—'The' Motor-Action Controversy*, ed. by O.G. Meijer, K. Roth (North-Holland, Amsterdam 1988)
4. O.G. Meijer, K. Roth (eds.), *Complex Movement Behaviour—'The' Motor-Action Controversy* (North-Holland, Amsterdam 1988)
5. H. Haken, A. Wunderlin: In *The Natural-Physical Approach to Movement Control*, ed. by H.T.A. Whiting, O.G. Meijer, P.C.W. van Wieringen (VU University Press, Amsterdam 1990)
6. P.N. Kugler: In *Motor Development in Children—Aspects of Coordination and Control*, ed. by M.G. Wade, H.T.A. Whiting (Nijhoff, The Hague 1986)
7. R. Rosen: *Life Itself—A Comprehensive Inquiry Into the Nature, Origin, and Fabrication of Life* (Columbia University Press, New York 1991)
8. H. Haken: *Erfolgsgeheimnisse der Natur—Synergetik: Die Lehre vom Zusammenwirken* (Deutsche Verlags-Anstalt, Stuttgart 1981)
9. P.A. Weiss: In *Hierarchically Organized Systems in Theory and Practice*, ed. by P.A. Weiss (Hafner, New York 1971)
10. I. Prigogine, I. Stengers: *Order out of Chaos* (Fontana, London 1986)

11. A.T. Winfree: Scientific American **230**, 82 (1974)
12. M. Eigen: Naturwissenschaften **58**, 456 (1971)
13. A.G. Cairns-Smith: *Seven Clues to the Origin of Life* (Cambridge University Press, Cambridge 1985)
14. F.H. Crick: Journal of Molecular Biology **19**, 548 (1966)
15. L. von Bertalanffy: *General Systems Theory* (George Braziller, New York 1968)
16. H.H. Pattee: In *Hierarchy Theory—The Challenge of Complex Systems*, ed. by H.H. Pattee (George Braziller, New York 1973)
17. J. Kirchhoff: *Friedrich Wilhelm Joseph Schelling* (Rowohlt, Reinbek bei Hamburg 1982)
18. I. Ekeland: *Mathematics and the Unexpected* (The University of Chicago Press, Chicago 1988)
19. J. Romein: *Op het Breukvlak van Twee Eeuwen* (Querido, Amsterdam 1976)
20. L.D. Landau, E.M. Lifshitz: *Statistical Physics, Part I* (Pergamon Press, Oxford 1980)
21. E. Jantsch: *The Self-Organizing Universe—Scientific and Human Implications of the Emergent Paradigm of Evolution* (Pergamon Press, Oxford 1985)
22. P. Glansdorff, I. Prigogine: *Thermodynamic Theory of Structure, Stability, and Function* (Wiley, New York 1971)
23. R. Thom: *Stabilité Structurelle et Morphogénèse—Essai d'une Théorie Générale des Modèles* (Benjamin, Reading MA 1973)
24. H. Haken: Zeitschrift für Physik **219**, 246 (1969)
25. H. Haken: *Synergetics—An Introduction* (Springer, Berlin 1983)
26. H. Haken: *Advanced Synergetics—Instability Hierarchies of Self-Organizing Systems and Devices* (Springer, Berlin 1987)
27. H. Haken: *Information and Self-Organization* (Springer, Berlin 1988)
28. B.H. Weber, D.J. Depew, J.D. Smith, eds.: *Entropy, Information, and Evolution—New Perspectives on Physical and Biological Evolution* (MIT Press, Cambridge MA 1988)
29. R. Penrose: *The Emperor's New Mind—Computers, Minds and the Laws of Physics* (Vintage, London 1991)
30. H. Haken, A. Wunderlin: *Die Selbststrukturierung der Materie—Synergetik in der unbelebten Welt* (Vieweg, Braunschweig 1991)
31. V.I. Arnold: *Gewöhnliche Differentialgleichungen* (Springer, Berlin 1980)
32. G.E. Hinton, T.J. Sejnowski: In *Parallel Distributed Processing—Explorations in the Microstructure of Cognition*, ed. by D.E. Rummelhart, J.L. McClelland, the PDP Research Group (MIT Press, Cambridge MA 1987)
33. H.A. Simon: In *Hierarchy Theory—The Challenge of Complex Systems*, ed. by H.H. Pattee (George Braziller, New York 1973)
34. S.W. Keele: *Psychological Bulletin* **70**, 387 (1968)
35. R.A. Schmidt: *Psychological Review* **82**, 225 (1975)
36. M. Flesher: Treasure in the attic—The impact of the Weber's study of human locomotion on the development of mathematical physics. Lecture presented at the Autumn Meeting of the International Society for Ecological Psychology, Hartford CT, 19 October 1991.

37. B. van der Pol, J. van der Mark: *Philosophical Magazine* **6**, 763 (1928)
38. E.P. Adrian, F.J.J. Buytendijk: *Journal of Physiology* **71**, 121 (1931)
39. M.T. Turvey: In *Perceiving, Acting, and Knowing*, ed. by R. Shaw, J. Bransford (Erlbaum, Hillsdale NJ 1977)
40. O.G. Meijer, R.C. Wagenaar, A.C.M. Blankendaal: In *Complex Movement Behaviour—'The' Motor-Action Controversy*, ed. by O.G. Meijer, K. Roth (North-Holland, Amsterdam 1988)
41. P.N. Kugler, M.T. Turvey: *Information, Natural Law, and the Self-Assemby of Rhythmic Movement* (Lawrence Erlbaum, Hillsdale NJ, 1987)
42. P.J. Beek: *Juggling Dynamics* (Free University Press, Amsterdam 1989)
43. R.C. Wagenaar: *Functional Recovery after Stroke* (VU University Press, Amsterdam 1990)
44. H. Haken, J.A.S. Kelso, H. Bunz: *Biological Cybernetics* **51**, 347 (1985)
45. G. Schöner, J.A.S. Kelso: *Science* **239**, 1513 (1988)
46. A.G. Fel'dman: *Biophysics* **11**, 766 (1966)
47. J.A.S. Kelso: *Journal of Experimental Psychology* **3**, 529 (1977)
48. B. Tuller, M.T. Turvey, H.L. Fitch: In *Human Motor Behavior—An Introduction*, ed. by J.A.S. Kelso (Erlbaum, Hillsdale NJ 1982)
49. E. Bizzi, A. Pollit, P. Morasso: *Journal of Neurophysiology* **39**, 435 (1976)
50. E. Bizzi, N. Accornero, W. Chapple, N. Hogan: *Journal of Neuroscience* **4**, 2738 (1984)
51. F.A. Mussa-Ivaldi, S.F. Giszter, E. Bizzi: *Cold Spring Harbor Symposia on Quantitative Biology*, **55** 827 (1990)
52. S.F. Giszter, F.A. Mussa-Ivaldi, E. Bizzi: *Society for Neuroscience Abstracts* **16-1**, 117
53. D. Bullock, S. Grossberg: *Human Movement Science* **10**, 3 (1991)
54. R. Dubuc, J.-M. Cabelguen, S. Rossignol: *Journal of Neurophysiology* **60**, 2014 (1988)
55. A.H. Cohen, S. Rossignol, S. Grillner, eds.: *Neural Control of Rhythmic Movements in Vertebrates* (Wiley, New York 1988)
56. A.I. Selverston: *Behavioral and Brain Sciences* **3**, 535 (1980)
57. F. Libersat, A. Levy, J.M. Camhi: *Journal of Comparative Physiology A* **142**, 339 (1989)
58. F. Libersat, R.S. Goldstein, J.M. Camhi: *Proceedings of the National Academy of Sciences* **84**, 8150 (1987)
59. R.M. Harris-Warrick, E. Marder: *Annual Review of Neuroscience* **14**, 39 (1991)
60. S. Grillner, T. Matsushima: *Neuron* **7**, 1 (1991)
61. A.H. Cohen, P.J. Holmes, R.H. Rand: *Journal of Mathematical Biology* **13**, 345 (1982)
62. A.H. Cohen, T. Kiemel: *American Zoologist* (in press)
63. O. Sachs: *Awakenings* (Duckworth, London 1973)
64. H. Forssberg, B. Johnels, G. Steg: *Advances in Neurology* **40**, 375 (1984)
65. G.O. Eni: *International Journal of Rehabilitation Research* **11**, 272 (1988)
66. M. Frischer: *Neuropsychologia* **27**, 1261 (1989)

67. J.C. Fentress: In *Perspectives on the Coordination of Movement*, ed. by S.A. Wallace (North-Holland, Amsterdam 1989)
68. J.C. Fentress: In *Signal and Sense—Local and Global Order in Perceptual Maps*, ed. by G.E. Edelman, W.E. Gall, W.M. Cowan (Wiley, New York 1990)
69. V.J. Bolivar, E.M. Coscia, W. Danilchuk, J.C. Fentress, K. Manley: *Society for Neuroscience Abstracts* **17-1**, 123 (1991)
70. G. Hinton: In *Human Motor Actions—Bernstein Reassessed*, ed. by H.T.A. Whiting (North-Holland, Amsterdam 1984)

Part V

Artificial Intelligence and Synergetic Computers

Artificial Life: An Engineering Perspective

A. Mikhailov

Department of Theoretical Physics, University of Bielefeld, and
German AeroSpace Research Establishment,
W-7101 Hardthausen 3, Fed. Rep. of Germany

Abstract. Possible engineering applications of artificial life studies are surveyed. The relationship between the problems of artificial life and of artificial intelligence is discussed.

1. Introduction

Artificial life (AL) is intended to display the essential properties of living beings without repetition of their biochemical material basis. In this respect, it is similar to *artificial intelligence* (AI) which aims to reproduce the functional abilities of the brain while leaving aside its actual underlying physiology.

When an animal develops from a fertilized egg, this process is governed by information present in the genes. However, the total amount of this information is much smaller than what would have been required to specify the positions and the differentiation of all the descendant cells. It means that the process of the individual development should be largely self-organized.

A movement of an animal is often produced by joint action of thousands of muscle elements. To calculate the dynamics of such a system and then to supply the detailed instructions to each individual element is beyond any reasonable capacity. Hence, generation of movements must also involve a large degree of self-organization.

Collective behavior of social insects, such as bees or ants, demonstrates extremely high complexity. However, each of the insects possesses in its memory only a set of simple, genetically inherited, laws of social interactions. When these laws are executed by all the members of the insect society, this results in the complex purposeful behavior required from the population as a whole.

Clearly, *self-organization* represents a characteristic functional property of natural living systems which is employed at all levels, from the processes inside a single living cell to the collective behavior of large populations and societies.

The situation in engineering is different. Modern machines and technological devices are based rather on the principle of externally enforced organization. To make an object of a required shape, we cast the melt into a mould or use a press of a suitable form. A complex

Springer Proceedings in Physics, Vol. 69
Evolution of Dynamical Structures in Complex Systems
Editors: R. Friedrich · A. Wunderlin

production task is frequently decomposed into a sequence of elementary operations executed according to an externally prescribed order. Movements are usually performed by dividing them into a series of primitive motions. When coordinated work of many individual subsystems is desirable, this is commonly done by providing them with external detailed instructions.

Theoretical studies of the last decades (see [1-4]) have revealed, however, that phenomena of self-organization are not an exceptional property of biological systems. Similar processes, which can be even described by almost identical mathematical models, have been found in inorganic nature, i.e. in physical or chemical systems. Once the general laws of self-organization are well understood, they can be used in engineering to construct machines and devices with the desired properties.

2. Artificial Life vs. Artificial Intelligence

The process of individual manual production consists of a sequence of operations with tools, each tool being a passive object independent of the others. A human formulates a logical program of actions with these tools and executes them in a required order. Execution of a production task is preceded by a rational analysis which results in decomposition of the task into a sequence of elementary operations.

The presence of a central organizing agent is also crucial for industrial production lines. There the operations of individual workers are performed sequentially in accordance with a certain externally imposed plan, and are themselves independent. Even in fully automatic machines one can discern a program of sequential operations which is recorded in their design.

This kind of production process differs in principle from the synergetic forms of behavior that are natural for living beings. The usual behavior of a living system is based on cooperative interactions between many units which result in complex self-organized behavior. Such a mode of operation does not presume the existence of an external agent who would have a complete plan of actions and would supply orders to individual units. Instead, the behavior of a living system follows entirely from interactions between the units.

Therefore, a synergetic "program of operation" would consist simply in laws of interaction that govern the dynamics. To change the behavior of a system, we have to modify these laws.

When a man performs computations he acts in effectively the same manner as in the process of manual labor. Numbers are taken from the memory (either internal or external) and then returned to it after completion of each individual arithmetical operation. The sequence of such operations is determined by a mental plan, or a program, which has been formulated before the computation started.

The abacus and other simple computing devices can be seen as tools which allow the mechanization of individual arithmetic operations, but still require the active participation of a human. Modern computing machines automatically execute a sequence of mathematical operations recorded in the program which was fed into it by a human. Actually, they act in the same way as fully automatized industrial production machines. Both presume the presence of some external agent who has already logically decomposed the problem into a sequence of elementary operations.

But this mode of information processing is not a natural one for living beings. When a cat jumps to catch a mouse, it is hardly imaginable that its brain solves by computation a complicated set of differential equations which describe the trajectory of motion. It is equally absurd to assume that a goalkeeper computes mentally the trajectory of his jump to intercept the flying ball. An alternative to computation is provided by *analog* information processing. This is based on the idea of mapping the problem onto the dynamics of some artificially constructed system, so that the answer can be obtained by following the evolution of this system.

In the middle of this century, when universal computers were not yet available, flight trajectories of missiles were often predicted by using specialized analog machines. These electronic devices included circuits in which the dynamics of the electric current obeyed effectively the same differential equations as those which describe the flight of a missile. Hence, it was possible to imitate any particular flight by adjusting the initial conditions and then observing the electric current dynamics in the circuit.

Instead of the dynamical systems with a few degrees of freedom, one can use much larger systems with a great number of interacting elements. Then more complicated physical processes can be directly simulated, by a proper choice of interactions between the elements in the model.

The advanced tasks of information processing, which are typical for intelligence, go beyond direct simulation of physical processes. For instance, the brain possesses the property of *associative memory*. When a new visual pattern is provided, it is very quickly classified and the most similar prototype stored in memory is retrieved. This property can be easily realized in an analog manner, using distributed dynamical systems which have several stable steady activity patterns. The relationship between associative memory and pattern formation in distributed dynamical systems, which was noted by H.Haken [5], is intensively studied today for particular models of artificial neural networks (see, e.g. [6]). Another class of problems, where the analog methods turn out to be very effective, involves *optimization* [3,6].

It seems plausible that many basic tasks of information processing, typical for human intelligence, can be realized by constructing an appropriate distributed dynamical system. In the AI studies these systems are usually called the "neural networks", since they intend to simulate the functional abilities of the brain. Despite their name, practical realizations of artificial neural networks have often very little in common with the actual biological neurons of the brain. They represent large systems of active elements which interact in a simple way one with another. The individual elements can be certain solid-state electronic devices, or optical cells, or macromolecular aggregates. These physical systems are far from thermal equilibrium and require a constant supply of energy to balance its dissipation. From a general point of view, the formation and evolution of activity patterns in these systems are self-organization phenomena.

Hence, engineering of self-organization becomes an important aspect of information processing. To employ the analog manner of information processing, one needs to know how the variations in properties of individual elements and in laws of interaction between them would influence the behaviour of a system as a whole.

As far as information processing is concerned, the actual material realization is of secondary importance. It does not matter whether we record information in electric potentials, optical images or conformations of macromolecules. If the dynamical laws of evolution of patterns are the same, all these distributed systems can be used equally well for the purposes of analog information processing.

However, the situation is different when one tries to employ self-organization phenomena in engineering devices. Here the physical nature of a system is of obvious importance. Depending on a particular task, we may need a system which would change its shape or configuration, execute some mechanical motions, vary its optical appearance, produce certain voltage distributions, etc. Therefore, the choice of a particular material system would be determined not only by its required functional properties but also by the desired manner of its physical operation.

3. Towards Living Machines

Today, when we need to produce an object of a given shape, this is achieved by cutting a solid material, by applying a press, or by pouring a liquid melt into a suitable mould. In all these technological operations the material itself plays only a passive role. Contrary to this, biological objects (such as plants) acquire their rich forms in a process of an autonomous development, as a result of interactions between the elements which constitute the living matter.

An application of AL in engineering may consist in creating "artificial living matter" which would be able to settle itself into a certain shape and maintain it, while energy is being supplied. By varying the flow of energy or other external parameters, it would be possible to gradually change the form of this living object or to induce its transitions from one stable steady form to another. Moreover, the object can be endowed with the property of associative memory, i.e. it can be designed to possess several distinct steady forms and to choose a particular form depending on the initial conditions.

Since the actual shape of an artificial living object is an outcome of interactions between its constituents, it can also be engineered to perform persistent self-oscillations, periodically changing its form with time. In such a property it would behave as a "motor" which is made from a single piece of material.

Many natural living objects, even those as simple as bacteria, are able to propel themselves through a passive medium. There are different opportunities to realize self-motion for artificial living objects too.

The electronic devices which are used today to generate and transform various electrical signals represent large ensembles of highly specialized passive elements whose functions are prescribed by the initial design. Using the effects of self-organization, it might be possible to introduce "living" electronic devices where the generation and transformation of signals would be achieved as a result of pattern formation and pattern interactions in an electrically active distributed medium.

The above examples give only the simplest illustrations of how self-organization phenomena can be used in engineering applications. Having in mind these incentives, we can now discuss the available repertoire of mechanisms and processes of pattern formation in physical systems far from thermal equilibrium.

The most thoroughly investigated class of distributed active systems are the so called *reaction-diffusion models*. They are most easily realized in chemistry, as chemical reactions which proceed in a dilute liquid solution. In this case, reactions between different molecules give rise to the local dynamics within a small volume element, whereas diffusion of reagents provides coupling between the dynamics in the neighbouring small volumes. Depending on the kind of local dynamics and on the relationship between the diffusion constants of different reacting components, the reaction-diffusion systems can demonstrate [3,7] a surprisingly rich variety of behaviour, from formation of stable stationary spatial patterns to self-supported propagation of pulses and emergence of complex wave structures.

The most spectacular experimental data is available for the famous *Belousov-Zhabotinskii (BZ) chemical reaction* (see, e.g., [8]). Here all basic kinds of patterns were observed and interactions between them were investigated. The BZ system is very convenient for experimental studies since the patterns are characterized by relatively large time and length scales, and can be easily observed by an eye.

However, when engineering applications are considered, the BZ reaction and other chemical reaction-diffusion systems are not of significant use. The emerging patterns are manifested here in distributions of reacting molecules which are present in only small concentrations. From this point of view, much more advantageous are other reaction-diffusion systems which represent solid-state or plasma devices. They can be realized in many different ways. For instance, one can use semiconductors (see [7,9]) where there are various kinds of charge carriers, i.e. electrons and holes in different energy bands, and where impurity energy levels may be also populated. Then interactions between these charge carriers (such as the electron-hole recombination process) and transitions between the bands and discrete energy levels may produce the desired local dynamics, whereas diffusion of these particles can provide the required spatial coupling. Note that the latter can also be realized by the process of heat conduction which obeys the same mathematical law as diffusion. In plasma reaction-diffusion systems (for instance, under the conditions of the electrical gas discharge [7,10]) the role of reacting particles is played by electrons and ions.

Experimental studies of pattern formation in semiconductor or plasma devices are more difficult than in the BZ reaction. However, a number of interesting effects have been recently observed in these systems [9,10]. In future, when greater experience in dealing with such systems is reached, it could become possible to purposefully design systems with the desired properties of pattern formation. These devices could then be used for generation and processing of electrical signals.

In engineering applications one might also require systems where patterns are manifested in mechanical motions or flows of a medium. Then the phenomena of pattern formation in liquids may be used. It is well known that very regular flow patterns (such as rolls or hexagonal cells in the Benard problem, see [11]) can spontaneously develop in liquid systems. However, the interval of existence of such coherent structures is narrow. They are observed at the threshold of the laminar-turbulent transition and even a small change of parameters is often sufficient to make the flow chaotic.

The hydrodynamical turbulence is suppressed in thin liquid films on solid substrates. Because of the small thickness of the film, its effective viscosity is significant and the flows have small Reynolds numbers. Now an important role in pattern formation can be played by capillary forces and by interactions between the film and the solid surface.

If the film covers only part of the solid, it can spread or retract on the substrate, depending on its wetting properties. When the surface is active, so that its properties change in time or in the plane, spreading can alternate with retraction [12]. When, the properties of the surface are locally influenced by the covering liquid, travelling films can be produced [13].

Above we considered continuous media. In this case one can speak about the local dynamics in a small volume element only in the figurative sense, since no distinct elements are actually present. In continuous media, both the local dynamics and the spatial interactions are described by the same set of partial differential equations. This condition imposes severe limitations on the class of admissible behaviour and makes the design of such systems more complicated.

A greater and more controllable variety of self-organization phenomena can be found if we turn our attention to *cellular* distributed systems which represent aggregations of distinct active elements (or "cells"), immersed into a continuous medium and interacting through it. Now the processes which determine the local dynamics take place within a separate active element and will therefore be described by their own dynamical equations, different from those specifying the interactions between the elements. The host medium in this case can be passive, i.e. energy may be continuosly supplied only to the active cells inside it. The role of the host medium is to transfer interactions between the cells.

The dynamics of the individual elements may be very simple. It can be even discrete, i.e. represent a sequence of quick transitions between several different states. Then an individual cell can be effectively modelled as a discrete automaton.

The interactions between the cells may be realized, for instance, by diffusion of certain chemical "mediators" through the intercellular medium (see [14-16] for further details). These mediator substances are released by the cells, depending on their current states, and then diffuse undirectionally through the medium. The mediator concentrations in a vicinity of a given cell element determine the transition to a subsequent state and thus influence the internal dynamics of the element. Investigations of this model [16] showed that it has a rich potential for pattern formation. When the proper laws of communication between the cells are employed, it can effectively imitate the behaviour typical of artificial "neural networks" and possesses the property of associative memory.

Another interesting realization of a cellular system can be based on use of an *elastic* host medium. Suppose that cells can change their form and/or volume, depending on their current states. If the cells are attached at certain locations to an elastic substrate, transitions between the states of the individual cells create deformations and strains in the substrate which are present even far from them. If we further assume that transitions between the states in individual cells are governed by the local strains in the substrate medium, that would give us a mechanism of interaction between the cells. The collective dynamics of the cells would result in elastic deformations and, hence, in changes of the shape of the substrate.

The elastic interactions between living cells are known to play an important role in the processes of biological pattern formation. Recently they were discussed in a detailed way by J.Murray (see [17]). Similar models may be realized in artificial systems, where the "cells" can be simple active physical elements which undergo certain transitions while receiving energy from an external source. In this manner one can, perhaps, design the "artificial living matter" hypothesized in the beginning of this section.

The artificial active elements can be also engineered to perform self-induced motions through the host medium. This behaviour is typical for biological organisms: even very simple bacteria are able to move themselves along a preferred direction.

One of the possibilities involves the use of capillary forces which are very sensitive to changes in temperature or in chemical composition of a solution. There is a popular experiment that shows how the capillary forces can be employed for generation of motion: Consider a match stick that floats on the surface of water. Let us take a tiny piece of soap (which is a surfactant) and attach it to the head of the stick. Now, instead of resting, the stick would start to glide over the water surface. It would rapidly move along a complicated trajectory, until the soap is spent.

The qualitative explanation of this behaviour is obvious: the dissolved soap changes the local value of the capillary coefficient. Since the soap concentration in the water is larger at the head of the stick than at its end, there is a net force applied to the stick which sets it in motion. This simple idea is further explored in [18] where a few other possibilities of engineering the self-supported motion are also discussed. The direction and the speed of self-motion may be determined by the internal state of a cell, which in turn can depend on the interactions between the cells. When the proper laws of interactions are chosen, very complicated forms of collective behaviour in the ensemble of active cells can thus be produced.

4. Complexity, Learning and Evolution

All natural living systems are very complex. How is this enormous complexity made compatible with the regular and predictable forms of behaviour? This question is far from trivial. When a system is produced simply by bringing together a large number of different elements, this would most probably result in a chaotic, unstable and unpredictable behaviour. Special measures should therefore be taken to match the complexity with the regular performance.

It turns out that living systems usually represent *hierarchies* of autonomously self-organized subsystems. As it was noted by H. Haken [1], the processes of pattern formation within any subsystem are controlled by a small set of variables which represent its *order parameters*. The patterns formed at a lower level can also serve as the elements of a higher level in the hierarchy.

Although the elements may have complicated internal dynamics which is described by many degrees of freedom, they usually enter into interactions with other elements as sufficiently simple units. Only a few internal variables are relevant for the processes of pattern formation and give rise to the order parameters of a subsystem. Other internal degrees of freedom are effectively *enslaved* by the order parameters.

The dynamics of the subsystems which belong to the higher levels is characterized by larger time scales. Therefore, the order parameters of these levels change only very slowly in time from the point of view of the lower-level systems. Because of this difference in the time scales, patterns at higher hierarchical levels do not directly interfere with the processes of self-organization at the lower levels. However, they may still execute certain control by effectively changing the parameters of the subsystems. In this way they can modify the properties of the formed patterns or induce transitions from one steady regime to another.

These principles of hierarchical organization can also be employed in the design of artificial living systems. Actually, the "cellular systems" discussed in the previous section can already be considered as the simplest implementation of such an approach. Indeed, we assumed there that the interactions between the cells were determined by their instantaneous discrete "states". The transitions between the states, influenced by changes in the environment of a cell, were described as sudden jumps of vanishing duration.

If we try to realize a cellular system in practice, the individual cells would be represented by some physical objects with a certain internal dynamics governed by a set of variables. This dynamics may be chosen in such a way that it would consist of quick transitions between a number of steady states. Then, from the point of view of the entire system with the larger characteristic time scales, an individual element can indeed be modelled as an automaton with a finite number of discrete states and abrupt transitions between them.

It should be noted in conclusion that, although the functional role of hierarchies is fairly well understood, no elaborate theoretical and experimental investigations of particular hierarchical self-organized systems has been performed until now. These studies must be carried out before the hierarchical approach can be consistently used in engineering applications.

An important property of all living creatures is that they can *reproduce* themselves, either sexually or by fission. Should not we try to imitate such a property in the artificial life devices? To answer this question, let us analyze what are the principal roles of reproduction.

The most obvious aspect of reproduction is that it allows a living system to maintain its existence within time spans that are much larger than the lifetime of an individual living organism, without any external supply of new elements.

Clearly, this property is of fundamental importance for the natural living systems. When the engineering applications are considered, it is not, however, very essential. Unless we need to colonize with the AL devices some distant and hardly accessible areas, we can always replace the worn out elements by the new ones that are taken from a stock or continuously produced elsewhere.

Another aspect of reproduction is that it provides a basis for the biological evolution. Although this evolution pursues no definite goals, it has led to gradual emergence of a very complex hierarchical structure which we know as *life*.

Recent mathematical studies [19-21] reveal that even simple hereditary algorithms may be sufficient to describe, for example, spontaneous splitting of a population into a set of distinct species, development of effective viruses and parasites, etc. These investigations can significantly improve our understanding of biological evolution, its functions and operation. They also show that a hierarchical structure can spontaneously start to develop without any external assistance. However, in practical applications of AL it still seems more wise to perform the main designing work ourselves, rather than delegating it to a slow evolutionary process.

Finally, evolution allows biological systems to adjust to gradual changes in their environment and to optimize their performance. In effect, it provides an opportunity to *learn* from the experience. This function may be desirable for the artificial living devices and therefore we should discuss it in more detail.

When the ideology of analog information processing is used, *learning* means the ability of a dynamical system to change intentionally its internal structure in such a way that its performance, for a certain class of tasks, becomes optimal. This can be achieved by changing the properties of individual elements and the laws of interaction between them.

When learning is realized without any external directives, it represents a slow dynamical process which goes on inside a special subsystem. The dynamics of such a subsystem is influenced by the overall efficiency of performance of the main system in a given environment. This subsystem controls the behaviour of the main system by slowly changing the parameters of individual elements and the laws of interactions between them; thus it effectively modifies the internal structure of the main system.

In biological species, the role of a learning subsystem is played by the genes, whereas the main system consists of the individual living organisms. Because of reproduction, the genes transcend the limits imposed by the finite span of individial lifes and are relayed from one generation to another. The characteristic dynamical time scales of the genetic subsystem are much larger than the lifetime of a single organism. The efficiency of a species is determined by its reproductive potential.

However, genetic reproduction is not the only mechanism which supports learning in natural living systems. At a higher level of the structural hierarchy, for human societies, the role of a learning subsystem is largely played by *culture* which is not genetically transmitted, but rather handed from one generation to another.

In the theory of artificial intelligence, various mathematical models of learning have been proposed. Some of them can be implemented (see [3]) as distributed dynamical systems. Learning is then treated as a special case of optimization. The learning subsystem is constructed in such a way that it tends to approach the state where the discrepancy between the desired and the actual performances of the main system becomes minimal.

This search for a minimum can be realized in a purely deterministic manner, by employing the dynamical systems which evolve to a minimum of a certain potential function (that specifies the error of performance). However, better results are obtained when the dynamics of learning is stochastic. Random fluctuations allow the system to overcome barriers in the potential landscape which might lie on the way to an optimal solution. An example of a learning distributed dynamical system that involves fluctuations and is known as the "Boltzmann machine" can be found in [3]. In the process of biological evolution, fluctuations are provided by mutations which represent the reproduction errors.

Hence, we see that the reproductive ability is not an obligatory prerequisite of AL systems. The distinguishing properties of AL devices should be rather an efficient use of the self-organization phenomena and, in more advanced forms, of a hierarchical internal structure. When it is required, a certain degree of intelligence can furthermore be achieved by adding a specialized slowly evolving dynamical subsystem.

References

1. H. Haken: *Synergetics: An Introduction* Springer Ser. Synergetics. Vol. 1 (Springer, Berlin, Heidelberg 1978)
2. H. Haken: *Advanced Synergetics* Springer Ser. Synergetics. Vol.20 (Springer, Berlin, Heidelberg 1987)
3. A. S. Mikhailov: *Foundations of Synergetics I. Distributed Active Systems* Springer Ser. Synergetics. Vol.51 (Springer, Berlin, Heidelberg 1990)
4. A. S. Mikhailov, A. Yu. Loskutov: *Foundations of Synergetics II. Complex Patterns* Springer Ser. Synergetics. Vol.52 (Springer, Berlin, Heidelberg 1991)
5. H. Haken: "Pattern formation and pattern recognition - an attempt at a synthesis" in *Pattern Formation by Dynamic Systems and Pattern Recognition*, ed. by H. Haken, Springer Ser. Synergetics, Vol.5 (Springer, Berlin, Heidelberg 1979)
6. H. Haken, ed.: *Neural and Synergetic Computers* Springer Ser. Synergetics, Vol.42 (Springer, Berlin, Heidelberg 1988)
7. B. S. Kerner, V. V. Osipov: Sov. Phys. Usp. **32**, 101-138 (1989)
8. A. V. Holden, M. Markus, H. G. Othmer, eds.: *Nonlinear Wave Processes in Excitable Media* (Plenum Press, New York 1991)
9. H.-G. Purwins, C. Radehaus, T. Dirkmeyer, R. Dohmen, R. Schmeling, H. Willebrand: "Application of the activator-inhibitor principle to physical systems" Phys. Lett. A **136**, 480-484(1989)
10. H. Willebrand, C. Radehaus, F.-J. Niedernostheide, R. Dohmen, H.-G. Purwins: "Observation of solitary filaments and spatially periodic patterns in a dc gas-discharge system" Phys. Lett. A **149**, 131-138 (1990)
11. H. Haken: "Spatial and temporal patterns formed by systems far from equilibrium", in *Nonequilibrium Dynamics in Chemical Systems*, ed. by C. Vidal, A. Pacault (Springer, Berlin, Heidelberg 1984) pp.37-46
12. P. G. de Gennes: "Wetting: statics and dynamics" Rev. Mod. Phys. **57**, 827-863 (1985)
13. D. Meinköhn, A. S. Mikhailov: "Liquid films on reactive solids", 1992, to be published
14. A. S. Mikhailov, I. V. Mit'kov, N. A. Sveshnikov: "Molecular associative memory" *BioSystems* **23**, 291-295 (1990)

15. E. M. Izhikevich, A. S. Mikhailov, N. A. Sveshnikov: "Memory, learning and neuromediators" *BioSystems* **25**, 219-229 (1991)
16. A. S. Mikhailov: "Information processing by systems with chemical communication", in *Rhythms in Physiological Systems*, eds. H. Haken and H. P. Koepchen (Springer, Berlin, Heidelberg 1991) pp.339-350
17. J. D. Murray: *Mathematical Biology* (Springer, Berlin, Heidelberg 1989), Ch.17
18. D. Meinköhn, A. S. Mikhailov: "Gliding bugs, crawling drops, and nonlinear dynamics of self-motion", 1992, to be published
19. M. Serva, L. Peliti: "A statistical model of an evolving population with sexual reproduction" J. Phys. A **24**, L705-L709 (1991)
20. P. G. Higgs, B. Derrida: "Stochastic models for species formation in evolving populations" J. Phys. A **24**, L985-L991 (1991)
21. H. Freund, R. Wolter: "Evolution of bit strings: some preliminary results" Complex Systems **5**, 359-370 (1991)

Path Finding with a Network of Oscillators

A. Babloyantz[1] *and J.A. Sepulchre*

Université Libre de Bruxelles, CP 231 – Campus de la Plaine, Boulevard du Triomphe, B-1050 Bruxelles, Belgium

The self-organizing properties of reaction-diffusion systems kept far from equilibrium have been the focus of much interest for several decades [1, 2, 3]. Experiments as well as numerical and analytical works have demonstrated the existence of self-organized collective behaviour in the form of spatial and spatio-temporal phenomena [4].

The spatial organization, refered commonly to Turing structures, although predicted analytically for decades have been confirmed experimentally only very recently [5]. However, target waves and spiral activity have been seen in many experimental systems. Spatio-temporal chaos is also seen in other experimental situations.

Recently spatio-temporal chaos, target waves and spiral waves were shown to appear also in networks of oscillating elements as well as in model cortices which are constructed of a network of biological neurons [6].

Haken and coworkers have also shown that a network of reacting and diffusing elements is endowed with pattern recognition and associative memory capabilities [7]. In this paper we want to show that the target waves propagating in a network of oscillating elements may be used for navigational problems. The latter capability is the result of unusual properties of the target waves in the presence of obstacles which has been observed recently [8].

In the first section we demonstrate the existence of target waves with unusual properties in a network of oscillators. The second section is devoted to propagation of target waves as well as spiral waves in the presence of obstacles and windows. It is shown that, contrary to the usual electromagnetic waves, the target waves do not show interference effects. Under appropriate conditions, they have the ability to inhibit each other. This latter property may be used to design systems with interesting navigational properties which is described in sect. 3.

1. Wave Propagation in a Network of Oscillators

Let us consider a two dimensional array of nonlinear oscillatory elements connected via nearest neighbour interactions. To be general, the system is described by the complex Ginzburg-Landau equation (CGL), which describes the generic form of a supercritical Hopf bifurcation. The network obeys to the following differential equations :

$$\begin{aligned}\frac{dX}{dt} &= X - \Delta\omega[k,j]Y - (X-\beta Y)(X^2+Y^2) + \mathcal{L}(X) - \alpha\mathcal{L}(Y)\\ \frac{dY}{dt} &= \Delta\omega[k,j]X + Y - (\beta X + Y)(X^2+Y^2) + \alpha\mathcal{L}(X) + \mathcal{L}(Y)\end{aligned} \qquad (1)$$

Springer Proceedings in Physics, Vol. 69
Evolution of Dynamical Structures in Complex Systems
Editors: R. Friedrich · A. Wunderlin

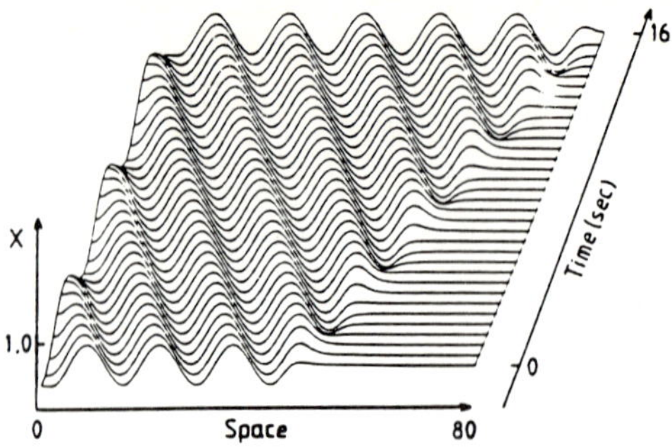

Figure 1: Temporal evolution of a spatial profile of the X variable computed from eqs. (1). A perturbation of one oscillator in the initially unstable state may produce phase waves whose velocity has opposite sign to the front velocity.

where $\mathcal{L}$ is a connectivity operator defined as:

$$\mathcal{L}(X)[k,j] = D\{X[k+1,j] + X[k-1,j] + X[k,j+1] + X[k,j-1] - 4X[k,j]\} \quad (2)$$

$$(k,j = 2,\ldots,N-1)$$

The boundary cells are subject to "zero flux" boundary conditions.

The unperturbed system may be in either one of the following states: All oscillators are held at the unstable steady state $(X,Y) = (0,0)$ or they are all oscillating with identical frequency.

Propagation in an unstable medium

Let us assume that the system of coupled oscillators has been prepared such that they are all in the absolutely unstable state. Thus a localized perturbation applied in the centre of the network, where the system is maintained in the uniform unstable state $(0,0)$, gives rise to a circular front propagating with a velocity v. Behind the front, according to the sign of coefficient β in eqs. (1), waves propagate either in the same direction as the front or in opposite direction. Figure 1 depicts the case where, as the front advances, the waves propagate inward behind the front. The explanation of such behaviour can be found in the original paper [9].

Propagation in an oscillatory medium

Let us now consider a network in an oscillatory state. Target waves may be generated under the action of a small perturbation of the parameters of a cluster of oscillators, which plays the role of a pacemaker. In this paper the waves were generated by a local frequency shift $\Delta\omega$ between the pacemaker region and the other oscillators. This centre initiates target waves in the system which again propagate in concentric rings starting from the pacemaker centre and gradually take over the entire network.

We also considered the case where two simultaneous pacemaker regions, with the same spatial extension and with different frequency shifts $\Delta\omega_1 < \Delta\omega_2$, were active.

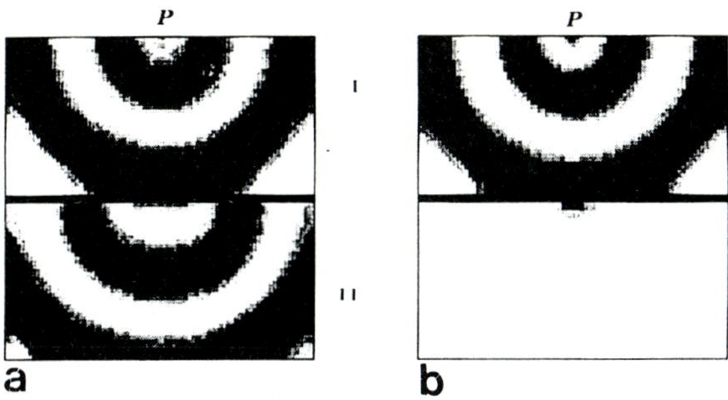

Figure 2: A square network of 80×80 oscillators is partitionned into two compartments by a barrier. Waves emitted in compartment I around a pacemaker located in P propagate through an aperture of size l in the compartment II. (a) $l > l_c$: Waves propagate in compartment II with the same frequency than in compartment I. (b) $l < l_c$: Frequency pulling occurs and the frequency of the waves in compartment II tends to 0 as $l \rightarrow 0$. The parameters of eqs. (1) and eq. (2) are $\Delta\omega = 1$, $\beta = 1$, $\alpha = 0$ and $D = 3.19$.

Target waves start to propagate from both centres, but after a while one sees that one of the centres remains active while the second one is prevented from emitting waves. The centre which wins the competition is the one whose frequency Ω of the emitted waves is the largest.

2. Waves and Obstacles

Let us now partition the network into two parts, communicating however via several openings. Waves are generated in the first compartment and we study their evolution in the second compartment.

Target waves and obstacles

As a first simulation we consider a partition between the compartments, with a single window size l taken as a control parameter. In this and in the three subsequent experiments, a single pacemaking region at point P is considered. When the first front starting from P in compartment I reaches the window, it generates a new set of similar target waves in the compartment II (Fig. 2).

The phase of each oscillator is defined as $\phi[k,j] = \arctan(Y[k,j]/X[k,j])$. Let us measure the phase difference $\phi_2 - \phi_1$ between two oscillators of compartment II and I. For a window length of $l > l_c$, this phase difference tends to a stationary value ϕ, therefore in this case, frequency locking between the two compartments occurs (Fig. 2a). However, as the size of the window l decreases, the stationary value ϕ tends to a limit point and the frequency-locked solution disappears at l_c. Below this value, a quasi-periodic solution is seen in the vicinity of the opening, and in the compartment II, target waves propagate with a lower frequency Ω_{II} compared to Ω_I in compartment I (Fig. 2b). A description of these phenomena in terms of numerical analysis and in the framework of bifurcation theory is given in the original paper [8].

Let us now consider the same network of oscillators as described above. We now introduce two openings l_1 and l_2 in the system (Fig. 3). Again a single local frequency shift $\Delta\omega$ is created at the point P. Target waves with frequency Ω_I start to propagate in the compartment I and reach successively the two openings l_1 and l_2. For $\Delta\omega = 3$, the two windows act as new pacemaker regions and generate in turn target waves in

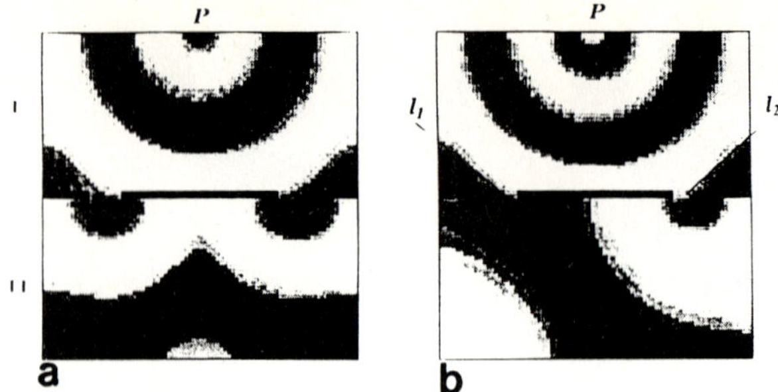

Figure 3: Wave propagation in the presence of two windows. The window sizes are $l_1 = 8$ and $l_2 = 10$ units in the network. (a) Waves propagate into compartment II from both windows and a cusplike structure is formed ($\Delta\omega = 3$). (b) Waves propagate from the largest aperture and inhibit the propagation from the small window ($\Delta\omega = 1$). Other parameters are the same as in Fig. 2

the second compartment with $\Omega_{II}(l_1) = \Omega_{II}(l_2) = \Omega_I$. After a while, the target waves propagating from the two centres collide and cusp-like structures are formed (Fig. 3a). In the next experiment, we decrease only $\Delta\omega$ at point P. A higher value of Ω_I, as compared with the preceding case, is obtained in the first compartment for the propagating target waves. In this experiment when the wavefront reaches the windows l_1 and l_2, new fronts are again generated. However, very soon the waves emerging from window l_2 take over the entire system and inhibit all propagation from l_1 (Fig. 3b). This unusual property distinguishes target waves from the usual electromagnetic phenomena.

Spiral waves and obstacles

Spiral waves may also propagate in coupled network of oscillators. Much work has been done on the characterization of spiral waves in an oscillating continuous medium [10]. These rotating patterns are also referred to vortices or defects. This latter name comes from the fact that at the centre of the spiral, the amplitude of the oscillator vanishes, and therefore the phase of the oscillator is not defined at this point and produces a defect in the phase field.

In the oscillatory network the target waves and spiral waves show a wealth of curious and unexpected behaviour. A few examples will be given here. A more complete catalogue of their properties has been established recently [11].

We also investigated the behaviour of spiral waves when they hit barriers with openings. Again the network considered is the same as in the previous section. Now, however, no frequency shift is imposed on any part of the system, but a spiral activity is generated in compartment I by the following initial conditions imposed on the network:

$$\begin{aligned} X[k,j] &= \gamma_1 \,(k - s_x) \\ Y[k,j] &= \gamma_2 \,(j - s_y) \end{aligned} \tag{3}$$

with $k, j = 1, \ldots, N$. The parameters γ_1 and γ_2 are chosen arbitrarily. With this initial condition, one sees the onset of a single spiral, the centre of which is located at the coordinate (s_x, s_y). The period of the spiral rotation is taken as the time unit. When the arm of the spiral reaches the opening a gradient is generated in the second compartment. In the set up of Fig. 4a the gradient emerges only from the larger opening. As the core of the spiral travel in the first compartment target waves emerge

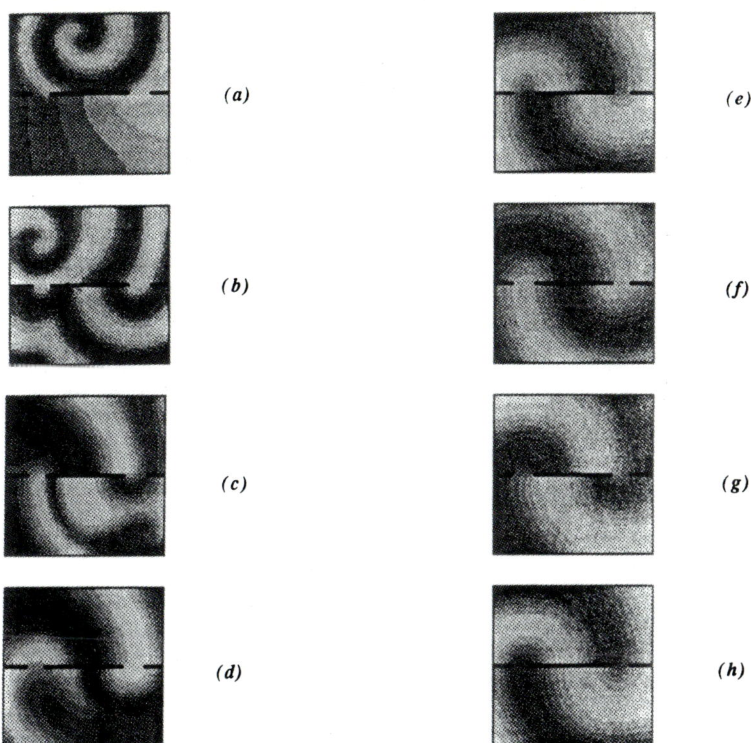

Figure 4: Evolution of a spiral in the same network as in Fig. 3. The initial condition is given by relation (3) with the parameter value $s_x = 40$, $s_y = 10$, $\gamma_1 = \gamma_2 = 1/40$. (a) to (d): Transient evolution of the oscillating network. (a),(b) The spiral in compartment I gives rise to a gradient in compartment II. (c),(d) less coherent transient structures. (e) to (h): Periodic activity of circulation waves between both compartments. The parameters of eqs. (1) and eq. (2) are $\Delta\omega = 0$, $\beta = 0.6$, $\alpha = -1$ and $D = 2.56$

Figure 5: Other initial conditions $s_x = 40$,$s_y = 20$, $\gamma_1 = 1/40$ and $\gamma_2 = 1/80$, gives a totally different asymptotic behaviour. The spiral remains in compartment I whereas target waves appear in compartment II. Waves propagate only from the largest aperture, inhibiting activity from the smaller opening.

from both openings and form a cusp-type structure (Fig. 4b). This behaviour can be paralled with cusp formation as seen in Fig. 3a. Gradually these well-defined patterns turn into an incoherent and less well-defined structures (Figs. 4c & d). However after about 16 time units, again a new organized spatio-temporal behaviour emerges.

From compartment I a wave crest enters into the second compartment, then another crest emerges from the latter and enters into the first compartment. Thus one sees a circulation of oscillatory activity. The spatio-temporal pattern itself shows a periodicity of 5 time units (see Figs. 4e to h).

For another set of initial conditions and for the same network, one sees radically different behaviour. After a brief initial transient behaviour the system settles down to a time and space pattern as depicted in Fig. 5. One sees only the propagation of target activity emerging from the larger opening. Here again as in the case of Fig. 3b all activity generated by small opening is inhibited.

Contrary to the preceding case, the spiral pattern seems to be stable and in our simulation was stable as long as for 40 time units.

3. Path Finding

The unusual properties of target waves, namely the fact that propagation from small openings are inhibited by waves emerging from the large windows may be used for path finding purposes.

To illustrate the path finding capacity of networks of oscillating elements, let us consider the following task. In Fig. 6 an object R of size d (e.g. a mobile robot) has to travel in different rooms through one of the possible doors m_1, m_2 or m_3 to reach a goal G. Some of the apertures are too small for R to pass through. The task is to direct the robot along the shortest path compatible with its size. All aspects of the situation are assumed to be dynamic. The location of the apertures, the position of the object R and the position of the goal G may be time dependent. To the physical space we associate a network described by eqs. (1). Points of the two-dimensional floor depicted in Fig. 6 are mapped into well-defined units in the network of oscillators. The internal walls are represented in the network by deleting the corresponding oscillators. All oscillators mapped to the internal as well as external boundaries obey zero-flux boundary conditions. To the point G there corresponds an element P in the network space. If this element is made to be a pacemaker centre, target waves will start to propagate from this centre and eventually reach the oscillator S which is mapped into the initial position of the object R in the physical space. Thus knowledge is created in the system, whereby the crest of these waves may become information carriers indicating a direction of propagation.

To see this we consider the discrete phase gradient of the waves. A finite-difference approximation of $\nabla\phi$ provides a vector field which is perpendicular to the crests and points to the direction of the pacemaker centre P. Thus if the vector field is computed in each cell as they are reached by the wave, a trajectory is constructed, which joins every point in the network to the point P. Due to the mapping of the network space into the physical space, a family of trajectories are also created in the physical space, relating every location of the room to the goal-point G. Such a mapping could be used for example for finding the shortest path between R and a point G where the trajectory starts. As a first experiment, the frequency shift $\Delta\omega$ in P is chosen such that the waves can propagate from all the openings m_1, m_2 and m_3. The vector field corresponding to this problem is depicted in Fig. 7a and it is seen that the shortest path is C_1 which passes through the door m_2.

However a more interesting case is the one in which for technical reasons, such as for

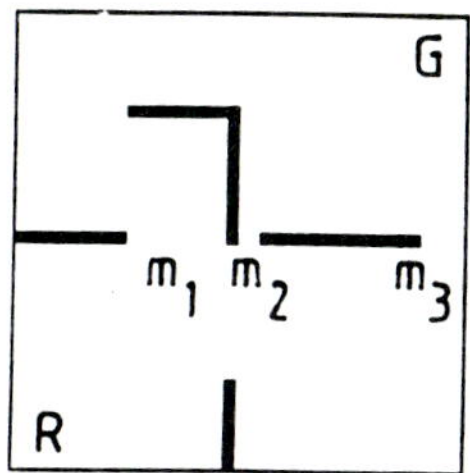

Figure 6: Physical space : The robot R of a given size d must reach goal G. The maze is formed from openings of various sizes. Robot size d exceeds the smallest value m_2

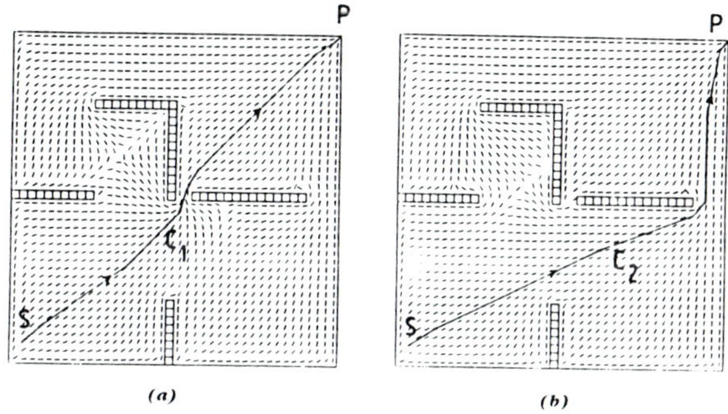

Figure 7: The vector field computed from eqs. (1) defines the trajectory between several compartments. (a) The shortest path with no size constraints. (b) if $m_2 < d < m_1$, C_2 is the shortest path.

example spatial extension of mobile object R, the door m_2 is too small and a suitable path must be found between point S and P. Such a situation is depicted in Fig. 7b. In this case the $\Delta\omega$ at P is chosen such that the waves only propagate through larger doors m_1 and m_3. These waves inhibit propagation from m_2. Again the vector field indicates that the path is the curve C_2 for every location S of mobile object R in the compartments.

4. Conclusion

In this paper we have described several unusual properties of target and spiral wave activity in a network of oscillators interacting with their nearest neighbours.

Let us note that in the limit of large numbers of units and small interunit distance the networks considered are a discretized version of reacting and diffusing systems. Thus it is expected that the properties seen here will be present also in non-stirred and nonlinear chemical media, and could be tested experimentally.

We believe such properties are not restricted to coupled oscillators only. Recently Destexhe and Babloyantz [6, 12] were able to show the presence of spiral as well as target waves in a model network of neurons. In these models the elements of the network are described by a linear process whereas all the nonlinearity results from interunit interactions. The network described in this paper is the other extreme case where the nonlinearity is due to the unit activity where the interunit interaction is linear.

We showed also with a simple example how the properties of target waves, notably their ability to distinguish openings of different sizes, could be used for path finding.

References

[1] G. Nicolis & I. Prigogine: *Self-organisation in Non-equilibrium Systems.* Wiley, New York (1977)

[2] H. Haken: *Advanced Synergetics*, Springer, Berlin (1983)

[3] A. Babloyantz: *Molecules, Dynamics and Life*, Wiley, New York (1986)

[4] For a recent overview, see e.g., Special Issue of Physica D, Vol. 49 (1991) Nos. 1 & 2, Selected papers from a meeting held in Pushchino, USSR, 28 May-1 June 1990, H.L. Swinney and V.I. Krinsky, (Eds)

[5] V. Castet, E. Dulos, J. Boissonade & P. De Kepper, Phys. Rev. Lett. 64 (1990) 2953

[6] A. Destexhe & A. Babloyantz, Deterministic chaos in a model of the thalamo-cortical system in: *Self-Organization, Emerging Properties and Learning*, A. Babloyantz (Ed.), New York, Plenum Press, ARW Series (1991)

[7] H. Haken in: *Neural and Synergetic Computers*, H. Haken (Ed.), Springer, Berlin (1988)

[8] J.A. Sepulchre & A. Babloyantz, Propagation of target waves in the Presence of Obstacles, Phys. Rev. Lett. 66 (1991) 1314

[9] J.A. Sepulchre, G. Dewel & A. Babloyantz, Propagation into Unstable States : The Direct Hopf Bifurcation, Phys. Lett. A 147 (1990) 380

[10] P.S. Hagan, Spiral waves in Reaction-Diffusion Equations, SIAM J. Appl. Math. 42 (1982) 762

[11] J.A. Sepulchre & A. Babloyantz, in preparation.

[12] A. Destexhe & A. Babloyantz, Pacemaker-induced coherence in cortical networks, Neural Comp. 3 (1991) 145

Synergetic Approach to Phenomena of Perception in Natural and Artificial Systems

T. Ditzinger, A. Fuchs, and H. Haken

Institute of Theoretical Physics and Synergetics,
University of Stuttgart, Pfaffenwaldring 57/4,
W-7000 Stuttgart 80, Fed. Rep. of Germany

1. Introduction

In this article we describe the synergetic approach to pattern recognition and associative memory as well as its extension to the modelling of phenomena observed in human perception. These are two examples of the application of phenomenological synergetics. In general, natural systems which show spontaneous self-organization can be divided into two classes: Systems for which the interaction at a microscopic level can be derived from first principles (like the laser or hydrodynamical systems), and systems where only the macroscopic phenomena are known. Examples of this class of systems which are treated with the methods of synergetics are the dynamics of spatio-temporal EEG patterns [2] and the so-called synergetic computers [1]. Synergetic computers are systems which we want to fulfill a special task, i.e. to act as an associative memory or to learn patterns which are offered from outside. Such systems can also be used to model phenomena in human perception, especially psychological experiments with ambiguous patterns as we shall describe below.

The article is organized as follows: First, we give a short survey of the standard model of pattern recognition and associative memory. This will lead us to one kind of universal dynamics which can also be observed in hydrodynamical systems. The application of the standard model to more complex patterns (scenes) shows that the algorithm can be generalized to be invariant with respect to shifts in real space, and that there are possibilities to control the dynamics. A further extension of this mechanism will lead us to the concept of time-dependent attention parameters in order to model the oscillations that occur in the human perception.

2. A Synergetic System for Associative Memory

2.1 The Basic Dynamical Process

The concept of associative memory [3] is one of the most powerful conceptual tools in the theory of pattern recognition. In 1987 Haken [4] introduced the standard model of the synergetic computer, a dynamical system which is able to perform the task of associative reconstruction of an input pattern to one of previously stored prototype patterns. The fundamental idea for the construction of such a synergetic computer is an energy-landscape where the temporal evolution from an initial pattern to the final prototype can be seen as an overdamped motion of a particle to one of the minima of the energy-function. The problem here, of course, is that these minima should correspond uniquely to the previously stored or learned patterns. It is well known that this is not the case for neural nets of the Hopfield-type [5] where so called spurious states occur – minima in the energy-

Springer Proceedings in Physics, Vol. 69
Evolution of Dynamical Structures in Complex Systems
Editors: R. Friedrich · A. Wunderlin

landscape which do not correspond to a prototype pattern. For the algorithm described below it can be shown that any stable fixed point represents one of the stored prototype patterns or its negative.

In order to obtain a mathematical formulation of this process, we consider a set of patterns as intrinsic in our system. These patterns are represented by vectors $\mathbf{v}^{(k)}$ of a high-dimensional space, where one component $v_i^{(k)}$ corresponds to the grey- or color-value of a specific pixel of a pattern. In the same way we code a starting or test-vector $\mathbf{q}$. Because the prototype vectors $\mathbf{v}^{(k)}$ will not be orthogonal in general, we introduce a set of adjoint vectors $\mathbf{v}^{(k)+}$ in such a way that the relations

$$\mathbf{v}^{(k)+} = \sum_{k'=1}^{M} A_{kk'}\, \mathbf{v}^{(k')} \tag{1}$$

and

$$(\mathbf{v}^{(k)+}\, \mathbf{v}^{(k')}) = \delta_{kk'} \tag{2}$$

are fulfilled.

The dynamics, i.e the temporal evolution of an initial pattern $\mathbf{q}(t{=}0)$, is given as an overdamped motion in the landscape of a potential and reads

$$\dot{\mathbf{q}} = -\frac{\partial V}{\partial \mathbf{q}^+} \tag{3}$$

with

$$\begin{aligned} V &= V_1 + V_2 + V_3 \\ V_1 &= -\frac{1}{2} \sum_{k=1}^{M} \lambda_k\, (\mathbf{q}^+\, \mathbf{v}^{(k)})^2 \\ V_2 &= \frac{1}{4}\, B {\sum_{k,k'=1}^{M}}{}' \, (\mathbf{q}^+\, \mathbf{v}^{(k)})^2\, (\mathbf{q}^+\, \mathbf{v}^{(k')})^2 \\ V_3 &= \frac{1}{4}\, C\, (\mathbf{q}^+\, \mathbf{q})^2 \end{aligned} \tag{4}$$

where the prime indicates $k \neq k'$ and $\mathbf{q}^+$ is defined by the relation

$$(\mathbf{q}^+\, \mathbf{v}^{(k)}) = (\mathbf{v}^{(k)+}\, \mathbf{q}) \quad . \tag{5}$$

If we assume $\lambda_k = \lambda > 0$, and $B > 0$, $C = \lambda > 0$ each minimum of the potential V corresponds to one prototype $\mathbf{v}^{(k)}$. The meaning of the single terms in the potential becomes clear if we plot V for only two prototypes as in Fig. 1. The first term, V_1, leads to the maximum in the middle and therefore to an unstable fixed point at $\mathbf{q} = \mathbf{0}$. V_3 is necessary for the global stability of the system and is called the 'saturation term'. A potential with only these two parts has the shape of a mexican hat. The term V_2, finally, leads to the saddle-points which separate the minima and is therefore called 'separation term'.

According to (3) the equation of motion for the test pattern reads

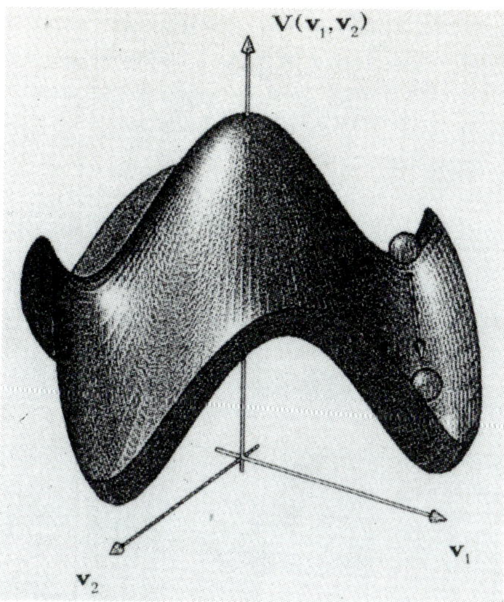

Figure 1: Shape of the potential for the case of two prototype patterns

$$\dot{\mathbf{q}} = \sum_{k=1}^{M} \lambda_k \left(\mathbf{v}^{(k)+}\,\mathbf{q}\right) \mathbf{v}^{(k)} - B \sum_{k,k'=1}^{M}{}' \left(\mathbf{v}^{(k)+}\,\mathbf{q}\right)^2 \left(\mathbf{v}^{(k')+}\,\mathbf{q}\right) \mathbf{v}^{(k')} - C \left(\mathbf{q}^{+}\,\mathbf{q}\right) \mathbf{q}. \qquad (6)$$

For a numerical simulation of this procedure we used a set of 10 digitized photographs of faces in a resolution of 60×60 pixels and 16 possible grey-values for each pixel. A capital letter as a symbolic name was added in the upper right corner of the patterns. From this set the adjoint patterns were calculated. The prototype patterns and their adjoint patterns are shown in Fig. 2.

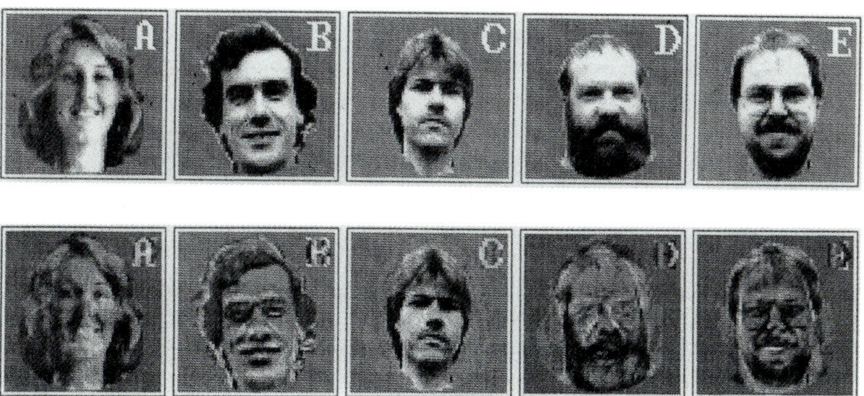

Figure 2: 5 prototypes from a set of 10 with their adjoint patterns

In Fig. 3 the associative property of the dynamical system (6) is shown using 3 different initial conditions: A part of a face, the name of a pattern, and a pattern that is disturbed by noise. In every case there is a complete restoration to the original prototype.

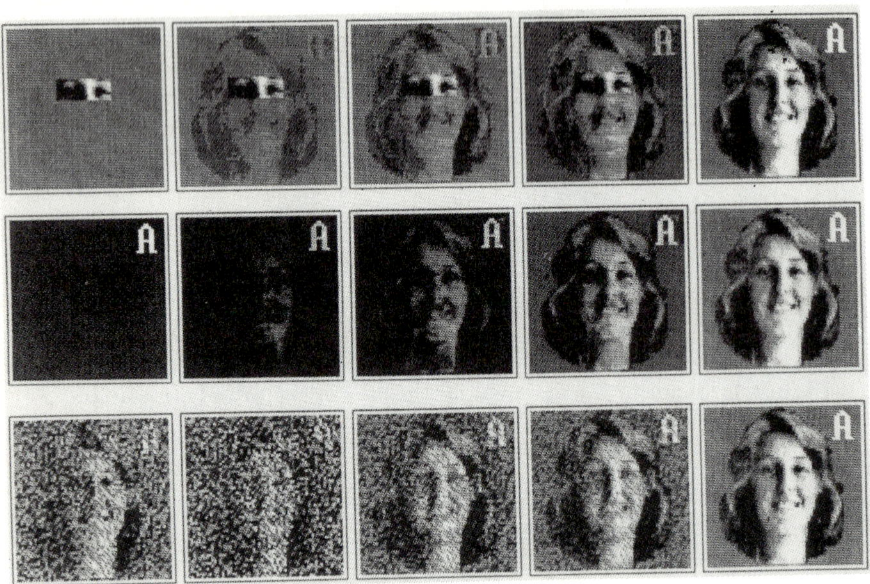

Figure 3: Associative recall of a pattern for 3 different initial conditions: A part of a face, the name of a pattern, and a pattern disturbed by noise

For the sake of completeness we show an example for the behavior of the system when λ is negative, i.e. below the critical point in Fig. 4. In this case every initially offered pattern disappears in time – the only stable state is $\mathbf{q} = 0$.

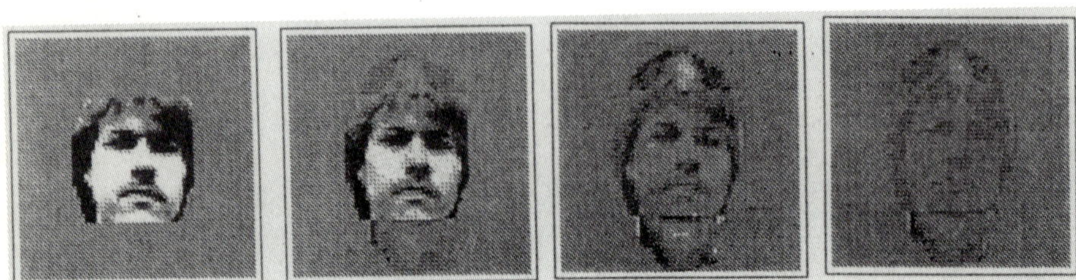

Figure 4: The system below threshold – the initial pattern disappears

2.2 Reduction of the Degrees of Freedom

So far we have treated the synergetic computer on a microscopic level where each pixel was represented by one component of a vector. This is a high-dimensional description and contains a lot of degrees of freedom which are not relevant for the dynamics from a macroscopic point of view. The synergetic theory for spontaneous pattern formation in nature tells us how a transformation can be performed in order to separate the relevant degrees of freedom, the order parameters, and the irrelevant degrees, the enslaved modes [6-8].

To this end we write the test pattern $\mathbf{q}(t=0)$ as a superposition of the prototypes $\mathbf{v}^{(k)}$ and a residual vector $\mathbf{w}(t)$ that collects the parts of $\mathbf{q}$ which do not lie in the subspace of the $\mathbf{v}^{(k)}$ by making the ansatz

$$\mathbf{q}(t) = \sum_{k'=1}^{M} \xi_{k'}(t)\, \mathbf{v}^{(k')} + \mathbf{w}(t) \tag{7}$$

Inserting (7) into (6) we obtain a system of equations for $\xi_k(t)$ and $\mathbf{w}(t)$. It turns out that the residual vector $\mathbf{w}(t)$ has no influence on the stationary state of the dynamics and can be neglected from the beginning. So we end up with a low-dimensional closed system of differential equations that describes the temporal evolution of the order parameters

$$\dot{\xi}_k = \lambda_k \xi_k - B \sum_{k' \neq k}^{M} \xi_{k'}^2\, \xi_k - C \sum_{k'=1}^{M} \xi_{k'}^2\, \xi_k \quad . \tag{8}$$

Fig. 5 shows the temporal behavior of the order parameters. As initial condition we used parts of two different faces. In the stationary state only one of the ξ_k has a nonvanishing value and corresponds to the identified pattern.

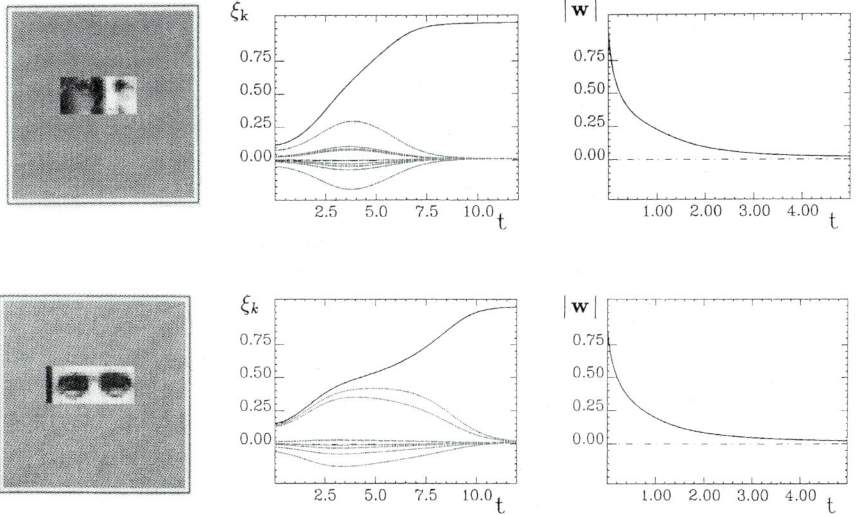

Figure 5: Competition of the order parameters for two different initial conditions

3. Pattern Formation and Pattern Recognition

In this section we want to point out that there exists a strong analogy between the spontaneous pattern formation in nature and the synergetic computer for pattern recognition described above. In order to show this connection we remind the reader of one of the most famous examples for spontaneous pattern formation – the Bénard experiment. If a fluid layer is heated from below and the temperature gradient inside the fluid exceeds a certain threshold, a convection pattern in form of horizontal rolls can be observed. If this experiment is performed in a rectangular vessel the orientation of the roll pattern is parallel to the shorter side. If we perform the same experiment in a circular vessel there are infinitely many directions which are degenerate, and one direction is selected by a spontaneous breaking of the circular symmetry of the system.

Mathematically, the dynamics of such a system is described by the Navier-Stokes equations, the equations of heat conduction and continuity. From these basic laws of hydrodynamics the so called Swift-Hohenberg equation [9] can be derived. That contains the interaction of rolls of different size and orientation and it is possible to obtain an equation that has the same structure as (8) and that describes the competition between rolls of different orientations (for details see [10]). For us the following points are important:

- The system has different internal states (rolls of different orientations) that correspond to the prototype patterns.
- If one direction is preferred due to a preparation from outside, the orientation of the rolls will be realized in that direction, i.e. one of the internal states (prototypes) is selected. This can be seen in the left and middle column of Fig. 6.
- If a superposition of two or more rolls with different orientations and different strengths is used as initial condition a competition takes place, and the initially strongest roll will eventually be realized (right column in Fig. 6).

So the fluid in this case acts as an associative memory that pulls an initial pattern into one of its internal states, in the present case rolls of coherent fluid motion. This shows that at the macroscopic level there is a unique description of the dynamics of systems of different kinds. Of course, equation (8) is only one example of a variety of equations which allow a classification of instabilities of different types.

4. Translational Invariance and Decomposition of Scenes

In this section we apply a preprocessing to our patterns in order to obtain a formulation of the algorithm that is invariant with respect to spatial shifts. To this end we apply a Fourier transformation to our prototype patterns to obtain a

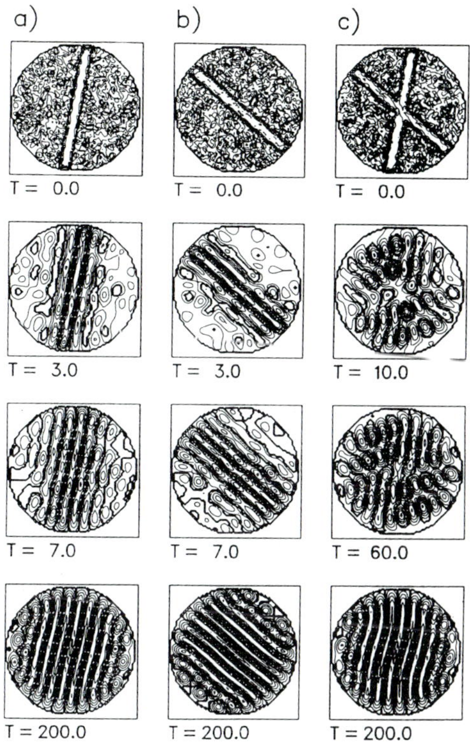

Figure 6: Competition of rolls with different orientations in a fluid layer. Due to its preparation, the system completes a roll pattern corresponding to the roll of the initial state (left and middle row). If a superposition of two orientations is used initially the dynamics leads to a selection and only one orientation survives

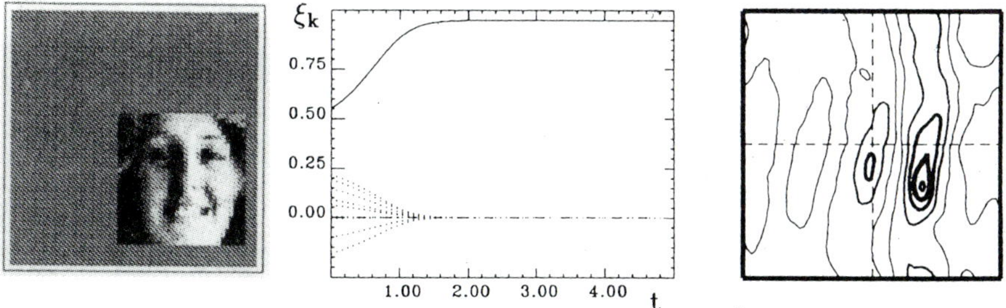

Figure 7: Simulation of the translational invariant algorithm.
left: Pattern in real-space – a part of a prototype shifted from its original location
middle: Temporal behavior of the order parameter ξ_k
right: Contour plot of the autocorrelation function

basis that consists of the magnitudes of the complex Fourier coefficients. These quantities do not depend on the spatial location of the patterns in real-space. From these vectors the adjoint basis is calculated in the same way as in Sect. 2.1. The vector that consists of the magnitudes of the Fourier coefficients from the initial test pattern is projected on the adjoint basis, leading to initial values for the order parameters ξ_k. Then the dynamics (8) is started and runs until it reaches a stationary state with only one nonvanishing ξ_l that corresponds to the identified pattern. An example showing a shifted input pattern and the behavior of the dynamics is given in Fig. 7. Now as the pattern is identified it is possible to calculate its original location which is determined by the absolute maximum of the correlation function $G(x_0, y_0)$

$$G(x_0, y_0) = \int\!\!\int_\Omega dx\, dy\; v^{(l)}(x - x_0, y - y_0)\; q(x, y, t{=}\infty) \quad . \tag{9}$$

A contour plot of $G(x_0, y_0)$ is shown in Fig. 7 (right).

Using this procedure it is possible to decompose complex scenes step by step. To this end a scene which consists for instance of one prototype in the foreground and one in the background as in Fig. 8 is offered to the system and the dynamics is started. When a stationary state is reached the most dominant pattern within the scene is identified (Phase I in Fig. 9). Then the eigenvalue λ_l that corresponds to this pattern is changed to 0 and the dynamics is restarted (Phase II in Fig. 9). Now the way to the former identified pattern is closed and the system identifies the pattern in the background (Phase III).

Figure 8: A foreground/background scene

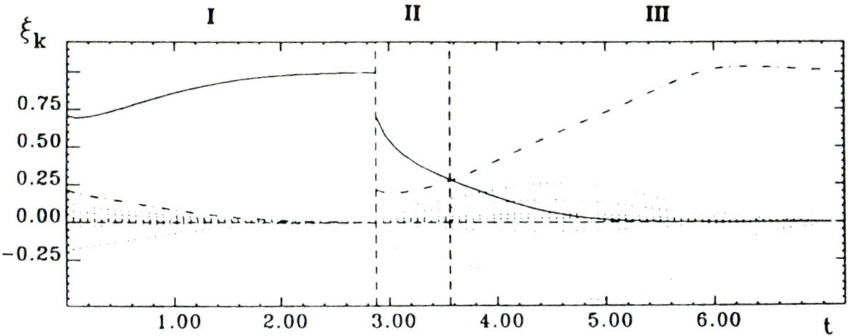

Figure 9: Decomposition of the scene of Fig. 8. The corresponding time dependence of the order parameters is depicted versus time

5. Time-Dependent Attention Parameters

So the human skill 'decomposition of scenes' is simulated with the idea of jumping parameters λ_k between the two values $\lambda_{low} = 0$ and $\lambda_{high} = \lambda = 1$. The parameters λ_k have a well known psychological correspondence: attention. If there is no attention to the foreground ($\lambda_{fore} = \lambda_{low}$), the background is recognized by means of a strong attention ($\lambda_{back} = \lambda_{high}$). Therefore the parameters λ_k are called 'attention parameters'.

In a natural further step to human-like pattern recognition we drop the assumption of the existence of only two allowed attention realizations λ_{low} and λ_{high} and introduce a continuous attention parameter range. Therefore, we profit by a psychological finding of Köhler [11]: he proposed an effect of time-dependent saturation of attention in the neural medium by means of a special physical model. Though not all psychologists accept Köhler's special model, there seems to be general agreement that the observed oscillations result from neuronal fatigue, inhibition or saturation. In accordance with this psychological model we establish an evolution dynamics for the attention parameters. The attention to the pattern k will saturate if its corresponding perception amplitude ξ_k increases. Thereby the term V_1 of our potential (4) becomes time dependent. Because of the invariance of our potential upon the replacement of ξ_k by $-\xi_k$, we wish to retain this property in the attention parameter equations:

$$\dot{\lambda}_k = \gamma \left(1 - \lambda_k - \xi_k^2\right) + F_k(t) \quad , \qquad k = 1, ...M \quad . \tag{10}$$

The attention parameters are subjected to fluctuating forces $F_k(t)$ which are assumed to have the following properties:

$$< F_i(t) >= 0, \tag{11}$$

and

$$< F_i(t)F_j(t') >= Q\delta(t - t')\delta_{ij} \qquad i, j = 1, ...M \quad . \tag{12}$$

In the following it is sufficient to restrict ourselves to two dimensions, M=2. An analytical treatment of our equations shows that an oscillation between the two fixed points

$\xi_1 = \frac{1}{\sqrt{1+C}}$, $\xi_2 = 0$, $\lambda_1 = \frac{C}{1+C}$, $\lambda_2 = 1$ and

$\xi_1 = 0$, $\xi_2 = \frac{1}{\sqrt{1+C}}$, $\lambda_1 = 1$, $\lambda_2 = \frac{C}{1+C}$

can occur. Oscillations will take place under the condition

$$B < 1 \quad . \tag{13}$$

Fig. 10 shows the time evolution of the order parameters ξ_1 and ξ_2 where the fluctuating forces have been taken into account. At this point we are able to compare the properties of our model with properties of pattern recognition by humans. Therefore we look at psychological measurements connected with so-called ambiguous patterns.

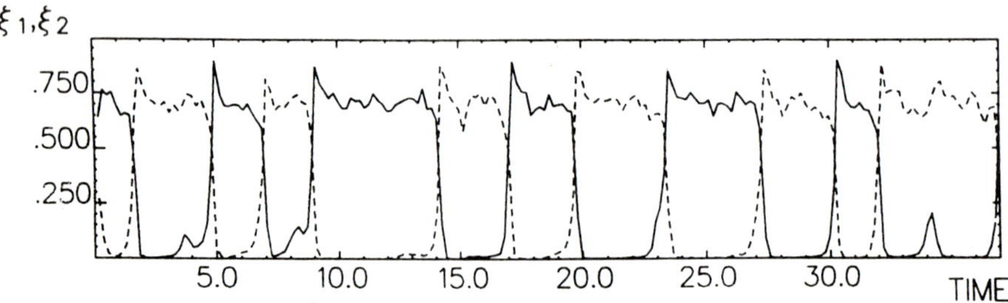

Figure 10: Temporal behaviour of the order parameters ξ_1 and ξ_2 with $B = 0.8$, $C = 1.0$, $\gamma = 0.1$, $Q = 0.001$. The solid and the dashed line refer to different interpretations of the ambiguous pattern. The oscillation of the perception can be clearly seen

6. Ambiguous Patterns

A famous example of an ambiguous pattern is provided by the Necker cube [12] (Fig.11) where a part of that figure is perceived either as front or back of a cube and the perception oscillates, i.e. the front is seen for a while and then suddenly changes to be perceived as the rear. These changes occur quite involuntarily and after a rather well-defined period of time. A good agreement of the measured characteristics with the order parameter characteristics of our model in Fig. 10 can be seen: stability followed by a sudden change of perception.

These phenomena of ambiguous patterns have aroused the interest of scientists for a long time. Therefore, a variety of experimental results have been obtained by different psychologists. An overview is given for example in [1,13-16] The measurements presented by different authors are very reproducible and reliable if they are obtained with respect to the same person, the same pattern, and the same influences of the surroundings, and if the method of measurement is the same.

Borsellino et al. [17] measured the fluctuations of the reversion time T, which is defined as the sum of the stable perception times $T_1 + T_2$ of the alternatives 1 and 2. The resulting characteristic probability distribution (frequency of occurrence versus reversion time) is shown in Fig. 12. The curve increases strongly for small reversion times and decreases slowly for larger times. In Fig. 13 the result of our model is depicted: we used the same parameter values as in Fig. 10 and plotted the frequency of occurrence of a specific reversion time for $N = 10000$ oscillations. The curve obtained corresponds very well, at least qualitatively, to the curve measured by Borsellino.

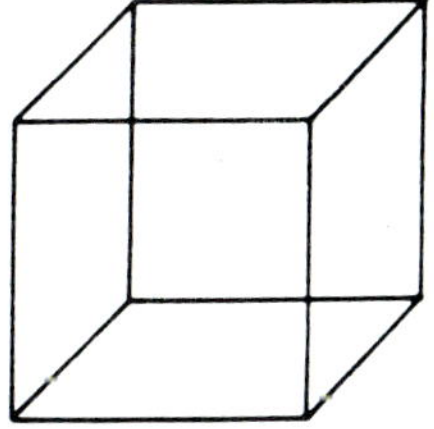

Figure 11: The Necker cube

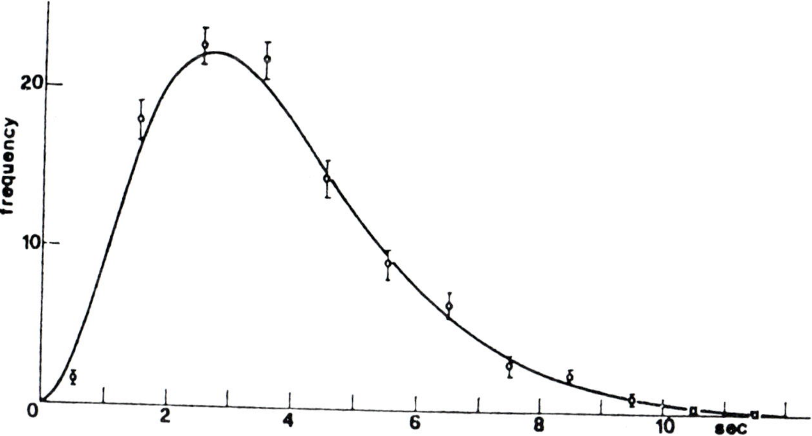

Figure 12: Distribution function of reversion times. The frequency of the occurrence of a specific reversion time is depicted versus that time (after [17])

7. Ambiguous Patterns with a Bias

In order to consider the full variety of ambiguous patterns a generalization has to be made: the alternatives of perception are usually not entirely equivalent. In most cases one alternative is preferred over the other. In order to model the intrinsic bias of a pattern we shift the ridge of the discrimination potential V_2 by an angle α. When we insert the explicit function of V_2 into (3) the order parameter equations in the case of two parameters read:

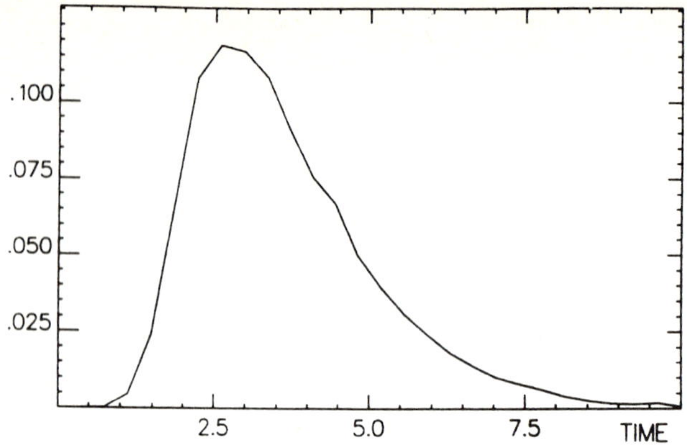

Figure 13: Theoretical distribution function derived from our model with the same parameter values as in Fig. 10. The meanings of ordinate and abscissa are the same as in Fig. 12

$$\dot{\xi}_1 = \xi_1 \left[\lambda_1 - C\xi_1^2 - (B+C)\xi_2^2 + 4B\alpha\xi_2^2 \left(1 - \frac{2\xi_2^4}{(\xi_1^2 + \xi_2^2)^2} \right) \right] \tag{14}$$

$$\dot{\xi}_2 = \xi_2 \left[\lambda_2 - C\xi_2^2 - (B+C)\xi_1^2 - 4B\alpha\xi_1^2 \left(1 - \frac{2\xi_1^4}{(\xi_1^2 + \xi_2^2)^2} \right) \right] \tag{15}$$

which allows us to simulate a large number of perception phenomena. Now oscillations will take place under the further condition:

$$-\alpha_{crit} < \alpha < \alpha_{crit} \tag{16}$$

where the value of α_{crit} is given by

$$\alpha_{crit} = \frac{1-B}{4B} \quad . \tag{17}$$

If the value of the bias parameter α reaches the critical limits $-\alpha_{crit}$ or α_{crit}, the bias is too extreme, so that the strong alternative is recognized with 100% certainty and an oscillation becomes impossible. The dependence of the reversion time on the bias parameter α is shown in Fig. 14, where the same parameter values as in Fig. 10 are used. In the upper part, the time evolution is shown for $\alpha = 0.02$, in the lower part for $\alpha = 0.04$. As can be seen, the individual reversion times $< T_1 >$ and $< T_2 >$ of the alternatives 1 and 2 are shifted in favour of the alternative 2 (dashed line) , i.e. $< T_2 >$ increases in agreement with psychophysical results. With increasing bias also the reversion time $< T > = < T_1 > + < T_2 >$ increases in agreement to the measurements. It can be shown that the following relationship between the average values and the bias parameter α exists:

$$\alpha = \alpha_{crit} \frac{< T_2 > - < T_1 >}{< T >} \tag{18}$$

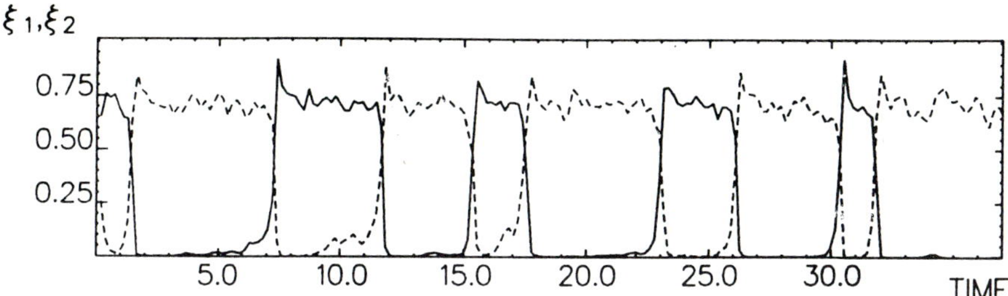

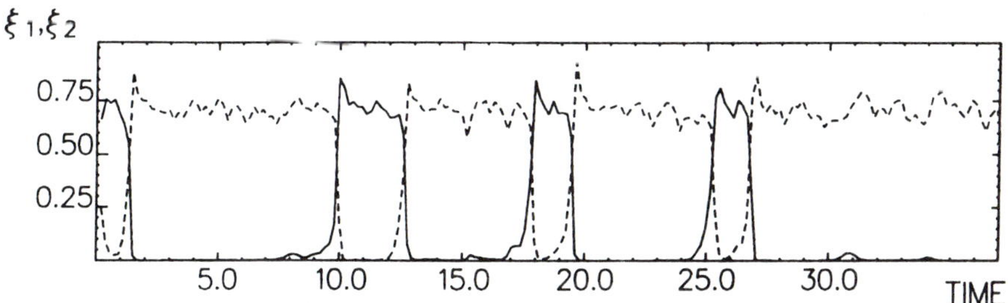

Figure 14: Temporal behaviour of the order parameters ξ_1 and ξ_2 for two different bias parameters (upper curve: $\alpha = 0.02$, lower curve: $\alpha = 0.04$)

and thus

$$< \frac{T_1}{T} >= \frac{1}{2} - \frac{\alpha}{2\alpha_{crit}} \quad , \qquad < \frac{T_2}{T} >= \frac{1}{2} + \frac{\alpha}{2\alpha_{crit}} \quad . \tag{19}$$

8. The Künnapas Experiment

In order to identify the bias parameter more precisely in connection with psychology we look at an experimental study by Künnapas [18]. He used a series of different Maltese-crosses as depicted in Fig. 15. This pattern can be either interpreted as a black cross on a white background or conversely as a white cross on a black background. Künnapas increased the angle of the black segment in small steps from the left to the right of Fig. 15. In this way the weight between the two interpretations can be changed continuously. The smaller area will be preferred, i.e. it will be interpreted as the pattern to be seen in the foreground (for details see [19]). Fig. 16 shows the results of the measurements by Künnapas for different subjects where the averaged reversion time of the alternative 'black cross on white background' is plotted versus the angle ω of the black segment from 5 to 85 degrees. The strong increase of the reversion time starting at about 85 degrees is particularly evident. In order to model this property we identify the angle ω of the black segments with the bias angle α. The different types of subjects are simulated by using different values of the parameter γ. The result of this simulation is depicted in Fig. 17: the averaged reversion time $< T_2 >$ is plotted versus α for three different values of γ with an excellent agreement with

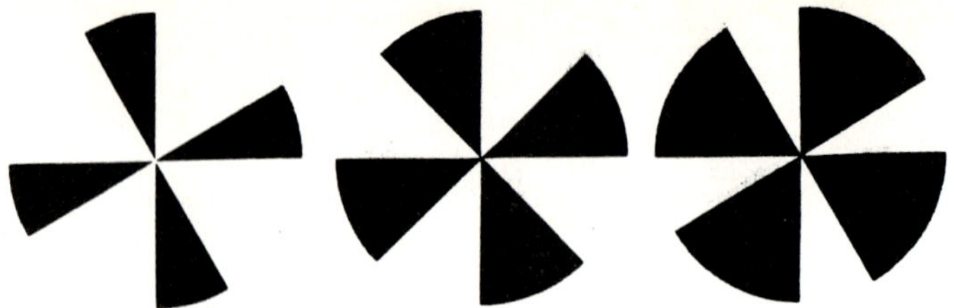

Figure 15: The Maltese crosses (from [13], see text)

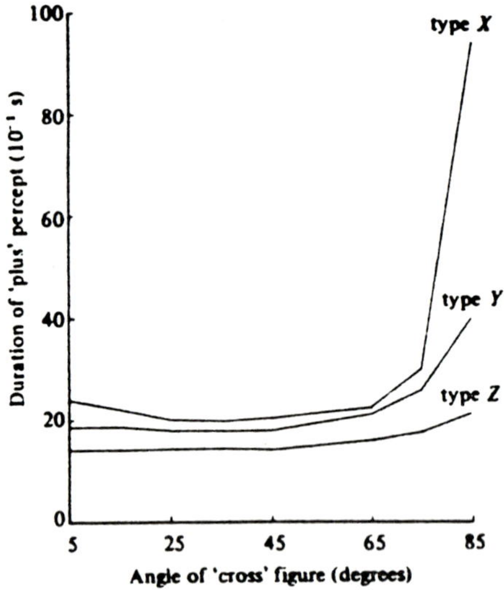

Figure 16: Mean reversion time of the 'black cross' percept versus angle of cross figures in degrees (after [18] and [13])

the results of Künnapas. A further result is the strong indication that there exists a linear relationship between the angles ω and α of the following kind:

$$\alpha = \alpha_{crit} \left(\frac{\omega - 45^o}{45^o} \right) \quad . \tag{20}$$

This relation can be obtained by gauging the values of α and ω at the points of strong increase of $< T_2 >$ in measurement (Fig. 16) and simulation (Fig. 17). Besides these results the role of the parameter γ can be identified as a rate of recognition speed of the individual subject, i.e. the higher the individual reversion rate, the bigger we shall choose γ.

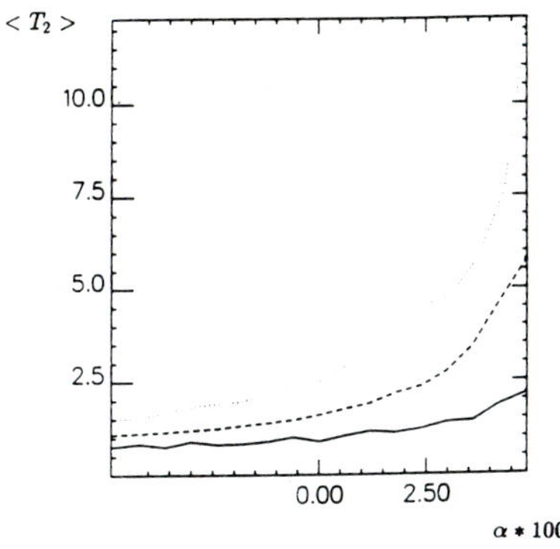

Figure 17: Simulation of Fig. 16 by means of our model with the following parameters: $B = 0.8$, $C = 1.0$, $Q = 0.0025$, $\gamma = 0.14$ (dotted), $\gamma = 0.2$ (dashed), $\gamma = 0.3$ (solid)

9. The Borsellino Experiment

In the next step we look at an experiment performed by Borsellino et al. [20] with a Necker cube as in Fig. 11. He varied the visual angle Θ, i.e., the size of the Necker cube shown to the subjects and studied the impact on the reversion times. The shape of the frequency distribution curve depends on the visual angle. While for small visual angles the distribution function has a narrow width, for increasing visual angles the width increases. This leads us directly to correlate the visual angle with the parameter Q of our model. This means in our interpretation that there are stronger fluctuations in the system with a bigger vision angle. This is certainly a rather plausible assumption: with a bigger vision angle there will be a greater probability that the surroundings will influence the attention parameters. With this psychological interpretation of the fluctuation parameter Q our model is fully characterized and is able to explain the measurements in dependence of the visual angle:

In Fig. 18 the averaged measured reversion times are plotted versus the visual angle for 10 different observers ([20]). There is a long plateau for small angles with relatively constant values for the averaged reversion times. As the vision angles increases in a higher range Borsellino was able to distinguish two types of observers according to their behaviour: for fast observers $< T >$ is little affected by Θ; for slow observers a strong increase of the measured time can be seen. This can be simulated by our model, at least qualitatively. In Fig. 19 the averaged reversion time $< T >$ is plotted versus the fluctuation strength. Again we simulated the different individual observers by the different values of the parameter γ. As in the measurement there is a relatively constant plateau for small values of Q. For higher values of Q we can again distinguish between two types of observers: for low values of γ, i.e. for slow observers, a strong increase of the reversion time can be seen, in contrast to higher values of γ, i.e. to faster observers. Comparing the position of strong increase of time in the simulation ($Q = 0.001$) to the measurements and

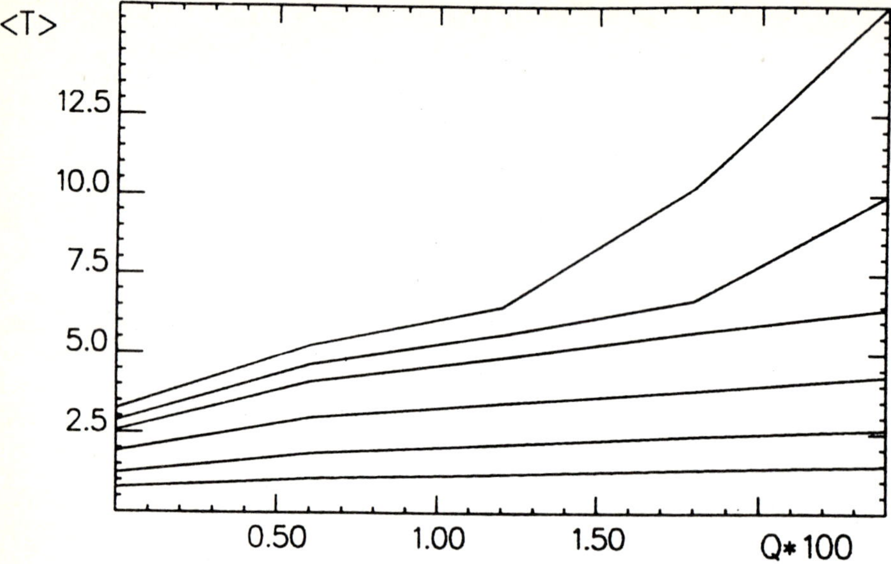

Figure 18: Average reversion times versus visual angle Θ for different observers (according to [20])

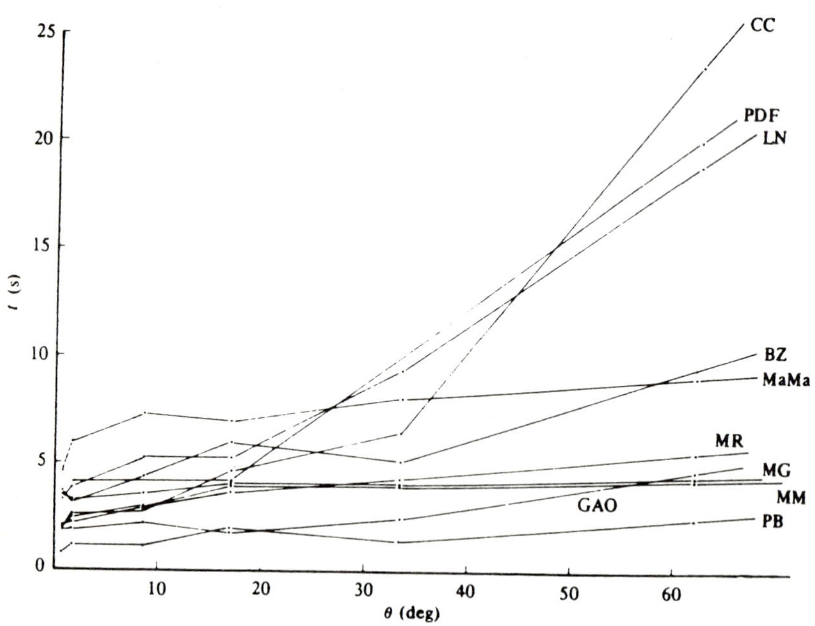

Figure 19: Theoretical curves of the average reversion times as a function of Q with $B = 0.8$, $C = 1.0$, $\alpha = 0.0$, $N = 10000$ oscillations and different sizes of γ: $\gamma_1 = 0.3$, $\gamma_2 = 0.2$, $\gamma_3 = 0.14$, $\gamma_5 = 0.1$, $\gamma_6 = 0.09$ in good qualitative agreement with Fig. 18

assuming a linear hypothesis we get a direct relationship between the fluctuation strength Q and the visual angle Θ: $\Theta = cQ$, with c about 3000 degrees.

By connecting the results of psychological measurements and the simulations of our model it was possible to assign the three parameters γ, α and Q psychological meanings:

- γ is a parameter specific for the individual persons and represents the frequency of the oscillations
- α is a parameter which depends on the pattern and measures the bias of the different alternatives
- Q is a parameter which here depends on the size of the pattern, irrespective of the alternatives.

There is a multitude of further psychological measurements which have been simulated by our model, such as hysteresis in perception, habituation and long scale time dependences, the probability of first-recognition of one special alternative, and the general case of multistable patterns with more than two alternatives.

10. Conclusion

In summary, we see that the ideas and methods of synergetics which are applied successfully in the investigation of pattern formation in systems far from thermal equilibrium are of fundamental importance in pattern recognition. Step by step we developed a synergetic system that models phenomena observed in the human visual system and its dynamics: We simulated the human ability of associative memory with a potential dynamics. Following the trail of natural pattern recognition, we made the system invariant with respect to translation, rotation and scaling of the patterns. We generalized the dynamics of our system to attention and introduced time- and recognition-strength-dependent attention parameters. By means of a connection of these generalizations it becomes possible to recognize complex scenes and to simulate many psycho-physical experiments on ambiguous patterns in great detail. Comparing the results of the model with the results of the measurements, we are able to supply the parameters γ, α and Q of the model with a psychological interpretation.

Acknowledgement

We thank R. Friedrich and R. Haas for valuable discussions and the Volkswagen foundation, Hannover, for financial support within the project on Synergetics.

References

[1] Haken H.: *Synergetic computers and cognition*, Springer, Berlin (1991)

[2] Friedrich R., Fuchs A., Haken H.: 'Modelling of spatio-temporal EEG patterns', in: *Mathematical approaches to brain functioning diagnostics*, I. Dvorak, A.V. Holden, Ed., Manchester University Press (1991)

[3] Kohonen T.: *Content-addressable Memories*, 2nd ed. 1988, Springer, Berlin

[4] Haken H.: 'Synergetic Computers for Pattern Recognition and Associative Memory', in: *Computational Systems, Natural and Artificial*, H. Haken, Ed., Springer, Berlin (1987)

[5] Hopfield J.J.: 'Neural Networks and Physical Systems with Emergent Computational Abilities', Proc. Natl. Acad. Sci. **79**: 2554-2558 (1982)

[6] Haken H.: *Synergetics. An Introduction*, 3rd ed., Springer, Berlin (1983)

[7] Haken H.: *Advanced Synergetics. An Introduction*, 2nd ed., Springer, Berlin (1987)

[8] Haken H.: *Information and Selforganization*, Springer, Berlin (1988)

[9] Swift J., Hohenberg P.C.: 'Hydrodynamic fluctuation at the convective instability', Phys. Rev. **A15**, 319-328 (1977)

[10] Bestehorn M., Haken H.: 'Associative memory of a dynamical system: The example of the convection instability', Z. Phys. B, **82**: 305-308 (1991)

[11] Köhler W.: *Dynamics in psychology*, Liveright, New York (1940)

[12] Necker L.A.: 'Observations on some remarkable phenomenon which occurs on viewing a figure of a crystal or geometrical solid', The London and Edinburgh Philosophical Magazine and Journal of Science **3**:329-337 (1832)

[13] Vickers D.: 'A cyclic decision model of perceptual alternation', Perception **1**:31-48 (1972)

[14] Attneave F.: 'Multistability in perception', Sci. Am. **225**:62-71 (1971)

[15] Kawamoto A.H., Anderson J.A.: 'A neural network model of multistable perception', Acta Psychol. **59**:35-65 (1985)

[16] Kruse P.: 'Stabilität, Instabilität, Multistabilität, Selbstorganisation und Selbstreferentialität in kognitiven Systemen', Delfin **11**:35-58 (1988)

[17] Borsellino A. et al.: 'Reversal time distribution in the perception of visual ambiguous stimuli', Kybernetik **10**:3, 139-144 (1972)

[18] Künnapas T.: 'Experiments on figural dominance', Journal of Experimental Psychology **53**:31-39 (1957)

[19] Metzger W.: *Gesetze des Sehens*, Kramer, 3rd ed., Frankfurt (1975)

[20] Borsellino A. et al.: 'Effects of visual angle on perspective reversal for ambiguous patterns', Perception **11**:263-273 (1982)

Part VI

Psychology and Social Sciences

Application of Synergetics to Psychology

G. Schiepek

University of Bamberg, Lehrstuhl für Klinische Psychologie, Markusplatz 3, W-8600 Bamberg, Fed. Rep. of Germany

Abstract

The present contribution illustrates some possible applications of synergetics to psychology. The success of Prof. Haken and his colleagues in explaining fundamental psychological functioning related to perception, pattern recognition and learning by means of synergetic modeling is readily apparent. In addition to these basic functions with which general psychology is concerned, examples will also be given of the application of synergetics to evolutionary patterns of schizophrenia and processes occurring within psychotherapy, research domains traditionally associated with psychiatry, abnormal and clinical psychology.

1. Introduction

There is no doubt that synergetics has become one of the most encompassing theoretical conceptualizations in modern science since it was introduced more than two decades ago (e.g. Haken 1970; Haken & Graham 1971). It represents a scientific concept which keeps its promise of offering an integrating and general approach. The precise mathematical formulation of its theoretical assumptions (Haken 1983a,b; 1990a; see Haken 1988 for macroscopic synergetics) forms the core of this approach. A large number of areas of application (domains), in addition to its original experimental paradigm, laser physics, were identified after these central formulations had been accordingly extended and specified, thus fulfilling the requirements of a structuralistic view of theorizing (e.g. Stegmüller 1973). Synergetics has been applied to such scientific fields as thermodynamics, fluid dynamics, and to phenomena of collective and dynamic order formation in chemistry, biochemistry, meteorology, biology (e.g. morphogenesis and population dynamics), neurobiology, cognitive science, sociology, and economics. Even the basic disciplines of psychology have already begun to model phenomena of perception, cognition and psychological development, all of which exhibit characteristics of complexity and discontinuous dynamics (phase transitions), using synergetic conceptualizations (see Haken 1992; contributions to Haken

Springer Proceedings in Physics, Vol. 69
Evolution of Dynamical Structures in Complex Systems
Editors: R. Friedrich · A. Wunderlin

& Stadler 1990, e.g. Stadler & Kruse 1990, Kruse & Stadler 1990). As has been pointed out (Stadler & Kruse 1990), there are obvious parallels to German Gestalt psychology (Köhler, Metzger, Wertheimer, Koffka and others, e.g. Köhler 1920; 1940; Wertheimer 1912; 1923).

Haken himself suggested the application of synergetics to psychology. It serves as a metaphor, tool and model of nonlinear phase transitions in human perception, experience and behavior.

2. The Application of Synergetics to General Psychology: Basic Motorics, Perception and Cognition

2.1 The Finger Movement Paradigm

Let us first consider a simple example, the finger movement paradigm, which is based on an experiment by Kelso (1984). In this experiment the subjects were asked to move their index fingers in parallel, which is easily possible when done slowly. However, when the subjects were asked to increase the speed of their finger movement, the parallel movement involuntarily turned into a symmetric one beyond a certain critical frequency. A new mode of behavior had developed. This phase transition can be precisely modeled by synergetics (see Haken, Kelso & Bunz 1985; Schöner, Haken & Kelso 1986).

2.2 Gestalt Perception and Autonomous Structuring of Perceptual Fields

Another experimental field for the application of synergetics is that of

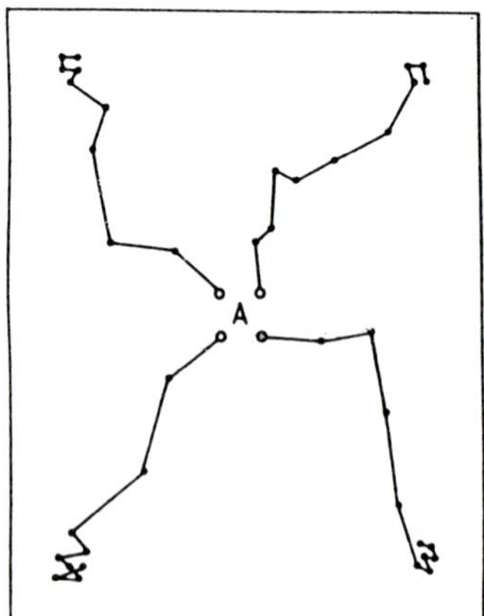

Fig. 1: The phenomenon of the wandering point (from Stadler et al. 1991).

structured perceptual fields in quasi-homogeneous stimulus areas. Evidence of such perceptual fields or attractors results from a simple but very creative experiment, the so-called Bartlett scenario (Stadler et al. 1991; Stadler & Kruse 1990). A subject is exposed to a white sheet of paper (without lines or lattice) with a single point near the intersection of the diagonals for a short time (less than one second). The subject is then asked to reproduce the perceived localization of the point on another white sheet of paper of the same size. This sheet is shown to the next subject, who is asked to repeat the same procedure; and so forth. After several iterations the point usually wanders to one of the four corners of the sheet of paper (fixed-point attractor), depending on minimal differences in the original position of the point (see Fig. 1). Different attractors would seem to exist, depending on minimal changes in the initial and boundary conditions. The results of this experiment are usually interpreted as evidence in favor of non-homogeneous perceptual fields as postulated by one of the best-known laws of Gestalt psychology, namely the tendency towards *Prägnanz*. "The changes in the structure from reproduction to reproduction, as observed in Bartlett scenarios, may be interpreted in Gestalt psychological terms as an effect of the tendency towards *Prägnanz*, which is nowadays sometimes identified with the tendency towards stability" (Stadler et al. 1991, p. 103; see also Kanizsa & Luccio 1990).

A recent elaboration of this experiment is described by Stadler, Kruse and colleagues (Stadler et al. 1991). To investigate the hidden psychological field structure of a homogeneous area, a sheet of paper was covered by an (invisible) square net with a mesh size of 10 mm. The dimensions of the net were such that $x=21$ multiplied by $y=29$ intersections marked the positions of the points on 609 different stimulus sheets (Fig. 2a). The 609 stimuli were exposed in random order for about one second each to the subjects, who had to reproduce the perceived position of the stimulus point immediately after its presentation on a pressure-sensitive table of the same size and orientation as the stimulus sheets. Here again, an effect of attraction similar to that observed in the Bartlett scenario was found. Altogether, ten subjects participated in the experiment conducted by Stadler et al. (1991), and Fig. 2b collects the positions of the 609 points reproduced by one subject. The effect became even more distinct when, after completion of the presentation of the 609 stimuli, the subjects were asked to draw on a white sheet of paper the place where in their opinion the stimulus points had been presented (Fig. 2c).

The results are interpreted in favor of a self-organization theory of perception (Köhler 1920), which "assumes the perceived order to be a product of the inner dynamics of the central nervous system, while the stimulus is seen as a rather meaningless initial condition" (Stadler et al. 1991, p. 103). In accordance with the assumptions of Gestalt psychology, the visual field is conceived of as a product of the intrinsic dynamics of the central nervous system, the visual field being defined as a *vector field*. "A vector field is a field in which every point is characterized by both a magnitude and a direction"

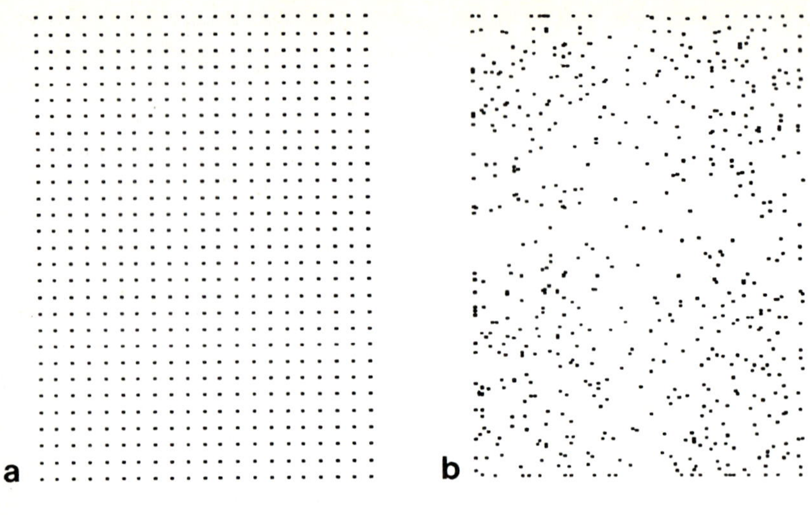

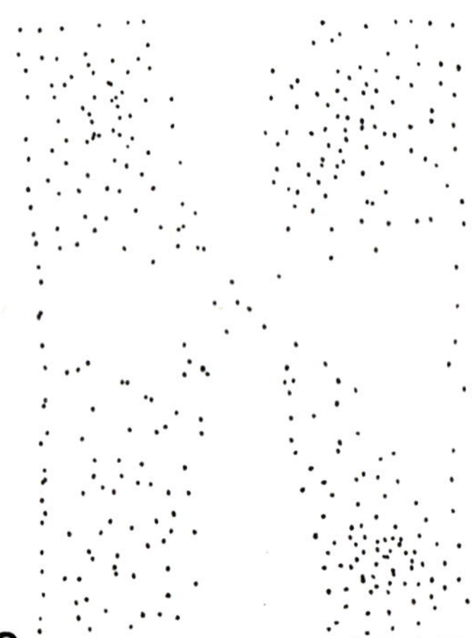

Fig. 2: (a) The stimulus matrix of the 609 points presented successively in random order to the subjects. (b) Reproductions of the stimuli by subject 9. (c) Reproductions of the supposed distribution of the stimuli on the DIN A4 sheet by the same subject (from Stadler et al. 1991).

(Brown & Voth 1937). The perceptual bias - or, rather, the perceptual organization of the subjects - can thus also be represented by means of a vector field. The mean vectors of all ten subjects result in the vector field given in Fig. 3. It is possible to reproduce the empirically obtained data by means of a mathematical analysis of the vector fields. A vector field can be represented as the sum of a gradient field and a rotational field, yet in the analysis carried out by Stadler et al. only the gradient components but not the circulation components, are taken into account. Figure 4 shows the gradient field calculated for the averaged data sets of all ten subjects, and the corresponding potential landscape is given in Fig. 5. The resulting potential landscape exhibits four minima near the four corners of the base area, which correspond precisely to the attractors of the moving points. One can easily imagine that such a minimum is reached following a double symmetry breaking (bifurcation): a fictitious ball at the top of the potential landscape has to "decide" whether to roll (a) to the front or the back and (b) to the left or the right side of the landscape. In this process, minimum fluctuations may have a decisive impact on the ball's direction at points of critical instability.

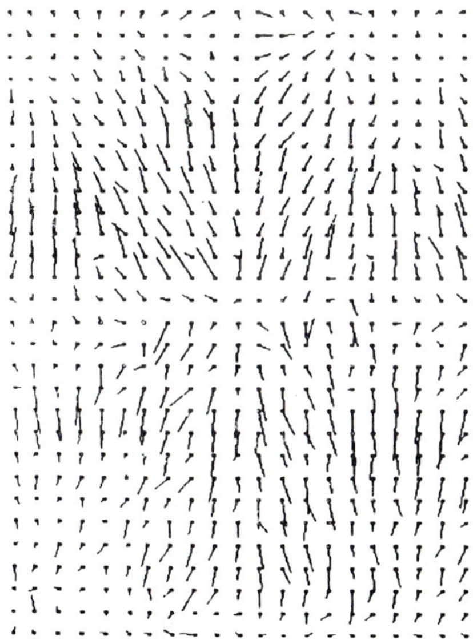

Fig. 3: The arithmetic mean of the displacement vector coordinates (10 subjects) (from Stadler et al. 1991).

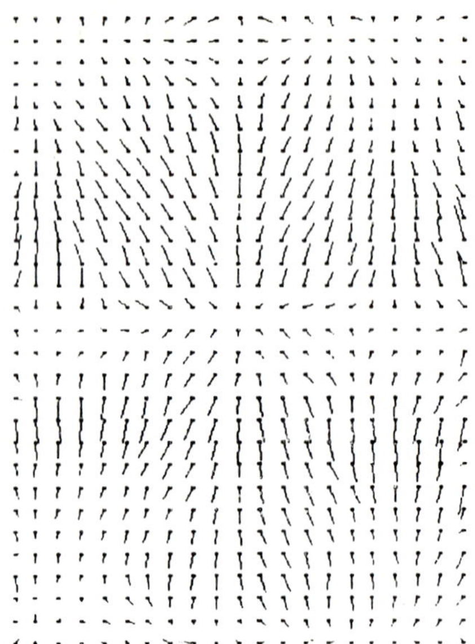

Fig. 4: The gradient fields calculated for the averaged data sets of all 10 subjects (see Fig. 5) (from Stadler et al. 1991).

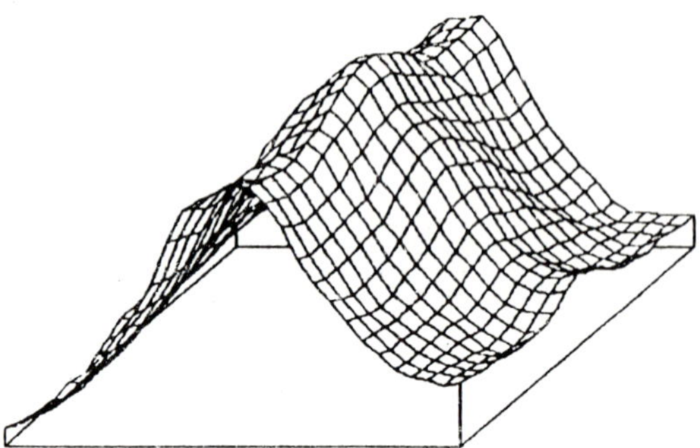

Fig. 5: The calculated gradient potential of the averaged empirical vector fields of all 10 subjects (from Stadler et al. 1991).

2.3 Recognition of Ambiguous Patterns

It can be assumed that, in addition to the empirical findings on perceptual field dynamics described here, every phenomenon of gestalt perception may well be explained in accordance with the concepts offered by synergetics (Stadler & Kruse 1986). Gestalt perception presupposes a self-organization theory of brain dynamics, as we are obviously concerned here with emergent order formation which can be explained neither by the physical properties of the given stimuli nor by the functioning of isolated neuronal networks (such as visual receptors) alone (compare Kanizsa & Luccio 1990; Kruse & Stadler 1990). It is especially with regard to questions concerning the recognition of ambiguous patterns - an old, central field of interest of perception research - that synergetics has contributed modeling approaches by means of which numerous phenomena arising from experimental research can be explained. The significance of the synergetic approach can hardly be overestimated, as it allows us to explain these phenomena without detailed knowledge of the precise rules governing the underlying processes on the neuronal, microscopic level, rules which cannot be easily modeled anyway due to their enormous complexity. It is the approach of macroscopic synergetics (Haken 1988) which allows a tremendous augmentation of the areas of application of synergetics, namely to systems whose functioning on the microscopic level is not known in detail.

2.4 The Paradigm of Pattern Recognition

There can be no doubt that one of the most interesting fields within perception psychology to which synergetics is applied is the synergetic computer (Haken 1987; 1991b, c). This mathematical model of human perception allows for an adequate simulation of processes of pattern recognition: that is, the predictions made by the model are in agreement with empirical findings. The computer model is capable of recognizing complex patterns such as human faces, even when they are noisy, distorted, partly covered or presented as seen from different angles.

I would like to point out a few remarkable similarities between the synergetic computer, on the one hand, and, on the other, connectionist models and the approach of Parallel Distributed Processing (McClelland & Rumelhart 1986; Rumelhart & McClelland 1986; Strohschneider & Schaub 1991; Schaub & Schiepek 1992), both of which are central paradigms of Artificial Intelligence research. In both cases, the pertinent computer models are capable of recognizing the "correct" figures starting from incomplete or mutilated patterns, based on processes of associative memory. In both cases, emergent psychological processes of self-organization on a macroscopic level, which can be interpreted in semantic terms, are simulated on the basis of processes on a neuronal, microscopic level, which cannot be interpreted in psychological terms. And finally, both approaches allow for learning processes by the "recognizing" systems, which are modeled in connectionist networks and synergetic computers via learning-dependent modifications of the activation rules of the neural net between the idealized neurons and via feedback effects of the perception process on the attention parameters, the degree of saturation, the learning matrix or the adaptation operation, respectively.

In the future, we may expect a fruitful exploitation of synergetic pattern recognition with regard to research into emotions. Since the synergetic computer is capable of differentiating not only between faces, but also between facial expressions (such as the expression of joy, anger or sorrow), it can be utilized to encode different emotional expressions. After a suitable examination of the validity of this procedure, it should allow for an on-line encoding of emotional states, for example during group discussions or psychotherapy sessions.

All these examples reveal the immense contribution made by Hermann Haken and colleagues to an improved understanding of highly complex psychological processes, especially with regard to human perception, cognition and information processing. Furthermore, the founder of synergetics has endowed psychology and psychiatry with an invaluable theoretical model and methodological tool, one which may be exploited in the investigation of the various different phenomena and problems with which these disciplines are concerned. In my opinion however, the mainstream of these disciplines has not yet fully recognized the significance of the synergetic approach. Synergetics is of such great importance for psychology and psychiatry because it (1) provides formal, mathematically elaborated, theoretical instruments, which at the same time do not impede but rather promote a qualitative understanding of the investigated phenomena, and (2) thus makes ambitious research possible, enabling a successful confrontation in a non-reductionist manner of the problems posed by high complexity and intricate dynamics. Last but not least, synergetics (3) enables us to identify phenomena of order formation and transformation (phase transitions), stability and instability, highly structured or chaotic dynamics, all of which represent the central difficulties of both a scientific and a practical nature with which these disciplines are faced. Such phenomena of self-organization can be identified at every possible level under consideration for healthy and pathological states, from neuronal and biochemical proceses within the individual to psychological and social systems and, finally, to processes encompassing society.

In the following I would like to discuss two possible areas of application which are of great practical significance for psychology, namely (a) the etiology of mental disorder, especially schizophrenia, and (b) psychotherapy.

3 Application of Synergetics to Abnormal Psychology and Psychotherapy

3.1 The Evolution of Dynamical Patterns in Schizophrenia

3.1.1 Mental Disorders as Dynamical Diseases

Mental suffering or "disorder" has been conceived of in terms of highly structured coherent states which impair and determine an individuals's mental and social being since as far back as the classical phenomenological approach to psychopathology; put differently, these states are seen as considerably reducing the individual's "degrees of freedom". Very often we can observe discontinuous transitions between different pathological states or between states of health and illness which resemble those nonlinear phase transitions observable within highly complex physical systems (compare Haken 1989). This has been demonstrated, for example, with regard to uni- or bipolar cyclic depression and schizophrenia. Strauss (1989) emphasizes the discontinuity in the evolution of patient functioning over time and describes several temporal patterns found in schizophrenia (e.g. "woodshedding", "the low turning point", "oscillating levels of function"; see also Schiepek, Schoppek & Tretter 1992). It therefore seems obvious to attempt the utilization of synergetic concepts with respect to etiological research and dynamic psychopathology. After all, synergetics presents an elaborate theory of disorder-to-order as well as order-to-order transitions within highly complex nonlinear systems. Moreover, this view is supported by the fact that the principle of strong causality obviously does not apply, with respect to etiological processes, something which is indicated by the unspecific effects of causal factors such as traumatic childhood experiences or life events.

Different initial conditions and different degrees of stress may result in similar pathological states (see for example Akiskal & McKinney 1975, regarding the concept of the "Final Common Pathway" of the evolution of depression), while only minimum fluctuations within an individual's intrapsychic or environmental conditions may also lead to quite different effects (compare the concept of "critical instability", Schiepek & Schaub 1991; Schaub & Schiepek 1992, see Fig. 6). Obviously, this represents two characteristic features of strange attractors: the postulate of convergence (attraction) and the postulate of divergence (expansion) (see Tschacher 1990).

Of special theoretical and therapeutic interest is the fact that, despite dramatic changes in the behavior of functional variables, the underlying control system may still be intact, with the exception of a single (or a few) parameter(s) being out of its (their) physiological ("healthy") range. This has already been shown to be the case in physiological and biochemical systems, (compare the concept of dynamical diseases, Mackey & an der Heiden 1982; an der Heiden & Mackey 1987; Glass 1987; for a comprehensive survey see Rensing, an der Heiden & Mackey (eds.) 1987). It seems worthwhile to attempt an application of this promising model to schizophrenic phenomena.

This could in fact be very rewarding, as various distinct longitudinal patterns characterized by discontinuous phase transitions between psychotic episodes, healthy functioning and chronic states have been detected in the long-term course of schizophrenia (cf. the findings of Bleuler 1972; Ciompi & Müller 1976; Huber, Gross & Schüttler 1979; Harding 1984). Investigations have shown that in the long run one can expect about 25% complete remissions, about 40-50% moderate residual states and only about 35% cases of permanent, severe, mental invalidism. There can be no doubt that an

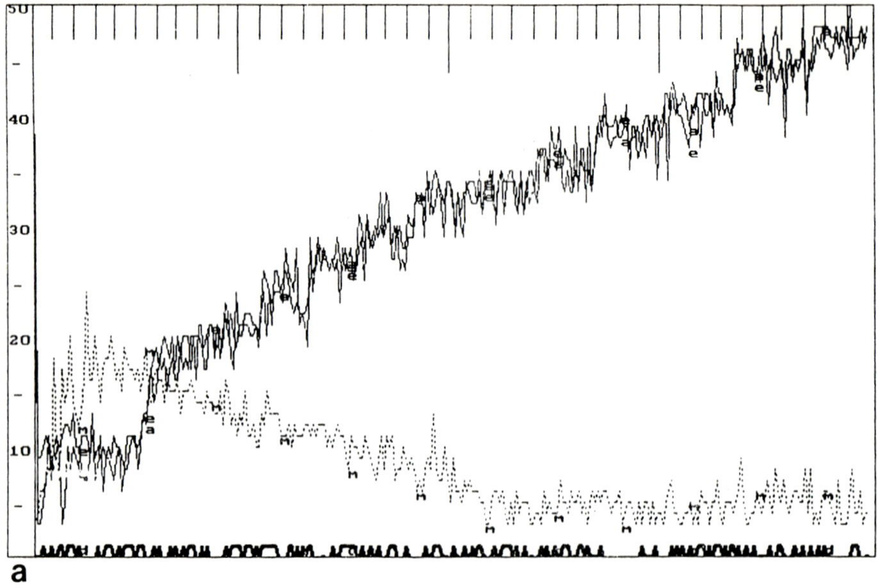

a

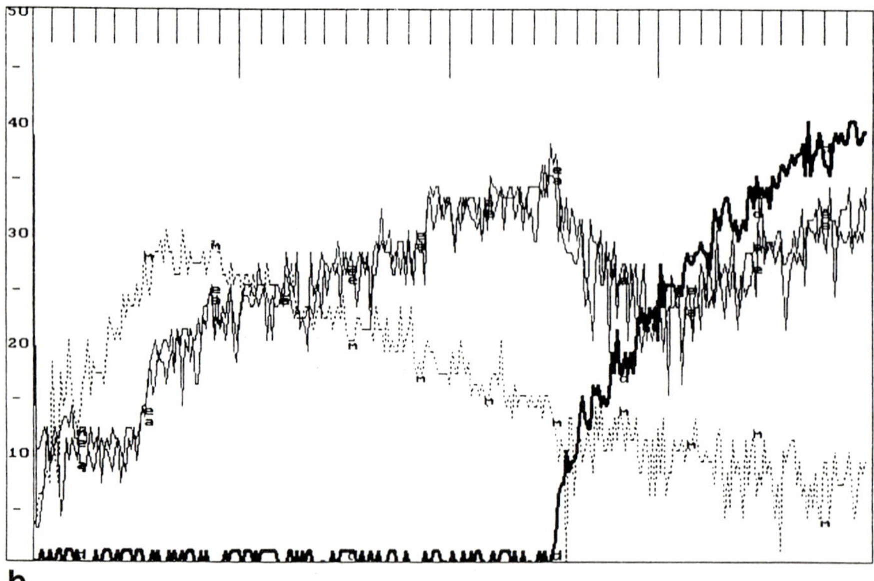

b

Fig. 6: Developments of the variables "demands"(a), "realization"(e), "motivation"(m) and "depression"(d). (a) The mean for "demands" was kept at value 10 for the first 500 time-steps. (b) The mean for "demands" was kept at value 11 for the first 500 time-steps. The consequences of this minimal starting difference for the further system development become apparent (critical instability).

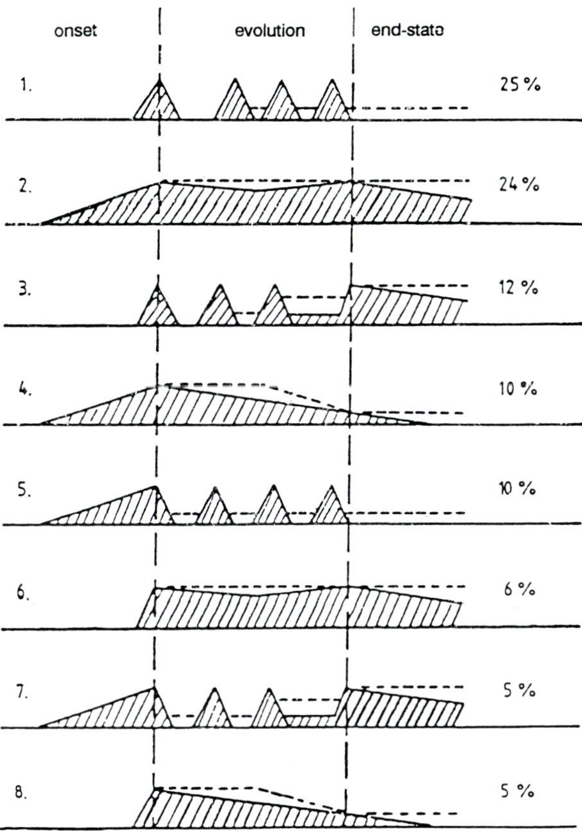

Fig. 7: Different patterns in the long-term evolution of schizophrenia. N = 228. Catamnestic time: 36.0 years (mean). (Ciompi & Müller 1976; Fig. from Ciompi 1989, p. 16).

improved care system and more differentiated therapeutic and rehabilitative facilities contribute to these results. The different patterns of the long-term evolution of schizophrenia as found by Ciompi & Müller (1976), together with their frequencies, are shown in Fig. 7 (compare Ciompi 1989). These patterns are the result of various combinations of slow vs. acute onset, acute episodes vs. progressive deterioration, and remission vs. chronic end-state. These patterns are the empirical basis, which will be explained (explanandum) by the simulation model presented in this section.

3.1.2 Development of a Structural Model

The etiology of schizophrenia has been the subject of research in various quite disparate disciplines: phenomenological and experimental psychopathology, psychology, physiology, biochemistry, neurobiology, communications and family research, psychoanalysis, sociology and ethnology, to name only a few. Each of these disciplines focuses on different aspects of the highly complex biopsychosocial system of "schizophrenia", so that it is hardly possible any longer to maintain an overall perspective. Accordingly, the selection of variables and parameters given below does not claim completeness. At the same time, however, it forms the basis for a structural model which does not postulate the status of a theory. The variables have been selected on the basis of reviews of psychiatric and psychological research from recent years (e.g. Böker & Brenner 1986; 1989), as well as the presentation by Ciompi (1989). Each of the constructs included could be regarded as a system of its own and analyzed at a still finer level of resolution (see also the concept of the microlevel approach as a virtual horizon of the high resolution of a system in psychological synergetics, Schiepek & Tschacher 1992).

The model interrelates five different variables or constructs (for details see Schiepek, Schoppek & Tretter 1992):

1) Cognitive disorders (c)
2) Stress (s)
3) Withdrawal (w)
4) Expressed Emotion (e)
5) Delusion (d)

Constructs which are invariable or subject to very slow modifications over the course of time (as opposed to the temporal dynamics of variables) take on the function of parameters in the equations proposed below. Their purpose is to mediate in specific ways the interactions of the variables. The following parameters are included:

1) *Diffuseness of affective-cognitive schemata* (σ). The construct of affective-cognitive frames of reference (or schemata) lies at the core of Ciompi's conceptualization of the "Affektlogik" (Ciompi 1982; 1986a, b). Programs for thinking, feeling and behaving are internalized as a condensation of the entire experience of an individual: external dynamics become internal structure (Ciompi 1986a, pp. 51ff.). Due to the diffuseness and contradictions experienced in interpersonal encounters with close relatives, the affective-cognitive systems of reference which mediate emotional experience and interpersonal behavior seem to be extremely diffuse and unclear. Interestingly, the diffuseness of these systems of reference evidently corresponds to certain phenomena which can be discovered in the neuronal substratum and are explained by the concept of "neural plasticity" (Haracz 1984;1985).

2) *Dopamine and serotonin metabolism* (δ). An important mediator in the evolution of schizophrenia seems to be the metabolism of neurotransmitters such as dopamine, serotonin and norepinephrin. Findings indicate an overactivity of the dopaminergic and serotonergic system, evidence for this including the blocking of dopamine receptors by neuroleptics. Recently developed drugs (e.g. pipamperone, setoperone, ritanserin, risperidone) affect the serotonin metabolism in an antagonistic manner (Gelders 1989). Dopamine metabolism is activated by amphetamine-like drugs, which can provoke psychoses. Modifications in the plasma level of the dopamine metabolite homovanillin acid parallel the intensity of psychotic symptoms, and increased levels of serotonin have been found in chronic schizophrenics. A recent finding is the increased number of dopamine receptors in schizophrenics post-mortem. Also, observations have shown that dopamine metabolism is greatly affected by stress (Haracz 1984;1985; Trulson & Preussler 1984). On the basis of these findings, dopamine may be assumed to be an important mediator between environmental stimuli and the cerebral substratum, thus gaining theoretical significance particularly with respect to concepts emphasizing the relationship between emotional reactions on the one hand and cognitive performances on the other (Ciompi 1982).

3) *Social (in)competence* (κ). As in the case of σ and δ, the high values of the parameter social competence κ (indicating a high degree of incompetence) suggest an increased vulnerability to schizophrenia. Prior to the onset of acute psychosis, vulnerable individuals seem to be characterized by deficiencies regarding their social skills. This is especially true in cases of poor premorbid adjustment. Social skills can hardly be developed to a sufficient degree within stressful family environments and contradictory communication structures accompanied by diffuse cognitive schemata for social perception and interpersonal interaction. Therefore, various treatment programs are directed at improving the social perceptual and interactional skills of schizophrenic patients.

4) β_c 5) β_s 6) β_w 7) β_e 8) β_d: Within the mixed feedback model proposed here, the extent of negative feedback is mediated by means of these five β-parameters: that is, increased values for β indicate stronger damping effects on the pertinent variable. Similar parameters are employed within models of population dynamics (cf. Krebs 1985; Odum 1983; Kriz 1990) to determine the effects of limited resources (resulting from competition or high population density, for example) on the decrease of population density. With regard to the dynamics of schizophrenia, high values of β_c, β_w and β_d, referring to cognitive disorders, withdrawal and delusions respectively, could be interpreted as an antipsychotic tendency on the way to recovery, whereas high values of β_s and β_e, referring to stress and high expressed emotion respectively, would indicate a natural relaxing tendency.

9) *Genetic risk* (γ). Our model does not provide for specific and direct influence by this parameter. Rather, it is assumed that certain genetic factors constitute the background for the vulnerability expressed by the parameters σ, δ and κ to evolve.

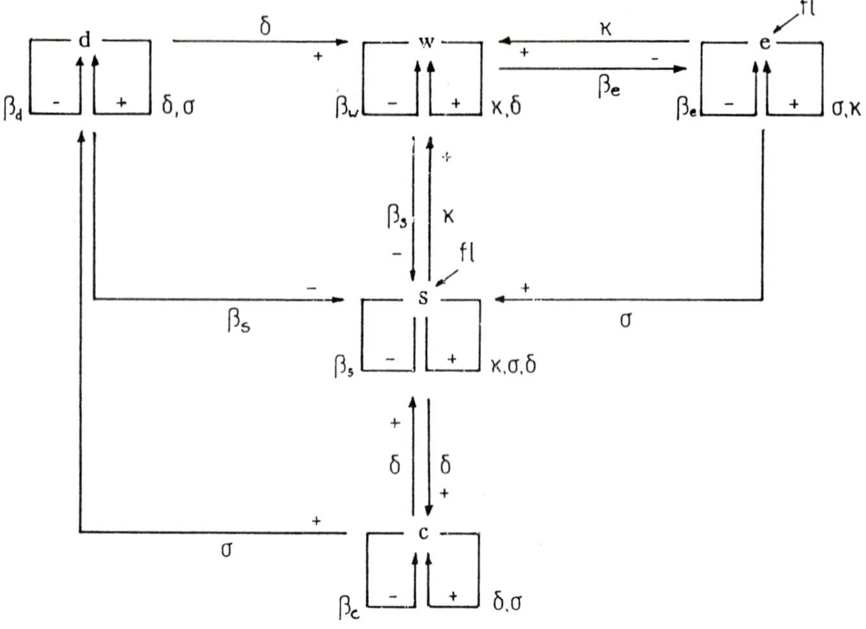

Fig. 8: The structural model as a basis for simulation.

The following description of the interactions between the variables can be compared directly with the difference equations (2) through (6) given in section 3.1.4, and also with Fig.8.

1) It is proposed that an increase of *cognitive disorders* (*c*) depends linearly on the prevailing intensity of cognitive disorders, and is mediated by the parameter σ. This is to say that the higher the degree of disturbance of affective-cognitive schemata for the perception and processing of internal (intrapsychic) and external (environmental) events, the smaller the possibility that the patient will comprehend and cope with perceived anomalies of cognitive functioning, and the greater the chance for an intensification and escalation of the cognitive disorders. In addition, cognitive disorders interact nonlinearly with stress. On the one hand, they evoke anxiety and insecurity and are thus stress-inducing; on the other hand, they are activated by stressful events. The interaction is mediated by the dopamine and serotonin metabolism (parameter δ). This is in agreement with findings indicating that medication reducing serotonin and dopamine activity has a positive effect on cognitive disorders and may also reduce the impact of stress on cognitive disorders. Finally, the model includes a nonlinear negative feedback loop with respect to cognitive disorders (parameter β_c).

2) An increase of *stress* (*s*) depends linearly on the degree of prevailing stress. The parameter κ mediates this linear dependency. Stress leads to more

stress if social skills of regulating interpersonal closeness, coping with difficult situations and stress management are only poorly developed. A nonlinear decline of stress also depends on the current degree of stress. Thus, there is a large decline in stress (e.g. through exhaustion or habituation), outbalancing further increase, for very high degrees of stress. Furthermore, the decline of stress is regulated by the interactive effects of stress and withdrawal, as well as stress and delusions. Intrapsychic or actual withdrawal means avoiding stress, whereas delusions reduce stress induced by experiences such as loss of coherent thinking, perceptual overtaxation or the dissolution of ego boundaries (represented by the variable of cognitive disorders). It is exactly this nonlinear interaction between stress and cognitive disorders which has a stress-inducing effect mediated by the serotonergic and dopaminergic activity (δ). Additionally, social interactions of the high expressed emotion type are stress-inducing, the impact being greater on a patient already suffering from a high degree of stress (nonlinear interaction). This effect grows more intense with decreased abilities to comprehend, and cope with, stressful interpersonal experiences. Therefore, the parameter σ (diffuseness of affective-cognitive schemata) is formally employed for mediating the stress-inducing effects of the interaction $e \cdot s$.

Since the occurrence of stressful events is determined not only by the processes described above but also by random events in the patient's environment which are neither predictable nor possible to control, random fluctuations (fl) generating stress must also be included in the model.

3) The intensity of a patient's *withdrawal* (w) is considered to be dependent linearly on the current degree of withdrawal, since autistic encapsulation and spatial isolation are thought to have a self-reinforcing effect which becomes increasingly marked with reduced social skills, allowing the person to keep or get in contact with others (parameter κ). Furthermore, there is a nonlinear interaction between withdrawal and stress, especially for low levels of social and stress-management skills, resulting in withdrawal as an auto-protective effort in the case of stressful events, be they intrapsychic or environmentally caused. The experience of highly emotional and overly critical social interactions (e) leads to basically analogous results. Furthermore, delusions interact with preexisting withdrawal to effect further withdrawal. After all, delusions are defined as a process of "wrapping oneself up" in an idiosyncratic world of one's own. Parameter d (serotonergic and dopaminergic activity) was chosen as the mediator, since positive symptoms such as delusions and hallucinations can be effectively influenced by neuroleptic medication. The parameter β_w determines the impact of withdrawal on future withdrawal by quadratic negative feedback.

4) In our model the variable e (*Expressed Emotion*) not only represents the "actual" behavior of the patient's relatives but also, more importantly, describes the patient's subjective experience of EE attitudes. The change of e over the course of time is considered a linear function of e in two different ways. It is

mediated by the parameter σ, on the one hand, and the parameter κ, on the other, for difficulties in coping with highly emotional and overly critical interpersonal encounters are the result both of diffuse schemata and of poorly developed social skills. Additionally, patients' socially incompetent reactions to EE attitudes contribute to further increases in criticism and highly emotional statements on the part of the family members. The nonlinear decline of e (mediated by β_e) is regulated by the interaction between e and w, as withdrawal means interrupting face-to-face encounters and escaping emotionally from stressful communications, and by means of a quadratic term, e^2. The model allows for random fluctuations (fl) of e, since this variable also represents relatives' behavior as influenced by external conditions.

5) The temporal changes in the degree of *delusion* (d) are, first of all, considered to be a linear function of the preexisting degree of delusion; this represents the tendency of delusional processes to spread out, affecting an ever-increasing part of ideation. This growth function is related to neurochemical processes of the dopamine and serotonin metabolism via parameter δ. Intrapsychic processes resulting in delusion are represented by the nonlinear interaction between cognitive disorders, which may occasion delusional ideas by their very occurrence — at least when consciously experienced as such — and the tendency for delusional interpretations currently existing. The parameter σ is suggested as mediator of this kind of interaction, as the inability to perceive and react appropriately to complex or ambiguous intrapsychic but also social stimuli represents the basis for strange interpretations and paranoidal ideas. Finally, the equation governing the dynamic processes of delusions contains a quadratic feedback term, relating the degree of delusions to parameter β_d.

3.1.3 Production and Protection

Following differentiated descriptions of the dynamic processes of schizophrenia, it becomes evident that some symptoms may be regarded as efforts at coping with primary symptoms or previously unsuccessful coping mechanisms (see also the concept of auto-protective strategies by Böker 1986; Gross 1986). This is apparently true with regard to withdrawal following perceptual overtaxation, stressful requirements and interpersonal encounters.

Following this line of argument, even delusions may be regarded as self-healing or auto-protective strategies. In accordance with a theory of auto-protective strategies (Böker & Brenner 1983), the variables included in our model can be categorized as those contributing to the *production* of schizophrenia, such as cognitive disorders, stress, expressed emotions, and those serving self-healing or *protective* efforts, such as withdrawal and delusions (see Fig.9).

This view corresponds exactly to that put forward within the theory of dynamical diseases. Within this approach, the rate of change of a certain

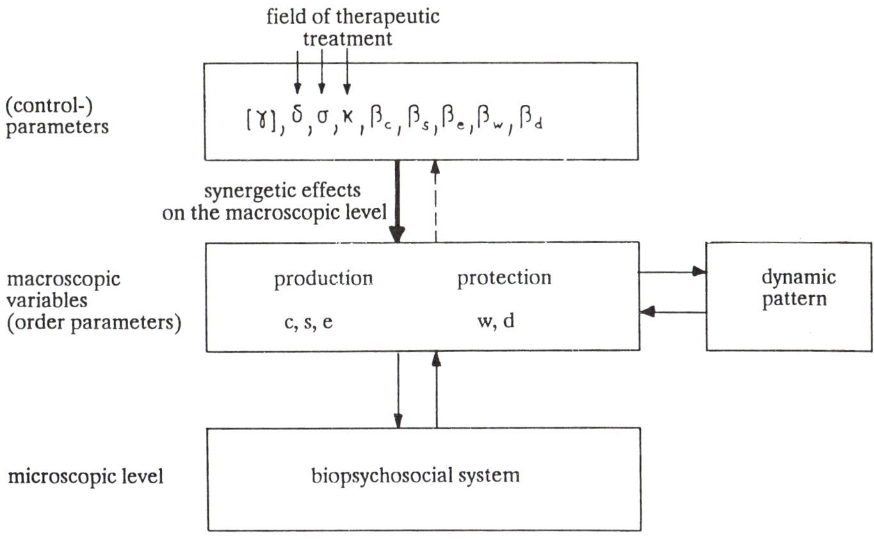

Fig. 9: The synergetic model of schizophrenia.

substance (here a variable) with time, *x(t)*, is considered as given by the difference between the production rate, *p*, and the destruction (or loss) rate, *d*:

$$\frac{dx}{dt} = p - d \tag{1}$$

We would like to emphasize that, in synergetic terms, the parameters introduced in section 3.1.2 can be considered the control parameters of a complex biopsychosocial system, which is described on the macroscopic level by the variables *c*, *s*, *w*, *e* and *d* (compare Haken 1988). It is even more reasonable to regard the parameters σ, δ and κ as control parameters, as they refer to three of the most important fields of the therapeutic treatment of schizophrenia, namely those of offering the patient clear and unambiguous communication, and training relatives to get acquainted with these principles, together with further treatment programs promoting the development of appropriate cognitive schemata aimed at improving the management of social situations (σ), neuroleptic medication (δ) and social skills training (κ) (see Fig.9).

3.1.4. Transformation of the Structural Model into a Set of Equations

There have been several attempts to date at creating a mathematical model of the dynamic patterns of the evolution of schizophrenia. For example, Cronin

(1977) describes the periodic patterns of psychomotor symptoms and biochemical processes occurring in periodic catatonic schizophrenia by means of a set of differential equations. The theory of dynamic systems has been applied in an analysis of dysfunctions within the feedback regulation of the dopamine metabolism by King and Barchas (1983). Elkaim, Goldbeter and Goldbeter (1980) have attempted an analysis of the intrafamiliar patterns of interactions in terms of symmetry breaking bifurcations.

Nonlinear differential or difference equations are especially suitable for a modelling of the dynamics of complex systems because of their ability to generate asymptotically stable, simple or complex periodic or even chaotic patterns, depending on the choice of parameters (see Mackey & an der Heiden 1982; Seifritz 1987; Schuster 1988).

In the following we would like to turn to an examination of discrete steps of change. Thus, the change of a variable x at time $t+1$ becomes a function of x at time t, which gives $x_{t+1} = f(x_t)$. With respect to the simulation of schizophrenia, one week has been chosen as a reasonable time interval $(t+1)$-t, where Δc, Δs, Δw, Δe and Δd represent the change in the level of the corresponding variables within the period of a single week. Δ denotes the corresponding difference operator. (For more detailed information on discrete maps and difference equations, see Goldberg 1968; Haken 1990a; 1983a, chapter 12.6; Seifritz 1987, chapter 2; Martienssen 1989.)

The above considerations present the basis which allows the transformation of the structural systems model of schizophrenia described in section 3.1.2 into a set of nonlinear difference equations (abbreviations correspond to those introduced in section 3.1.2):

$$\Delta c = \sigma c + \delta sc - \beta_c c^2 \quad (2)$$

$$\Delta s = \kappa s + \sigma es + \delta cs - \beta_s (w + d + s)s + fl \quad (3)$$

$$\Delta w = \kappa w + \kappa sw + \kappa ew + \delta dw - \beta_w w^2 \quad (4)$$

$$\Delta e = \sigma e + \kappa e - \beta_e (w + e)e + fl \quad (5)$$

$$\Delta d = \delta d + \sigma cd - \beta_d d^2 \; . \quad (6)$$

These equations combine the effects of symbiosis and competition between the variables, designating a model of mixed feedback in accordance with Fig.8 (cf. an der Heiden & Mackey 1987). Equations (3) and (5) provide for random fluctuations, since it must be assumed that stress and expressed emotion are also activated by external environmental influences.

3.1.5. System Dynamics

Depending on the parameter values chosen, each single equation describes a logistic rate function which takes the form of a sigmoid curve with a more or less steep slope representing the growth or the decline of the pertinent variable. The *y*-coordinates may be interpreted roughly as higher or lower values; precise quantitative interpretations are not possible. This applies to the following figures as well.

However, in the present simulation the system behavior of each single equation converges asymptopically towards a fixed point in accordance with the parameter values chosen, which corresponds to points "to the left of" the first bifurcation of the Feigenbaum route. Thus, the overall behavior of the system results not from the intrinsic dynamics of each single equation, but rather (a) from their interactions - that is, from the interconnectedness of the complete system - and (b) from the external impacts (input) via random fluctuations.

The system behavior resulting from the interactions of all five equations (still without random fluctuations) depicts one single schizophrenic episode (Fig.10).

The activation of schizophrenic processes by random fluctuations and their "dying out" without such external influence - that is, their tending towards latency (see Fig.10) - correspond exactly to the assumptions made by the diathesis-stress model of schizophrenia (Zubin & Spring 1977; Nuechterlein & Dawson 1984). This model maintains that the physiological and intrapsychic

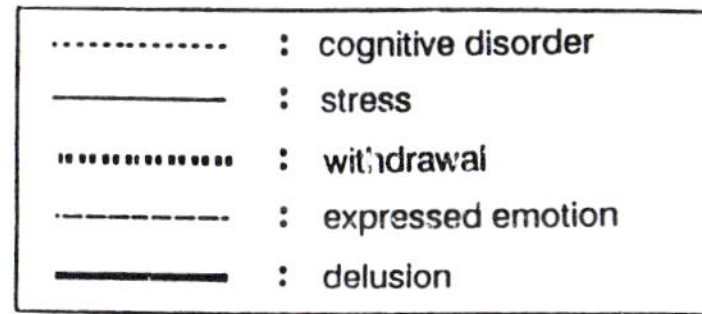

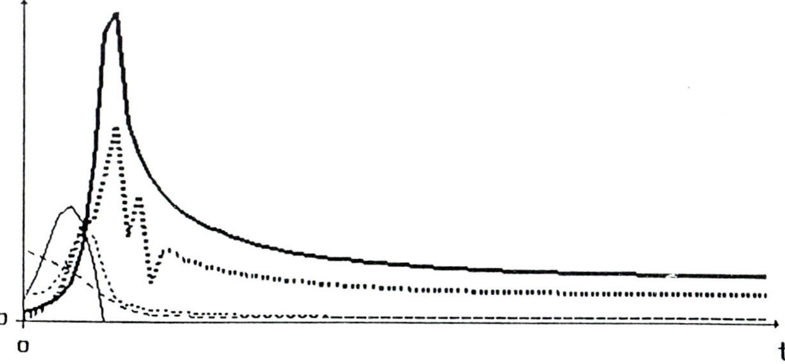

Fig. 10: Evolutionary patterns of variables *c*, *s*, *w*, *e* and *d* according to equations (2) to (6); without random fluctuations.

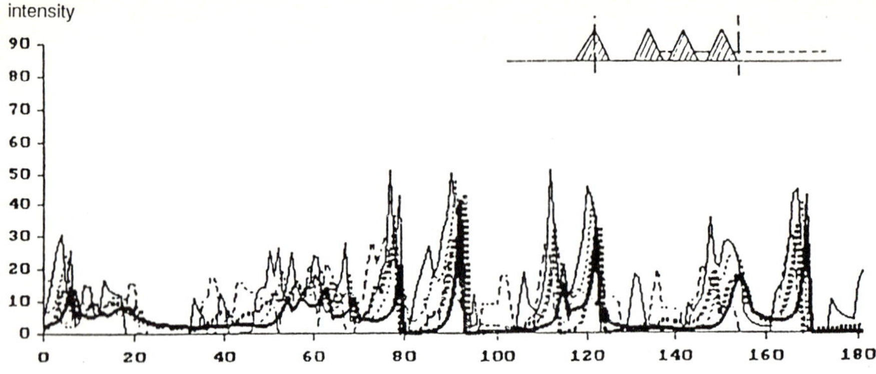

Fig. 11: Episodic evolutionary pattern with prodromal symptoms, acute onset. The parameter values are $\sigma = 0.055$; $\delta = 0.041$; $\kappa = 0.034$; $\beta_c = 0.050$; $\beta_s = 0.030$; $\beta_w = 0.100$; $\beta_e = 0.009$; $\beta_d = 0.080$.

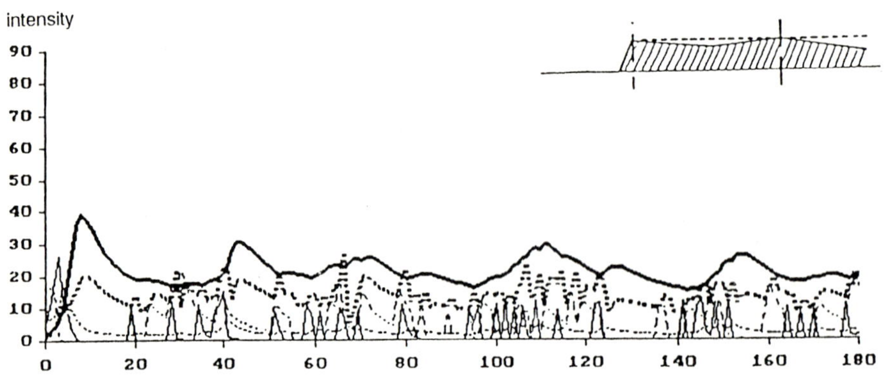

Fig. 12: Acute onset, chronic evolution, chronic end state. Same parameter values as in Fig. 11, except for: $\beta_w = 0.080$ and $\beta_d = 0.009$

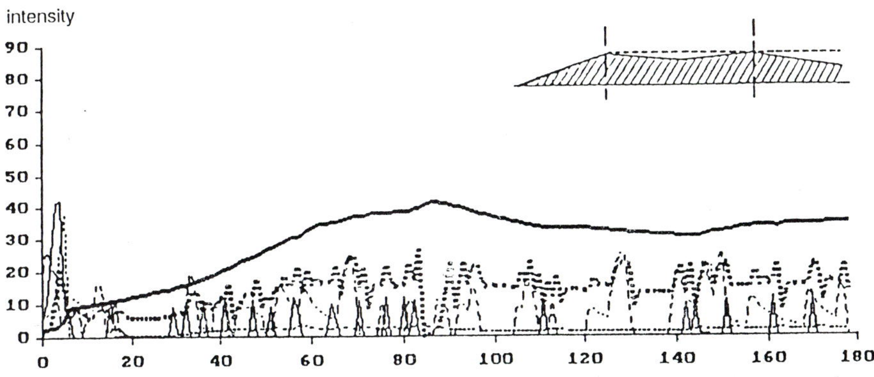

Fig. 13: Slow onset, chronic evolution. Same parameter values as in Fig. 11, except for: σ (only in equation (6)) = 0.015 and β_d = 0.002.

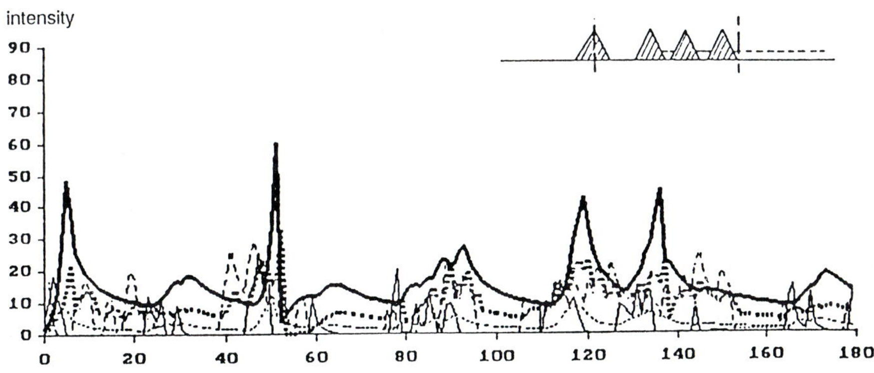

Fig. 14: Episodic evolution; acute onset without complete remission of symptoms between episodes. Same parameter values as in Fig. 11, except for: σ (only in equation (5)) = 0.130 and β_d = 0.028.

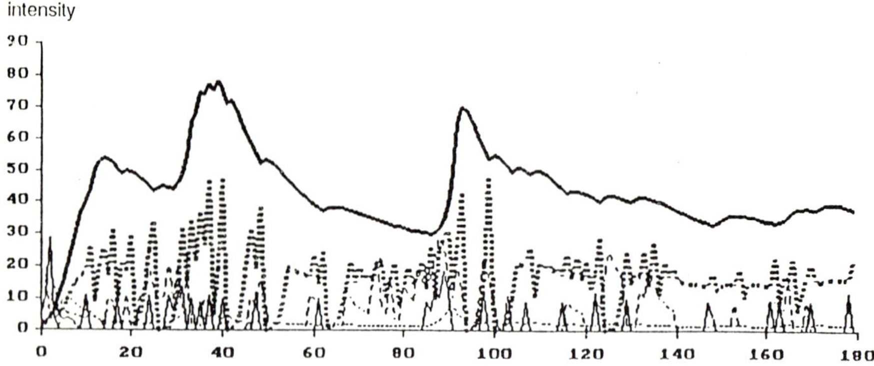

Fig. 15: Nontypical evolution. Same parameter values as in Fig. 11, except for: β_d = 0.004

systems and those pertaining to the regulation of interpersonal interaction are not pathological per se. Rather, they are very sensitive and vulnerable in their response to external irritations and demands, tending to overreact in particular ways. This view, in turn, corresponds to that of the concept of dynamical diseases, which holds that the systems' functions remain intact, while particular (pathological) - albeit principally reversible - changes in the parameters result in patterns of system behavior differing qualitatively from those of a "healthy" system behavior (cf. Mackey & an der Heiden 1982; Hesch 1990).

This very phenomenon is demonstrated by the different simulations presented in Figures 11 through 15, for each single simulation was generated by means of the same mechanisms contained in the structural model (Fig.8) and the set of equations. Additionally, random fluctuations were effective with respect to "stress" and "expressed emotion". They differ with respect to only a few parameter values, which are stated for each figure. Moreover, the law of parsimony of scientific explanations (Occam's razor) has been very closely observed.

One might argue that it is merely the coincidence of two marked random fluctuations which causes the peaks (or schizophrenic episodes) such as those shown in Figure 11. However, this objection is overruled by the fact that such an episodic pattern did not occur in every simulation and, moreover, that the patterns produced vary systematically in accordance with the parameter values chosen, largely independent of the flickering of external noise. This would certainly be different for parameter values within the chaotic regime, where there is a sensitive dependence on initial conditions.

The astonishing similarity between the evolutionary patterns obtained via simulation and those found by empirical research into the long-term evolution of schizophrenia can be regarded as the essential result of this simulation (see Fig 7).

3.1.6 Parameter Dynamics

The parameters σ, δ and κ may be considered to indicate temporally relatively stable states of intrapsychic, psychosocial and biochemical systems. However, their relatively slow rate of change, as opposed to that of the variables, does not mean that they are invariable. It is precisely the development of cognitive-affective schemata (in German terms: majorisierende Äquilibration, Affektlogik) and social skills, together with the evolution of the neuro-peptide system, upon which current etiological theories focus. Thus, a separate model should describe the evolutionary dynamics of the parameters, taking into account their interactions with the dynamics of the variables. The implication here is that we are concerned with coupled dynamic systems (cf. Glass 1987; an der Heiden 1992), with the only constant factor being the genetic disposition, γ.

The vulnerability of the neurotransmitter system may be interpreted as a consequence of certain genetic features regulating the processes of protein synthesis and responsible for the interactive effects of the agonistic and antagonistic properties of neurotransmitters as well as for receptor characteristics ($\gamma\ \delta$), i.e. internal processes of the system, together with their interaction with the development of intrapsychic structures ($\sigma\ \delta$).

A specific constitutional disposition, as well as prolonged periods of stress, s, and inconsistent interpersonal experience, e, the effects of which become apparent only after some delay ($t-\tau$), are responsible for the development of diffuse intrapsychic schemata. Furthermore, findings concerning neuronal plasticity (Haracz 1984;1985) indicate an interaction between the intrapsychic systems and the biochemistry of neurotransmitters ($\delta\ \sigma$)(see sect. 3.1.2).

Social (in)competence (κ) results from the interaction between κ and the diffuseness of intrapsychic schemata guiding perception and behavior within social contexts σ, as well as from prolonged periods of stress and impairment due to cognitive disorders.

These hypotheses lead to the formulation of the following difference equations (7) to (10):

$$\gamma = \text{const.} \tag{7}$$

$$\Delta\delta = \gamma\delta + \sigma\delta - \beta_\delta \delta^2 \tag{8}$$

$$\Delta\sigma = \gamma\sigma + \delta\sigma + \rho_s \bar{s}_{t-\tau} + \rho_e \bar{e}_{t-\tau} - \beta_\sigma \sigma^2 \tag{9}$$

$$\Delta\kappa = \sigma\kappa + \rho_s \bar{s}_{t-\tau} + \rho_c \bar{c}_{t-\tau} - \beta_\kappa \kappa^2 \tag{10}$$

where $\bar{s}$, $\bar{e}$ and $\bar{c}$ denote the mean values of the corresponding variables obtained within a certain time interval following the current point in time t by τ time intervals, ρ_s, ρ_e and ρ_c represent damping constants, and β_δ,

β_σ and β_κ denote feedback parameters. It is undoubtedly rather unusual in mathematics to treat the parameters of one set of equations as variables in the next. However, findings are available indicating a differentiation of variables into slowly and quickly varying variables in oscillating systems (cf. Tyson 1976, pp.54ff.). Haken (1990b, p.11) also proposes possible interactions between order and control parameters: "More generally we may expect an interplay between control parameters and order parameters where for instance an order parameter at one hierarchical level acts as control parameter at another one, etc."

Establishing a system of parameter dynamics means adhering even more closely to the frames of reference given by the vulnerability approaches to schizophrenia. For specific vulnerabilities to evolve, these models hypothesize (a) specific interactions of a biopsychological system with itself, (b) specific interactions with the system's social and biological environment, and finally, (c) specific interactions with illness and health care processes (Fig.16).

Furthermore, it is possible for the parameter values to enter a regime within which they cause the set of equations (2)-(6) to generate chaotic behavior. Whether they remain within this chaotic region, or begin to oscillate between asymptotically stable, periodic or chaotic regimes on their own, all these phenomena open up new perspectives for an analysis of schizophrenia as a dynamical disease by means of the framework provided by the theory of dynamic systems and chaos theory (Schmid 1991).

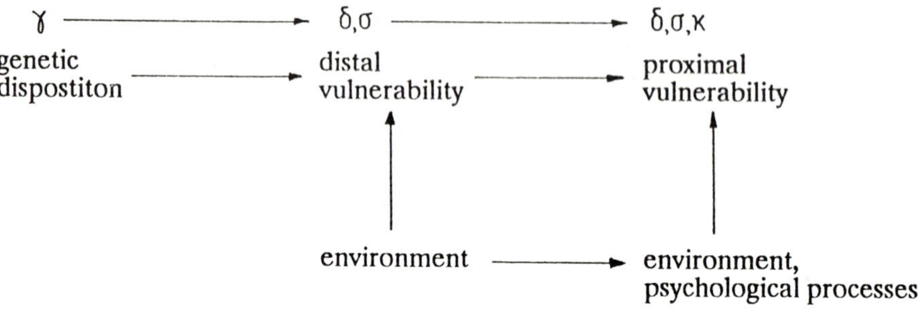

Fig. 16: The vulnerability model interpreted as evolutionary model of relevant parameters.

3.2 Psychotherapy

3.2.1 Basic Considerations

Another important application of the concept of nonlinear phase transitions within psychology are processes taking place within psychotherapy. They may be considered to represent nonlinear phase transitions between different

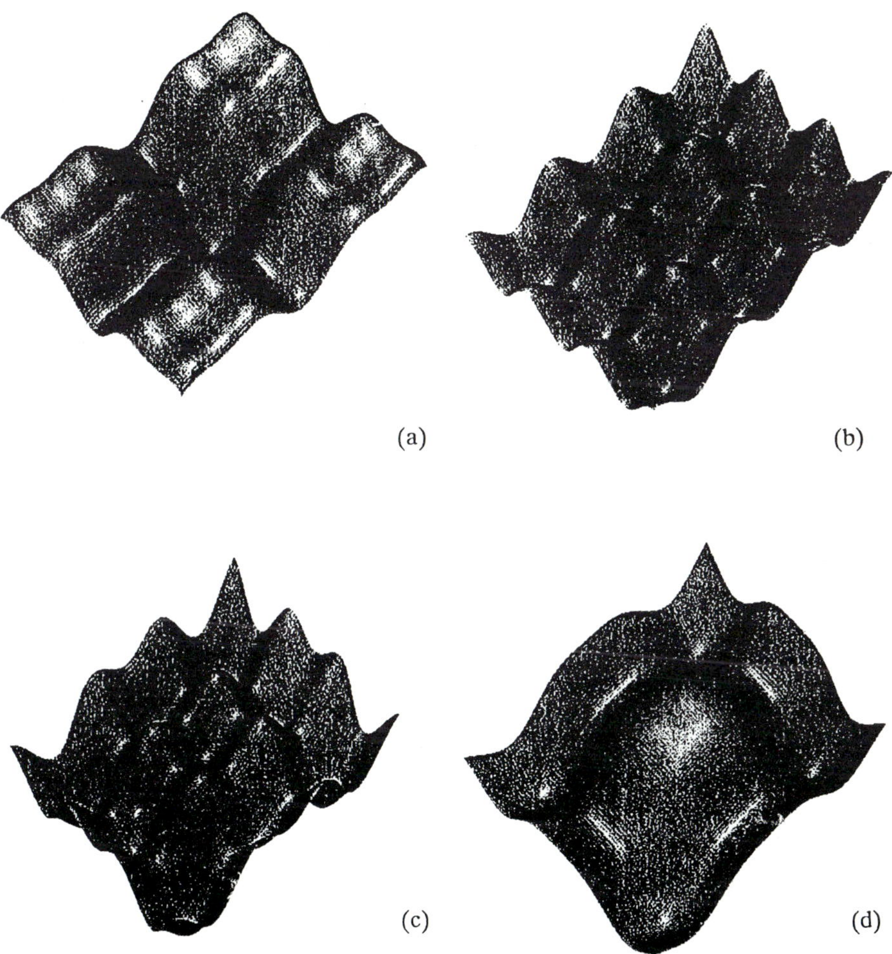

Fig. 17 (a-d): The temporal evolution of a potential landscape (representing a client's lifestyle scenario): A central potential valley is changing its shape in favor of several other potential valleys.

Technical remarks for the production of these figures: Each potential valley shown in figures (a)-(d) is a function of two variables z(x,y) over the x-y plane:

$z(x,y) = a \sin(x/4) + b(\cos(x) + \cos(y)) + c (\cos(2x) + \cos(2y))$

The plane is constituted by $-1.5\,\pi \leq x \leq 1.5\,\pi$ and $-1.5\,\pi \leq y \leq 1.5\,\pi$. The parameter values a, b, c of the individual figures are: (a) $a = 0, b = -1, c = -0.5$; (b) $a = 0, b = 0, c = -1$; (c) $a = -0.5, b = 1, c = -1$; (d) $a = -1, b = 1, c = -0.25$. (We are indebted to W. Lorenz of the University of Stuttgart, Institut für Theoretische Physik und Synergetik, for providing the figures presented here.)

biopsychosocial states of order. This view is supported by both practical experience and detailed analysis, indicating that therapeutic change obviously does not take place in the form of a continuous, incremental process, as may be suggested for example by the model of Skinner's shaping. On the contrary, we can observe the occurrence of sudden inspiration, "aha-experience" or the restructuring of problems similar to the cognitive bistable perception of ambigous patterns (Stadler & Kruse 1990; Kruse et al. 1992; Mahoney 1980). Therapists are acquainted with the phenomenon of casual remarks or ideas being picked up by clients only after a substantial delay - and it often seem to be these fluctuations which then trigger the jump into a modified pattern of experience and behavior. This may be considered an effect similar to that described by Zeigarnik, who postulates a period of latency or distraction to be necessary after several trials of learning in order that the individual be enabled to spontaneously achieve a higher level of functioning than before. It is the occurrence of this kind of phenomena which supports the postulate of nonlinear processes being effective within psychotherapy. The transfer of methods applied in chaos research to the analysis of psychotherapeutic processes is therefore obvious (Fuhriman & Burlingame 1991; Burlingame et al. 1991).

Kelso (1990) and co-workers have carried out numerous studies aimed at investigating the phase transitions between different rhythmic movements. He concludes that in the case of intended behavior not corresponding to the intrinsic dynamics of the system - which is the usual case as far as learning processes are concerned - the question of which attractor will determine overt behavior depends on the intensity of the behavioral information given. "1. If behavioral information is weak or absent, the system's behavior is determined primarily by the intrinsic dynamics; 2. If behavioral information specifying a required behavioral pattern is strong, the system will possess a stable state at that required pattern" (Kelso 1990, p. 262).

Yet it is questionable with regard to therapeutic processes whether goal-oriented behavioral information alone is sufficient to effect a phase transition of the behavior of a biopsychosocial system into a desired state of attraction. Rather, one has to expect bistable or even multistable system behavior once the system begins to develop one or more attractors in addition to the attractor determining behavior so far (Stadler & Kruse 1990; Kruse et al. 1992). We may also recognize such multistabilities during practical therapeutic work. Clients sometimes seem to process their insights while experiencing different mental and emotional states ("states of mind", see Horowitz 1987). They may also repress or dissociate insights they have already gained, or reverse behavioral changes and pick them up again at a later date to work on them at an even more profound level. Planned, goal-oriented development is therefore rather unlikely during periods of psychosocial instability, because new attractors which determine behavior have not yet evolved; also, behavioral information determining emotional and social processes is still being worked out during a process of co-evolution on the basis of the intrinsic emotional dynamics of the client. It may be assumed that these processes, exspecially at instability points, are very sensitive to external fluctuations and take place at an unconscious rather than a conscious level (see Revenstorf 1985; 1990; Kruse et al. 1992). At any rate, the synergetic concept of phase transitions between biopsychosocial states of order encourages researchers to take a closer look at characteristic features and processes of stability and instability within psychotherapy. The identification and empirical description of patterns or states of order within biopsychosocial systems is a fundamental prerequisite of this kind of research. In the following we wish to propose a state-of-order concept suitable for the field of psychothcrapy from a theoretical, but also a practical, point of view. This concept is designated by the term "lifestyle".

3.2.2 The Biopsychosocial Lifestyle of the Client

"Lifestyle" refers to the state of order governing the individual's present way of life. The state of order may be characterized by specific enduring negative emotional schemata, together with their complementary positive self-schemata. The activation of certain schemata may cause specific patterns of perception to be activated, especially in the field of interpersonal and emotional perception. These patterns in turn suggest corresponding patterns of behavior which may serve to confirm the corresponding patterns of perception. Thus we find typical emotion-behavior-perception cycles, into which significant others may be intricately integrated. These cycles may also be repeated with people not yet entangled in close relationships (compare Freud's principle of repetition compulsion). It is not usually by chance that others appear in an individual's lifespace. Rather, they are selected consciously or unconsciously so that they fit in as partner, friend, authority, colleague, idol, etc.

However, it is not only intrapsychic schemata which determine the process of experience and the patterns of interaction. Various other psychological, social or even biological processes and their recursive interaction have an impact on the reproduction of life-themes, success and failure, the course taken by one's life with respect to career or health, the selection of partners, conflicts, living conditions etc., all of which may be typical for a certain lifestyle scenario. However, the individual should not be considered solely responsible for the creation of his or her lifestyle. Significant others and their characteristics, as well as social conditions or material circumstances (e.g. air pollution, food pollution, the shortage of housing, structural unemployment), also play an important part. Nonetheless, the daily reproduction of a person's lifestyle can be considered an active achievement, partly conscious but for the greater part unconscious. Alternatives to the actual lifestyle do exist, at least in principle, yet they are neither perceived nor actively taken into consideration. The individual therefore perceives his or her lifestyle to be normal or necessary, or even characterizes it by saying "there is no way out".

The concept of the lifestyle scenario introduced here is similar to the appreciation of psychological functioning as presented in the schema theoretical framework (see Ciompi 1986b; Grawe 1987; Zingg 1988). It goes beyond that appreciation, however, as we include a complex system of different - not only psychological - interrelated processes constituting relatively stable dynamics. In particular, immunological, endocrinological and physiological processes must also be taken into account, as they are fundamental to physical and mental health and illness (compare Hesch 1990).

In synergetic terms, it describes the fact that a large number of processes on a microscopic level combine to form a coherent, dynamic pattern (mode) representing a fundamental attractor of the dynamics predominant in a person's life. Attractors can not only be identified in the lifestyle of one individual, but may also depict the pattern of cooperation in systems made up of several individuals (couples, families, teams).

When potential landscapes are utilized as a means of representation, an attractor can be recognized by a potential valley. It can be assumed that the biopsychosocial *potential landscape* is the result of the cooperation of several order parameters. The potential valleys describe the amount of energy contained in a system which may potentially carry out work. All forces acting in the system are mathematical derivatives of the potential function (see Haken 1983a, pp.105; Haken 1983b). The depth and diameter of a potential valley may vary in accordance with the phases in actual life. Two or more potential valleys may exist at the same time, transitions between them being possible. The ability to make such transitions may be regarded as a criterion of mental health (Fig. 17).

It is to be expected that scale-invariant phenomena (Großmann 1989) can be observed in the lifestyle of an individual: similar behavioral patterns on different levels of behavioral organization as well as on different time-scales (self-similarity). Examples of this are nonverbal signals conveying vagueness and detachment; vagueness of content; discussion of difficulties in decision-making; long-term avoidance of decision-making and acknowledgement of responsibility (Ciompi, personal communication).

3.2.3 The Therapeutic System

The therapeutic system comprises all events within the scope of the relationship between client and therapist which are defined as therapy. It involves one or more clients (or one client and one or more related persons not defined as clients) and one or more therapists. The existence of the therapeutic system is temporally and spatially confined. Empirical analysis is possible at different levels. An example is the analysis of immunological and endocrinological parameters before, during and after therapy. However, the methods normally used within psychotherapy research aim at an analysis of data on two different levels. These are (a) the level of the psychological systems of client and therapist, which are autonomous, operationally closed, but structurally coupled (for details, see Schiepek 1991, Chapter IV); and (b) the level of the social or - following Luhmann (1984) - communication system constituted through therapy. The communication system evolves on the basis resulting from the functioning of the participating psychological systems; however, it evolves over and above these to constitute an emergent reality.

Self-similarity can also be expected when we compare an individual's lifestyle scenario and the therapeutic system in which he or she is involved. As far as clients are concerned, this phenomenon has been termed "transference" in psychoanalysis (for a recent conceptualization, see the plan-concept (in German: Test-Konzept) proposed by the Mount Zion Psychotherapy Research Group; cf. Silberschatz, Curtis & Nathans 1989). Such phenomena also occur with respect to therapists (counter-transference) and are made the subject of encounter groups and supervision.

Finally, all attempts at describing the field of psychotherapy remain incomplete if they do not take into account the context in which therapy takes place. The context can result for example from the fact that therapists are supervised, that therapy is carried out within an institutional frame of varying complexity or in cooperation with a team of therapists, as envisaged by many so-called systemic approaches. The systems involved are thus interrelated so as to create more or less useful or problematic ways of cooperation. Interrelational patterns can be produced within each of the systems, or between the systems involved, similar to those produced by the clients and significant others, or to those observable in the therapeutic system (self-similarity). The recursive interactions between the various institutions, teams and therapists involved on a particular case may close to form a hypercycle, for which the client - theoretically the central figure - constitutes merely the trigger enabling significant eigendynamics to occur (for details see Schweitzer 1989; Imber-Black 1990; Schiepek & Reicherts 1992).

The therapeutic system, together with the interacting systems forming the context of a therapy, may take on more or less significance for the lifestyle of a client (a couple, a family). For clients who define themselves as part of a therapeutic system over a long period of time, the therapeutic system may become a constitutive part of their lifestyle. For people living in a mental health care unit (a children's home, a psychiatric clinic), the institution becomes nearly identical with the "lifespace" and will determine their lifestyle in a fundamental way.

Fig. 18 shows the systems described above which are basic to the understanding of self-organization in psychotherapy.

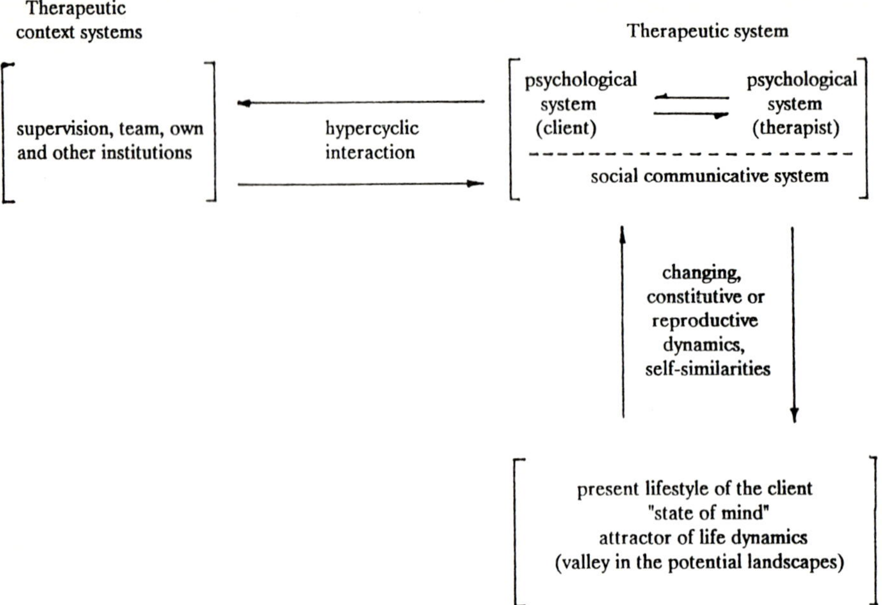

Fig. 18: Systems constituting the process of psychotherapy.

3.2.4 Self-Organization: The Deformation of Potential Landscapes

How is the process of self-organization in psychotherapy to be conceived?

Usually the reason for seeking therapy lies in the continuing existence of a specific state of order experienced by the client or his/her significant others as unpleasant, distressing or intolerable. Furthermore, the state of order tends to be relatively stable; attempts to overcome it on one's own have failed repeatedly. The client finally seeks professional assistance, which he/she hopes will help towards the achievement of a different state of order.

Thus, successful therapy can be conceived as an order-to-order transition (phase transition). In the terminology introduced above, a state of order is represented by a certain present lifestyle scenario (or "states of mind" which dominate the client's lifestyle, or at least part of it, see Horowitz 1987).

If we describe this phenomenon in terms of potential landscapes, a dominant lifestyle is represented by a deep potential valley, while the system can be viewed as a particle resting at the bottom of it. The fluctuations of the system are not sufficient to allow it to overcome the potential barriers and leave the potential valley.

Psychotherapy can therefore follow two different courses, which will usually be combined:

(a) It may contribute to the deformation of the potential landscape by flattening the dominant potential valley (i.e. by decreasing the dominance of the present state of mind), by lowering the potential barriers, and by forming alternative potential valleys (see Fig. 19).

(b) It can encourage more intensive fluctuations of the system, i.e. of the particle resting in the potential valley. From the viewpoint of the client, this means trying to act differently than before, experimenting with new ways of

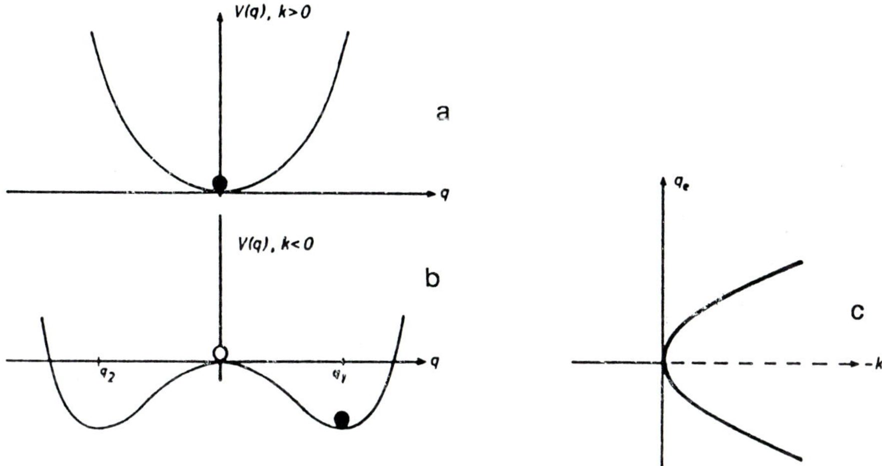

Fig. 19: The transition from a symmetrically stable state (a) to two equiprobable potential valleys (b). In this process, the particle rolls into one of two potential valleys following a symmetry-breaking instability. The potential function is: $F(q) = -kq - kq^3$, where q is the elongation from the equilibrium position and k is the bifurcation parameter (control parameter); with $k > 0$ (a) and $k < 0$ (b). This is the so-called anharmonic oscillator which contains a cubic term besides a linear term in its force F. (c) represents the bifurcation diagram of the system (anharmonic oscillator), with the equilibrium coordinate q_e as function of k. For $k > 0$, $q_e = 0$, but for $k < 0$, $q_e = 0$ becomes unstable (dashed line) and is replaced by two stable positions (solid fork).

behavior and new ideas in certain situations, or confronting oneself with situations or emotions so far avoided.

When the previously deep potential valley becomes flatter, phenomena characteristic to instability will occur, such as critical fluctuations and critical slowing down: feedback processes do not take the system back to the stable state as quickly as before. If certain parameters defining the potential function become less than zero, then a symmetry-breaking instability occurs: minimal fluctuations are sufficient to allow the particle to roll into one of two new equiprobable potential valleys (Fig. 19). It is at this point that chance and nonpredictability come into play. With regard to psychotherapeutic practice, analysis of the stability of the system dynamics would be desirable, allowing an assessment of the degree to which the system dynamics have approached the boundaries of instability, whether creative or dangerous.

Following this model, two different approaches are possible in psychotherapy. Either an amplification of the fluctuations of a system is attempted, so that it moves on from one potential valley into another one, existing within the lifespace but currently not in use (such "jumps", which are often not precisely aimed, are turbulent in nature); or efforts are made to deform the potential landscape by using minimum irregularities of the surface to create new potential valleys via processes of deviation-amplifying feedback. As time passes, continued use of the new potential valleys may further shape them. These two approaches are complementary, rather than mutually exclusive. However, the client's cooperation and active participation are fundamental prerequisites.

It is the function of therapy not only to transfer the system dynamics into the regime of a different attractor but also generally to increase accessibility to alternative attractors (states of mind, lifestyles). Thus, the focus of attention is not merely reliability within another fixed lifestyle, leading once again to a low degree of fluctuations and deep potential valleys, but adaptability and ease of switching, requiring large fluctuations and flat potential curves (compare Haken 1983a).

Of course, this is quite a general description of the principles of self-organizational processes occurring withing psychotherapy. It is both possible and necessary, however, to specify these principles with respect to different therapeutic approaches (see Kruse et al. 1992 for considerations regarding hypnotherapy; Schiepek, Fricke & Kaimer (1992) for solution-oriented brief therapy).

An interesting question, one which is still to be answered, concerns which of the *order parameters* are in fact those governing the system's states of order. In view of the enormous complexity of biopsychosocial systems, the determination of order parameters proves to be much more difficult here than in physical systems (compare Stadler & Kruse 1990). For lack of a mathematical model for the synergetics of psychotherapy, the states of order of a system (here: lifestyles) will be characterized not by a quantifiable value of one or more order parameters but by the description of the respective lifestyle patterns. The method to be applied here is the construction of an idiographic model of the system, one which recursively interrelates the relevant processes and variables constituting a specific, dynamic lifestyle pattern (Schiepek 1986; 1991; Schiepek & Tschacher 1992).

The *control parameters* of therapeutic order-to-order transitions are assumed to be inherent in the processing of the therapeutic system, which means that a change in the value of the relevant control parameter is a result of the interaction between therapist and client. Empirical results concerning clients' receptiveness (Ambühl & Grawe 1988; 1989; Orlinsky & Howard 1986) show that the techniques employed by the therapist, and even the therapist's behavior generally, exert no significant influence on therapeutic change unless precisely adjusted to the client and his specific receptiveness.

One possible control parameter might be the amount of motivational energy invested by the client, influencing the degree of fluctuations and the deviation-amplifying feedback processes effective in the deformation of the potential landscape. Everything that happens during therapist-client interaction which effectively raises the level of motivational energy also raises the value of this control parameter.

Another possibility for the identification of relevant control parameters could be the search for either one or a small number of variables typical for the client's lifestyle which would respond to treatment. These characteristics are symbolized by the core variables in the idiographic systems model of the client's lifestyle. It is also possible to rely on the client's mentioning his/her central complaints and goals, as contained in the approach put forward by the Brief Family Therapy Center (BFTC) (de Shazer 1985).

A word of caution should be voiced, however. The term "control parameter" should not be misunderstood to mean that there is a parameter independent of the client which can be manipulated by technical means, that a therapist need only make use of his/her professional techniques to modify this parameter and effect a phase transition in the client's lifestyle. The concept of control parameters refers to psychological or social processes taking place within the frame of the therapeutic system or outside - at any rate, however, produced by the processing of the therapeutic system and thus stimulated by the cooperation of therapist and client.

The concept of self-organization in psychotherapy as proposed here represents a qualitative analysis. Thus it becomes clear thta the formal model provided by synergetics also promotes an improved qualitative understanding of highly complex psychological processes. Nevertheless, quantitative empirical analysis is necessary and efforts in this direction are already being undertaken. Empirical methods corresponding to the individual aspects of the therapeutic system are illustrated in Fig. 20.

Concerning the level of psychological - i.e. cognitive-emotional - processes, a method is used which allows the modification of topics focused on during therapy and their semantic association to be traced over the course of therapy. Therapy transcripts are used to demonstrate how the structures of the topics introduced change during a continuous process of co-creation (Reconstruction of Cognitive Representation, RCR). This can be shown for the therapist as well as for the client (Schiepek, Fricke & Kaimer 1992).

Another method aims at analyzing, at a level of high temporal resolution, the sequence of activation of interactional plans (Caspar 1989) of each

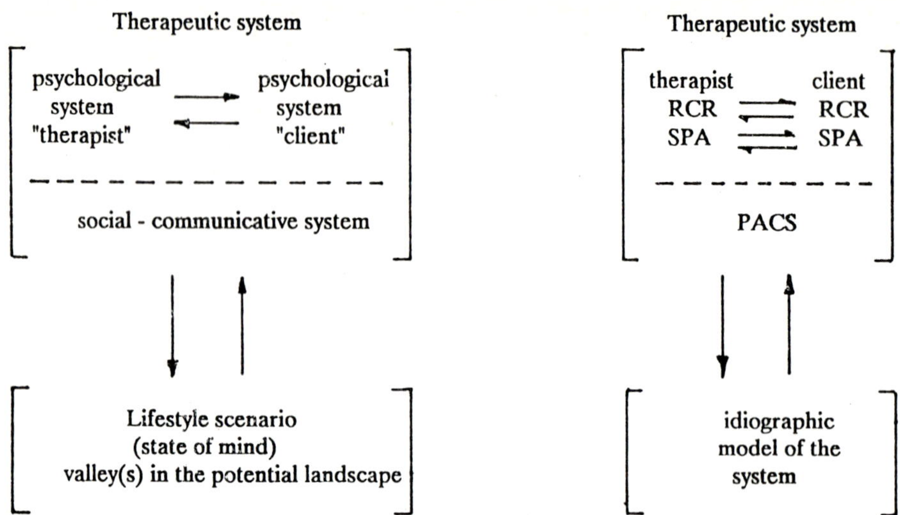

Fig. 20: Synopsis of the proposed empirical methods for system-process research on psychotherapy, related to the theoretical model (compare Fig. 18).

individual system in the course of their structural coupling (Sequential Plan Analysis, SPA). Using this method, our research group at the University of Bamberg is preparing long time-series by subdividing video recordings of two completed therapies (9 and 13 sessions) into ten-second units and coding the interactional plans of both client and therapist. The dimension analysis of these data (e.g. Grassberger & Procaccia 1983) will allow us to ascertain their correlational dimensions, and thus obtain empirical evidence for the chaotic dynamics effective within therapeutic processes.

The method to be applied in the process analysis of the communicative therapeutic system abstracts from the psychological systems of the participants and is directed at analyzing how the self-referential sequence of communicative proceedings constitutes and modifies the social system "therapy" with its corresponding structures, limits, identities etc., that is to say, how this social system self-organizes and creates disorder-to-order transitions (for similar considerations concerning self-organization of the legal system, compare Teubner 1987; 1989). The coding scheme for relevant communicative proceedings is called "Process Analysis of Communicative Systems" (PACS).

The recursive interrelations of the macroscopic variables constituting a client's lifestyle scenario are established and recorded by means of idiographic systems modeling (Schiepek 1986).

Of course, the examples described here illustrate only a few possible applications of synergetics to psychology and psychiatry. Further examples are to be seen in research on structure formation in groups (Tschacher, Brunner & Schiepek 1992) or in more complex social systems such as institutional networks of the psychosocial care system (Bergold 1992), or system games (Schiepek & Reicherts 1992). System games are live simulations of the social dynamics occurring within or between various groups and institutions, representing an adequate research paradigm for the investigation of self-organizational processes within complex social systems, as they allow an assessment under experimental conditions which are nevertheless characterized by great realism. A survey of numerous other examples for the application of synergetics to psychology is given by two volumes published in the Springer Series in Synergetics, namely "Synergetics of Cognition" (Haken & Stadler 1990) and "Self-Organization and Clinical Psychology" (Tschacher, Schiepek & Brunner 1992).

It is no surprise that synergetics itself is generating synergistic effects, with research in such different fields as neurobiology, psychiatry, sociology, psychology, and also physics becoming interrelated. The resulting fruitful interdisciplinary exchange and cooperation, one that can be seen as a form of comprehensive human-science synergetics, may be considered the order parameter for creative research across the boundaries of these various disciplines.

Acknowledgement. I would like to thank Dipl.-Psych. Martina Weisensel (Staatl. gepr. Übersetzerin) for the translation of the present contribution.

References

Akiskal HS, McKinney WT (1975) Overview of Recent Research in Depression: Integration of ten Conceptual Models into a Comprehensive Clinical Frame. Arch Gen Psychiatry 32: 285-295

Ambühl H, Grawe K (1988) Die Wirkungen von Psychotherapien als Ergebnis der Wechselwirkungen zwischen therapeutischem Angebot und Aufnahmebereitschaft der Klient/inn/en. Z Klin Psychol, Psychopath Psychother 36: 308-327

Ambühl H, Grawe K (1989) Psychotherapeutisches Handeln als Verwirklichung therapeutischer Heuristiken. Psychother med Psychol 39: 1-10

Bergold J (1992) The Systemic Character of the Psychosocial and Psychiatric Health Services. In: Tschacher W, Schiepek G, Brunner EJ (Eds) Self-Organization and Clinical Psychology. Springer, Berlin (in press)

Bleuler M (1972) Die schizophrenen Geistesstörungen im Lichte langjähriger Kranken- und Familiengeschichten. Thieme, Stuttgart

Böker W (1986) Zur Selbsthilfe Schizophrener: Problemanalyse und eigene empirische Untersuchungen. In: Böker W, Brenner HD (Hrsg) Bewältigung der Schizophrenie. Huber, Bern, S 176-188

Böker W, Brenner HD (1983) Selbstheilungsversuche Schizophrener: Psychopathologische Befunde und Folgerungen für Forschung und Therapie. Nervenarzt 54: 578-589

Böker W, Brenner HD (Hrsg) (1986) Bewältigung der Schizophrenie. Huber, Bern

Böker W, Brenner HD (Hrsg) (1989) Schizophrenie als systemische Störung. Huber, Bern

Brown JF, Voth AC (1937) The Path of Seen Movement as a Function of the Vector Field. Am J Psychol 49: 543-563

Burlingame GM, Fuhriman A, Barnum K, Lyman R (1991) Deterministic Chaos in Group Psychotherapy Interaction: An Illustration. Paper presented at the Society for Psychotherapy Research 22nd Annual Meeting, July, 1991, Lyon, France

Caspar F (1989) Beziehungen und Probleme verstehen. Eine Einführung in die psychotherapeutische Plananalyse. Huber, Bern

Ciompi L (1982) Affektlogik. Klett-Cotta, Stuttgart

Ciompi L (1986a) Auf dem Weg zu einem kohärenten multidimensionalen Krankheits- und Therapieverständnis der Schizophrenie: Konvergierende neue Konzepte. In: Böker W, Brenner HD (Hrsg) Bewältigung der Schizophrenie. Huber, Bern, S 47-61

Ciompi L (1986b) Zur Integration von Fühlen und Denken im Licht der "Affektlogik". Die Psyche als Teil eines autopoietischen Systems. In: Kisker KP et al (Hrsg) Psychiatrie der Gegenwart Bd 1. Springer, Berlin, S 373-410

Ciompi L (1989) Zur Dynamik komplexer biologisch-psychosozialer Systeme: Vier fundamentale Mediatoren in der Langzeitentwicklung der Schizophrenie. In: Böker W, Brenner HD (Hrsg) Schizophrenie als systemische Störung. Huber, Bern, S 27-38

Ciompi L, Müller C (1976) Lebensweg und Alter der Schizophrenen. Eine katamnestische Langzeitstudie. Springer, Berlin

Cronin J (1977) Mathematical Aspects of Periodic Catatonic Schizophrenia. Bull Math Biol 39: 187-197

Elkaim M, Goldbeter A, Goldbeter E (1980) Analyse des transactions de comportement dans un système familial en termes de bifurcation. Cahiers Critiques de Thérapie Familiale et de Pratique des Réseaux 3: 18-34

Fuhriman A, Burlingame GM (1991) Chaos: Emerging Developmental Patterns in Process Research. Paper presented at the Society for Psychotherapy Research 22nd Annual Meeting, July, 1991, Lyon, France

Gelders YG (1989) Thymosthenische Wirkstoffe, ein neuartiger Ansatz in der Behandlung der Schizophrenie. In: Böker W, Brenner HD (Hrsg) Schizophrenie als systemische Störung. Huber, Bern, S 65-74

Glass L (1987) Coupled Oscillators in Health and Disease. In: Rensing L, an der Heiden U, Mackey MC (Eds) Temporal Disorders in Human Oscillatory Systems. Springer, Berlin, pp 8-14

Goldberg S (1968) Differenzengleichungen und ihre Anwendung in Wirtschaftswissenschaften, Psychologie und Soziologie. Oldenbourg, München

Grassberger P, Procaccia I (1983) Measuring the Strangeness of Strange Attractors. Physica 9D: 189-208

Grawe K (1987) Psychotherapie als Entwicklungsstimulation von Schemata. Ein Prozeß mit nicht vorhersehbarem Ausgang. In: Caspar F (Hrsg) Problemanalyse in der Psychotherapie. Forum 13, dgvt, Tübingen, S 72-87

Gross G (1986) Basissymptome und Coping Behavior bei Schizophrenen. In: Böker W, Brenner HD (Hrsg) Bewältigung der Schizophrenie. Huber, Bern, S 132-141

Großmann S (1989) Selbstähnlichkeit: Das Strukturgesetz im und vor dem Chaos. In: Gerok W (Hrsg) Ordnung und Chaos in der unbelebten und belebten Natur. S Hirzel, Wissenschaftliche Verlagsgesellschaft, Stuttgart, S 101-122

Haken H (1970) Laser Theory. Handbuch der Physik/Encyclopedia of Physics, Vol. 25/20. Springer, Berlin

Haken H (1983a) Synergetics. An Introduction. Springer, Berlin (3rd edition)

Haken H (1983b) Advanced Synergetics. Instability Hierarchies of Selforganizing Systems and Devices. Springer, Berlin (3rd edition)

Haken H (1987) Synergetic Computers for Pattern Recognition and Associative Memory. In: Haken H (Ed) Computational Systems, Natural and Artificial. Springer, Berlin

Haken H (1988) Information and Self-Organization: A Macroscopic Approach to Complex Systems. Springer, Berlin

Haken H (1989) Synergetik: Vom Chaos zur Ordnung und weiter ins Chaos. In: Gerok W (Hrsg) Ordnung und Chaos in der unbelebten und belebten Natur. S Hirzel, Wissenschaftliche Verlagsgesellschaft, Stuttgart, S 65-75

Haken H (1990a) Synergetik. Eine Einführung. Springer, Berlin

Haken H (1990b) Synergetics as a Tool for the Conceptualization and Mathematization of Cognition and Behavior - How Far Can We Go? In: Haken H, Stadler M (Eds) Synergetics in Cognition. Springer, Berlin, pp 2-31

Haken H (1991c) Synergetic Computers and Cognition. Springer, Berlin.

Haken H (1992) Synergetics in Psychology. In: Tschacher W, Schiepek G, Brunner EJ (Eds) Self-Organization and Clinical Psychology. Springer, Berlin (in press)

Haken H, Graham R (1971) Synergetik - Die Lehre vom Zusammenwirken. Umschau 6: 191

Haken H, Kelso J, Bunz H (1985) A Theoretical Model of Phase Transitions in Human Hand Movements. Biol Cybern 51: 347

Haken H, Stadler M (Eds) (1990) Synergetics of Cognition. Springer, Berlin

Haracz JL (1984) A Neural Plasticity Hypothesis of Schizophrenia. Neuroscience and Biobehavioral Reviews 8: 55-71

Haracz JL (1985) Neural Plasticity in Schizophrenia. Schizophrenia Bulletin 11: 191-229

Harding CM (1984) Long-term Outcome Functioning of Subjects rediagnosed as meeting the DSM III Criteria for Schizophrenia. Doctoral Dissertation, University of Vermont

an der Heiden U (1992) Chaos in Health and Disease. In: Tschacher W, Schiepek G, Brunner EJ (Eds) (1992) Self-Organization and Clinical Psychology. Springer, Berlin (in press)

an der Heiden U, Mackey MC (1987) Mixed Feedback. A Paradigm for Regular and Irregular Oscillations. In: Rensing L, an der Heiden U, Mackey MC (Eds) Temporal Disorder in Human Oscillatory Systems. Springer, Berlin, pp 30-46

Hesch RD (1990) Sein im deterministischen Chaos: Ansätze einer neuen Systemtheorie der Medizin. Vortragsmanuskript, Hannover

Horowitz MJ (1987) States of Mind. Plenum Press, New York

Huber G, Gross G, Schüttler R (1979) Schizophrenie. Eine Verlaufs- und sozialpsychiatrische Studie. Springer, Berlin

Imber-Black E (1990) Familien und größere Systeme. Im Gestrüpp der Institutionen. Auer, Heidelberg

Kanizsa G, Luccio R (1990) The Phenomenology of Autonomous Order Formation in Perception. In: Haken H, Stadler M (Eds) Synergetics of Cognition. Springer, Berlin, pp 186-200

Kelso J (1984) Phase Transitions and Critical Behavior in Human Bimanual Coordination. Am J Physiol: Reg Integ Comp 15, R1000-R1004

Kelso J (1990) Phase Transitions: Foundation of Behavior. In: Haken H, Stadler M (Eds) Synergetic of Cognition. Springer, Berlin, pp. 249-268

King R, Barchas JD (1983) An Application of Dynamic Systems Theory to Human Behavior. In: Basar E, Flohr H, Haken H, Mandell AJ (Eds) Synergetics of the Brain. Springer, Berlin

Köhler W (1920) Die physischen Gestalten in Ruhe und im stationären Zustand. Vieweg, Braunschweig

Köhler W (1940) Dynamics in Psychology. Liveright, New York

Krebs CJ (1985) Ecology. Harper & Row, New York

Kriz J (1990) Synergetics in Clincal Psychology. In: Haken H, Stadler M (Eds) Synergetics of Cognition. Springer, Berlin, pp 393-404

Kruse P, Stadler M (1990) Stability and Instability in Cognitive Systems: Multistability, Suggestion, and Psychosomatic Interaction. In: Haken H, Stadler M (Eds) Synergetics of Cognition. Springer, Berlin, pp. 201-215

Kruse P, Stadler M, Pavlecovic B, Gheorghiu V (1992) Instability and Cognitive Order Formation: Self-Organization Principles, Psychological Experiments, and Psychotherapeutic Interventions. In: Tschacher W, Schiepek G, Brunner EJ (Eds) Self-Organization and Clinical Psychology. Springer, Berlin (in press)

Luhmann N (1984) Soziale Systeme. Grundriß einer allgemeinen Theorie. Suhrkamp, Frankfurt am Main

Mackey MC, an der Heiden U (1982) Dynamical Diseases and Bifurcations: Understanding Functional Disorders in Physiological Systems. Funkt Biol Med 1: 156-164

Mahoney MJ (Ed) (1980) Psychotherapy Process. Current Issues and future Directions. Plenum Press, New York

Martienssen W (1989) Gesetz und Zufall in der Natur. In: Gerok W (Hrsg) Ordnung und Chaos in der belebten und unbelebten Natur. Wissenschaftliche Verlagsgesellschaft, Stuttgart, S 77-99

McClelland JL, Rumelhart ED (1986) Parallel Distributed Processing: Explorations in the Microstructure of Cognition. Vol. 2: Psychological and Biological Models. MIT Press, Cambridge Mass.

Nuechterlein KH, Dawson ME (1984) A Heuristic Vulnerability-Stress Model of Schizophrenic Episodes. Schizophrenia Bulletin 10: 300-312

Odum EP (1983) Grundlagen der Ökologie. Bd 1: Grundlagen. Bd 2: Standorte und Anwendungen. Thieme, Stuttgart

Orlinsky DE, Howard KI (1986) The Relation of Process to Outcome in Psychotherapy. In: Garfield SL, Bergin AE (Eds) Handbook of Psychotherapy and Behavior Change. Wiley, New York, pp 311-381

Rensing L, an der Heiden U, Mackey MC (1987) (Eds) Temporal Disorder in Human Oscillatory Systems. Springer, Berlin

Revenstorf D (1985) Nonverbale und verbale Informationsverarbeitung als Grundlage psychotherapeutischer Intervention. Hypnose und Kognition 2(2): 13-35

Revenstorf D (1990) Zur Theorie der Hypnose. In: Revenstorf D (Hrsg) Klinische Hypnose. Springer, Berlin, S 79-99

Rumelhart DE, McClelland JL (1986) Parallel Distributed Processing: Explorations in the Microstructure of Cognition. Vol 1: Foundations. MIT Press, Cambridge, Mass.

Schaub H, Schiepek G (1991) Simulation of Psychological Processes - Basic Issues and Illustration within the Etiology of Depressive Disorder. In: Tschacher W, Schiepek G, Brunner EJ (Hrsg) Self-Organization and Clinical Psychology. Springer, Berlin (in press)

Schiepek G (1986) Systemische Diagnostik in der Klinischen Psychologie. Psychologie Verlags Union, München

Schiepek G (1991) Systemtheorie der Klinischen Psychologie. Vieweg, Braunschweig

Schiepek G, Schaub H (1991) Als die Theorien laufen lernten... Ein Simulationsmodell zur Depressionsentwicklung. In: Schiepek G Systemtheorie der Klinischen Psychologie. Vieweg, Braunschweig, S 221-305

Schiepek G, Fricke B, Kaimer P (1992) Synergetics of Psychotherapy. In: Tschacher W, Schiepek G, Brunner EJ (Eds) Self-Organization and Clinical Psychology. Springer, Berlin (in press)

Schiepek G, Reicherts M (1992) The Concept of System-Games as Research Paradigm for the Self-Organization of Complex Social Systems. In: Tschacher W, Schiepek G, Brunner EJ (Eds) Self-Organization and Clinical Psychology. Springer, Berlin (in press)

Schiepek G, Schoppek W, Tretter F (1992) Synergetics in Psychiatry: Simulation of Evolutionary Patterns of Schizophrenia on the Basis of Nonlinear Difference Equations. In: Tschacher W, Schiepek G, Brunner EJ (Eds) Self-Organization and Clinical Psychology. Springer, Berlin (in press)

Schiepek G, Tschacher W (1992) Application of Synergetics to Clinical Psychology. In: Tschacher W, Schiepek G, Brunner EJ (Eds) Self-Organization and Clinical Psychology. Springer, Berlin (in press)

Schmid GB (1991) Chaos Theory and Schizophrenia: Elementary Aspects. Psychopathology 24: 185-198

Schöner H, Haken H, Kelso J (1986) A Stochastic Theory of Phase Transitions in Human Hand Movements. Biol Cybern 53: 247

Schuster HG (1988) Deterministic Chaos. An Introduction. VCH, Weinheim

Schwegler H (1978) Sind Katastrophen vorhersehbar? - Geometrische Modelle für sprunghafte Veränderungen in Natur und Gesellschaft. Jahrbuch der Wittheit zu Bremen XXII: 181-198

Schweitzer J (1989) Professionelle (Nicht-)Kooperation: Ihr Beitrag zur Eskalation dissozialer Karrieren. Z system Ther 7(4): 247-254

Seifritz W (1987) Wachstum, Rückkoppelung und Chaos. Hanser, München

Shazer S de (1985) Keys to Solution in Brief Therapy. Norton, New York

Silberschatz G, Curtis JT, Nathans S (1989) Using the Patient's Plan to Assess Progress in Psychotherapy. Psychotherapy 26: 40-46

Stadler M, Kruse P (1986) Gestalttheorie und die Theorie der Selbstorganisation. Gestalttheory 8: 75-98

Stadler M, Kruse P (1990) The Self-Organization Perspective in Cognition Research: Historical Remarks and New Experimental Approaches. In: Haken H, Stadler M (Eds) Synergetics of Cognition. Springer, Berlin, pp 32-52

Stadler M, Richter PH, Pfaff S, Kruse P (1991) Attractors and Perceptual Field Dynamics of Homogeneous Stimulus Areas. Psychol Research 53: 102-112

Stegmüller W (1973) Theorie und Erfahrung, Zweiter Halbband: Theorienstrukturen und Theoriendynamik. Springer, Berlin

Strauss JS (1989) Intermediäre Prozesse in der Schizophrenie: zu einer neuen dynamisch orientierten Psychiatrie. In: Böker W, Brenner HD (Hrsg) Schizophrenie als systemische Störung. Huber, Bern, S 35-50

Strohschneider S, Schaub H (1991) Konnektionismus und Kognitive Psychologie. Systeme 5(2) (in press)

Teubner G (1987) Episodenverknüpfung: Zur Steigerung von Selbstreferenz im Recht. In: Baecker D, Markowitz J, Stichweh R, Tyrell H, Willke H (Hrsg) Theorie als Passion. Suhrkamp, Frankfurt am Main, S 423-446

Teubner G (1990) Hyperzyklus in Recht und Organisation. Zum Verhältnis von Selbstbeobachtung, Selbstkonstitution und Autopoiese. In: Krohn W, Küppers G (Hrsg) Selbstorganisation: Aspekte einer wissenschaftlichen Revolution. Vieweg, Braunschweig, S 231-264

Trulson ME, Preussler DW (1984) Dopamine-containing Ventral Tegmental Area Neurons in freely Moving Cats: Activity during the Sleep-waking Cycle and Effects of Stress. Experimental Neurology 83: 367-377

Tschacher W (1990) Interaktion in selbstorganisierenden Systemen. Asanger, Heidelberg

Tschacher W, Brunner EJ, Schiepek G (1992) Self-Organization in Social Groups. In: Tschacher W, Schiepek G, Brunner EJ (Eds) (1992) Self-Organization and Clinical Psychology. Springer, Berlin (in press)

Tschacher W, Schiepek G, Brunner EJ (Eds) (1992) Self-Organization and Clinical Psychology. Springer, Berlin (in press)

Tyson JJ (1976) The Belousov-Zhabotinski-Reaction. Springer, Berlin

Wertheimer M (1912) Experimentelle Studien über das Sehen von Bewegung. Z Psychol 61: 165-292

Wertheimer M (1923) Untersuchungen zur Lehre von der Gestalt II. Psychol Forsch 4: 301-350

Zingg MB (1988) Schematheoretische Sichtweise psychologischer Störungen. Inauguraldissertation, Universität Bern

Zubin J, Spring B (1977) Vulnerability: A New View of Schizophrenia. J.Abnorm. Psychol. 86: 103-123

Organization and Self-Organization

W. Tschacher[1] *and E.J. Brunner*[2]

[1]Sozialpsychiatrische Universitätsklinik, Murtenstr. 21, CH-3010 Bern, Switzerland
[2]Institut für Erziehungswissenschaft, Universität Tübingen, Münzgasse 22–30, W-7400 Tübingen, Fed. Rep. of Germany

Abstract. A systemic approach to organizational theory is discussed on the basis of self-organization theory. Organizations are conceived of as nonlinear systems characterized by microscopic complexity, circular causality and openness to their psycho-social environments. Finally, consequences of the self-organizational view for the field of management and organizational theory are discussed. As a synthesis of interventionist and non-interventionist arguments as well a position of indirect evolutionary management is inferred from synergetics and recent trends in the field of industrial organization.

1. Organization and Self-Organization in Industrial Management Theory

The play on the words in the title is intentional: when we speak of larger social systems, that is to say of organizations, it seems natural, as we should like to show, to view them from the standpoint of the theory of self-organization.

To do this we shall first of all turn our attention to publications in the field of industrial management theory. How are industrial organizations defined there?

As regards the concept "organization" a distinction is made in industrial management theory between a sociological concept of organization and one which is business-centered (Hahn, 1990). In the sociological concept a company is seen as a social complex arising from the division of labor in which business is done. The industrial economists express it in this way: "A company *is* an organization"; here organization is used as an umbrella term for the full range of goal-directed social systems (Bühler, 1987, p. 4). In contrast, the business-centered concept speaks of a firm as an organization because it has an organization (structure). "A company *has* an organization" (Hahn, 1990, p. 391 f.)

In the latter definition what is meant by organization is the total of all measures with the help of which the management tries to set up an order, previously conceived by planners, according to which the formal process of on-going business is permanently orientated. "These measures form the concrete content of the firm's organization" (Fries, 1987, p. 61).

Formulations such as these suggest an understanding of systems theory, which is indeed to be found in industrial management theory. Borrmann (1984, p. 338) holds that any enterprise (organization) can be seen as a system "in which the financial, technical and human resources are combined with one another in such a way that its aims are effectively (and as a rule economically) achieved. This process

Springer Proceedings in Physics, Vol. 69
Evolution of Dynamical Structures in Complex Systems
Editors: R. Friedrich · A. Wunderlin

of combination is controlled by the "central nervous system" of the firm: the management system. Rall & Hagemann (1984, p. 329) view organizational structure as an important instrument of management. What decides the success of a firm, however, is that organization is not applied in isolation, "but as one of the elements, which together form the system of management, the ordering framework for the day's work and which act in combination as in a system of communicating pipes".

In the theory of organization in industrial management studies the systems character is, as a rule, seen in the formal structures. The formal structure is described as the consciously created, rationally formed structure for the fulfilment of the firm's objectives. In contrast, by informal organization we are to understand "the social structures determined by the personal aims, wishes, sympathies and behavior of the employees" (Bühler, 1987, p. 6). These are consistently referred to as sources of "disturbance".

A contrary opinion is held on this point by Ulrich (1985). For him conscious, planned organizing represents at most a necessary complement to or correction of informal structures in a firm. A basic understanding of organization ought rather to be of an informal structure constantly re-forming and adjusting by virtue of its self-organizing abilities. In this light the problem of organization lies in strengthening the human faculty of self-organization through forward-looking outlining of structure that also fosters economically viable action.

Similarly in Malik's view (1986) the basic problem for management is seen in the control of complexity. According to Staehle (1989) one may distinguish a directive-technomorphous theory type from a systemic-evolutionary theory type. The two types are condensed into tabular form in Table 1.

After Staehle systems theory and above all cybernetic approaches have exerted a considerable influence particularly in management research. The author sees here a development in the area of industrial management theory and takes up the distinction between cybernetics of the first and of the second order. Whereas in the first phase of cybernetics management research was mostly concerned with processes maintaining balance with the aim of stability it was, according to cybernetics of the second order, interested in problems of instability, flexibility, change, learning, evolution, autonomy and self-reference. Seen in this light imbalance is no catastrophe but rather the normal case and prerequisite for change.

The development in the theory of industrial economics would be an appropriate paradigm – in our view – for a long-needed revision of the concept of organization in the social sciences. A self-organization viewpoint, which we consider unavoidable, is, however, far from a matter of course as yet. Wiedl and Graf (1991, p. 394), from the field of work and industrial psychology, describe an organization structure "as the total of process relations between members of an organization and the tasks specific to the organization." This perception is systemic in the sense that organizations are seen as complex structures in which many elements stand in relation to one another. The conception, however, does not get beyond the position of cybernetics of the first order outlined above.

Directive-technomorphous theory type		Systemic-evolutionary theory type
Management...		Management...
1.	... is leading people	...is shaping and steering of whole institutions in their environment
2.	... is leadership by the few	... is leadership by many
3.	... is the task of the few	... is the task of many
4.	... is direct influence	... is indirect influence
5.	... is geared to controllability	... is geared to achieving the optimum
6.	... has information on the whole	... never has enough information
7.	... has the aim of maximization of profits	... has the aim of the maximization of the ability to survive

Table 1. Contrasting two types of management

2. Self-Organization in Organizations

We start from the assumption that – in analogy to observable phenomena in other areas of the social sciences – in larger social organizations collective, ordered behavior of components occurs spontaneously in the form of macroscopic patterns. What we postulated for small groups (Tschacher, 1990; Brunner and Tschacher, 1991a; Tschacher et al., 1992) applies also to larger social systems:

In organizations (firms; schools; administrative bodies; etc.) we are concerned with "multi-component" systems as well. It is the communicated cognitions and emotions which represent the elementary components on the micro level. Organizations as self-organizing systems are complex, that is to say the number of their components and/or the connections between them is very large. Macroscopically visible order emerges here too – as in the small groups – from the "chaos" of very many degrees of freedom as a new kind of quality. Self-organized organizations are composed of a great many separate (micro-) components which form the constituents for structures on the phenomena level.

In order to be able to explain the structures and processes on the macro level only a few order parameters ("principle of enslavement") are required. The principle of openness continues to hold: self-organized systems – organizations themselves too after all – exist in a state of permanent exchange with their environment, which influences them by means of control parameters. And finally the connections between the components of organizations and between systems and control parameters are non-linear, e.g. circular.

On the macro level, i.e. on the level of visible structures, there are many single (social) groups in every organization, out of which the over-all system of the institution in question is constituted. The (formal and informal) small groups are characterized in accordance with self-organization theory on the one hand by external boundary conditions (for instance functional duties); on the other hand through the internal formation of boundaries they give themselves the rules by which

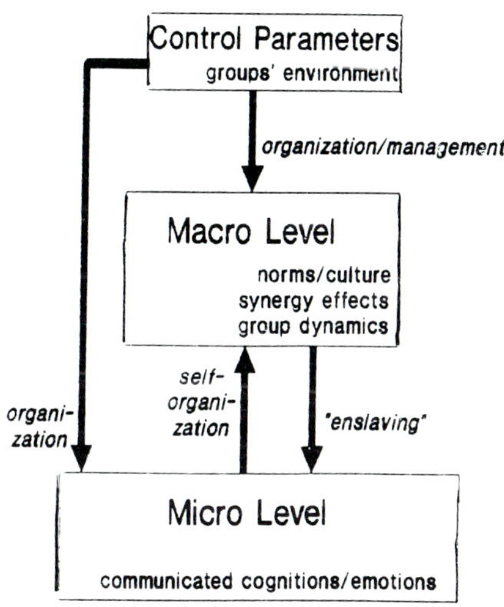

Fig. 1. Schema of organization and self-organization in a complex social system

specific results are obtained.

The degrees of freedom of these "self-organized organizations" are limited in two respects: in the first place these systems exist in an environment, from which they draw energy and/or matter. In the second place organizations are – as indicated above – composed of different kinds of small groups which interact in a multiplicity of ways (Brunner and Tschacher, 1991b). All the groups are connected to one another by virtue of having members in common and/or they have the same social network and are bound together in the hierarchy of the organization in question. Industrial firms can be interpreted for example as miniature societies composed of many interdependent subgroups. Each of these groups will tend to form norms, standards and traditions which are decisive in creating their own particular "work climate".

On the level of the over-all organization too a certain "way of life" will establish itself, creating a company's identity ("corporate identity") which comprises the norms and values characteristic of the organization as a whole, right down to rites and codes of behavior of single persons. In the management of a firm the improvement or creation of an organization climate that seems appropriate is seen as an important aspect of successful organizations. From the field of industrial organization it is known well that "trust, effective communication, prompt feedback and creativity" are among the prerequisites for positive synergy effects (Sprüngli, 1981).

3. Intervention in Complex Systems

How can systems be managed? – this is an aspect central not only in regard to any theory of industrial organization but also especially to self-organizational processes within an organization. There have been several quite controversial debates on this subject so that it seems close at hand to dichotomize between the advocates of interventionism and of non-interventionism. We concede that this confrontation is incomplete and somewhat exaggerated; therefore we will utilize it only in order to reach a synthesis, which will then be based on the perspective of self-organization.

On the side of *interventionism* the tradition of behavior theory must be mentioned; behaviorists claim that behavior can be shaped almost freely either by the management of contingencies and/or by applying appropriate stimuli. To the present day this optimistic preconception can be sensed in modern developments and applications of behavior theory like in the variants of cognitive behavior therapy or – maybe to a lesser degree – in behavioral medicine.

A corresponding interventionist tendency is found in the established organizational theories of industrial economics (the directive-technomorph type of theory), which is based on models of action theory and decision theory. Leadership is emphasized to be a central function in the organization's hierarchy. Without leadership no effective and reasonable action is possible; problems of control in systems are explained by leadership being implemented in a suboptimal way (Ochsenbauer, 1989).

In sum, interventionist approaches assume that even complex problems may be reduced to simple ones, which are then accessible to fixed strategies of problem solving.

In contrast to this stand arguments for *non-interventionism*. Von Foerster's example of the "nontrivial machine" can be listed here; the machine is a simple system with input, output and some state variable (a kind of S-O-R concept). By combinatorial considerations von Foerster is lead to the conclusion that the machine is de facto unpredictable if its interior state is not known. This argument says that even simple systems may show unpredictable behavior the laws of which may never be extracted from overt behavior.

Another argument pointing to an impossibility of goal-oriented interventions stems from language philosophy and constructivism: the argument of self-reference. From a social systems point of view the manager is herself part of the system she tries to manage. Thus, management is self-management of a system, planning becomes planning plannings (Krohn & Küppers, 1991; cf. Schiepek & Tschacher, 1992); in a cascade of meta-levels the firm ground necessary to control and direct a system is lost. If a system is closed self-referentially in this way, and there is no demon or transcendence left out, then intervention is merely what the system does anyway, i.e. its eigen-behavior.

From a different strain of research, namely cognitive psychology, we know of findings that also cast doubt on the possibility of successful interventions in complex systems. In this respect Dörner (1989) speaks of a "logic of failure".

All anti-interventionist arguments have a basic statement of this kind in common: complex systems are not reducible and generate complex behavior which can hardly be prognosticated, if at all.

Starting from these preparatory remarks we will try to reach a synthesis by differentiating between different aspects of the problem of management and intervention. It is a well-known fact in synergetics (Haken, 1990; 1991) that complex systems under certain conditions bring forth rather simple behavior: one single mode "enslaves" all the others. From the phenomenology of self-organized systems we may deduce two hypotheses:

i) A system is sometimes less than the sum of its parts;

ii) control by management can be broken down into influencing and directing events; there is also need for prognosis; these components must be viewed separately.

Therefore, when considering the question of intervention we have to differentiate depending on what kind of dynamical regime is realized. For elucidation a simple bifurcation diagram showing a system in parameter space may be used (Fig. 2). If the system is located close to a critical symmetry-breaking point

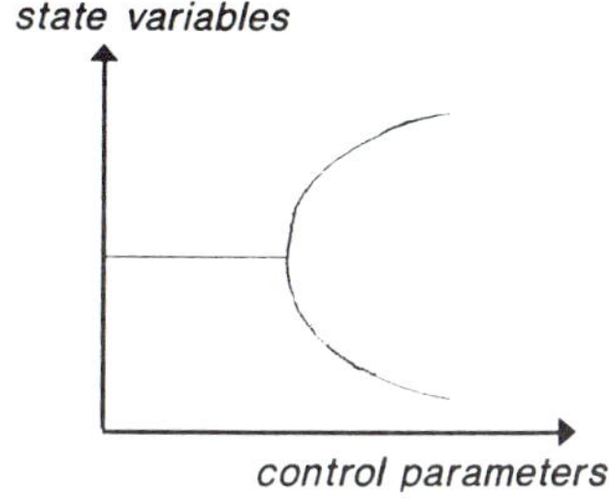

Fig. 2. Bifurcation diagram of group evolution

we can give no deterministic prognosis of the system's further development. The attractor (i.e. the representation of the system's homeostasis) is shallow, so that even small fluctuations at the appropriate time can decide into which attractor the system will settle down in the future. At the same time our chance at influencing the system is obviously great: we could well add the decisive bias ourselves. The situation in other words: prognosis is unreliable, while influence is easy. But there are restrictions to this influence; controlling the system is not possible in just any direction since our choice is limited to the paths offered by the system. Attractors are not modelled by the manager or therapist in any direct way. As a rule, we will not even know about the new attractor. Thus, this kind of control is more like an influence with eyes closed.

The opposite situation occurs in areas of parameter space well away from critical points. Here prognosis is easy as the system is stable (stability need not mean stationarity, but may also refer to stable, i.e. attracting, oscillations). This is why intervention may meet with the system's "resistence".

Another aspect worth considering in regard to the control of dynamical systems is given by the arrow of time that is linked to dissipation. At least in social and mental systems symmetry breakings are *irreversible*. One may infer intuitively that irreversibility by dissipation (Prigogine, 1980) corresponds to axioms of communication pragmatics (Watzlawick et al., 1967). Communications cannot be taken back nor be inverted, because the system (if sensitive to exterior influence at all) is changed by any event immediately.

How can we manage a chaotic system? In some respect, dealing with deterministic *chaos* resembles the situation close to critical points as we already mentioned: small causes may have great effects. Simultaneously this means that the system is influenced easily but cannot be predicted in the long run. It seems important for the functionality of chaotic attractors that they undergo qualitative changes when parameters of the system's environment vary but slightly; this applies to chaotic dynamics in cognition (Nicolis, 1986) – probably also in economics?

In order to apply our phenomenological descriptions of dynamical systems to the management of groups and organizations we may re-evaluate two points that were already mentioned above in the context of systemic management theory:

1) Management must necessarily be indirect because a self-organized system can be influenced and directed only insofar as attractors are generated by the system. This amounts to a rejection of both interventionism and non-interventionism. According to Malik and other systemic theoreticians of (industrial) organization, management is performed by presenting a framework for self-organization. Management is working on the system's control parameters; management cannot determine the system's equilibrium directly.

2) It is obviously a central point which equilibria systems in economy bring forth when allowed to be more or less self-controlled. In this respect the term "synergy" is used widely, expressing the belief that benefits can be expected to emerge from the interaction of previously independent parts (on the relation of synergy and synergetics cf. Ulrich & Probst, 1984). Doctoroff (1977), for instance, proposed to establish synergy effects by informal gatherings.

4. Trends of Development in Organizational Theory

We will now present some recent developments of organizational psychology and management theory in order to assess these trends from the perspective of synergetics.

During the last decade there has been growing emphasis on issues of *organizational culture* (Schein, 1985). It is accepted increasingly that the "soft" cultural aspects of an industrial company can be at least as essential as the "hard facts" of economy. Culture makes it possible for persons in an organization to identify with that organization. Organizational culture becomes visible in specific behavior patterns, ways of communicating, rituals, etc. Interestingly, these are emergent phenomena of an informal kind which we introduced as being connected to the self-organization of groups. Thus, emergent aspects seem to move into the focal area of organizational theory.

Theory-Z companies put special emphasis on organizational culture. The distinction of theory-Z and other types of companies was introduced by Ouchi (1981), who was the first of many authors to scrutinize the reasons for the success of Japanese firms. To state some of the topics that have been found to differentiate between western (i.e. mainly US-American) and Japanese attributes: A Theory-Z style of enterprise stresses human resources and the finding of consensual decisions more than the competition of individuals in reaching economical goals; the philosophy of the company is central; job rotation within a firm is common since one-sided specialization of employees is not encouraged.

Computer Integrated Manufacturing (CIM) is jargon for the informationally enmeshed company of the future. In this respect some observations of innovation motivated by computerized integration make clear that these issues are compatible with a systemic understanding of organization: the prototypical case of a computer-based organization is not the infamous "fully automated robot factory", but a flexible production in which work units are delegated to more autonomous groups and individuals. This is a basic claim of the so-called IZ-philosophy proposing an "Ideal Zero" of informational interfaces between stages of production. Production becomes object-oriented rather than organized by functional segmentation.

Participation which substitutes for linear directive leadership is a common denominator to the different trends. From the increased autonomy of teams and from the expansion of work segments flat hierarchy necessarily results. This is also economically efficient, as can be exemplified by the installation of quality circles in many organizations. Quality circles are teams recruited from members of the production sector which serve the function of the traditional department of quality control; division of labor among functional teams is decreased.

To summarize what has been said we may state that for managers it becomes increasingly important to know how group systems – or, more fundamentally, complex dynamical systems – evolve and maintain themselves. This follows mainly from the fact that the single groups and teams may and must act more independently in modern types of organization. Management theories of the systemic-evolutionary type (cf. especially the school of St. Gall; Ulrich, 1984; Ulrich & Probst, 1990) therefore advocate that the psycho-social self-organization of groups in organizations has to be considered with priority. There is growing attention to, and trust in, evolutionary mechanisms rather than directive leadership. The informal level of human interaction is seen as significant and is no longer categorized as dysfunctional only. Holistic work processes will in the long run replace the extreme Tayloristic division of labor. As far as management is concerned, knowledge about how to handle groups and their spontaneous activity is of vital interest. An organization is no longer viewed as waiting to be organized by some external manager but as a self-organized entity.

References

Brunner, E.J. & Tschacher, W. (1991a) Distanzregulierung und Gruppenstruktur beim Prozeß der Gruppenentwicklung. I.: Theoretische Grundlagen und methodische Überlegungen. Zeitschrift für Sozialpsychologie, **22**, 87-101.

Brunner, E.J. & Tschacher, W. (1991b) Selbstorganisation und die Dynamik von Gruppen – Die systemische Perspektive in der Sozial- und Organisationspsychologie. In: Niedersen, U. & Pohlmann L. (Eds.), Jahrbuch "Selbstorganisation", (Vol. 2); Berlin: Duncker & Humblot, 53-67.

Borrmann, W.A. (1984) Organisationsentwicklung. In: Management Enzyklopädie. Lexikon der modernen Wirtschaftspraxis. Vol. 7. Weinheim: Zweiburgen Verlag, 338-347.

Bühler, R. (1987) Betriebswirtschaftliche Organisationslehre. München: Oldenbourg. (3rd ed.)

Doctoroff, M. (1977) Synergistic Management: Creating the Climate for Superior Performance. New York: AMACOM.

Fries, H.-P. (1987) Betriebswirtschaftslehre des Industriebetriebes. München: Oldenbourg. (2nd ed.)

Hahn, O. (1990) Allgemeine Betriebswirtschaftslehre. München: Oldenbourg.

Haken, H. (1990) Synergetik: eine Einführung (Nichtgleichgewichts-Phasenübergänge und Selbstorganisation in Physik, Chemie und Biologie). Berlin: Springer (3rd ed.)

Haken, H. (1991) Synergetik im Management. In: Balck, H. & Kreibich, R., Evolutionäre Wege in die Zukunft. Weinheim: Beltz, 65-91.

Krohn, W., & Küppers, G. (1990). Selbstreferenz und Planung. In: U. Niedersen & L. Pohlmann (Eds.), Selbstorganisation und Determination. Berlin: Duncker & Humblot, 109-127.

Malik, F. (1986) Strategie des Managements komplexer Systeme. Bern: Haupt.

Nicolis, J.S. (1986) Dynamics of Hierarchical Systems. Berlin: Springer.

Ochsenbauer, C. (1989) Organisatorische Alternativen zur Hierarchie. München: GBI-Verlag.

Ouchi, W.G. (1981) Theory Z. How American Company Can Meet the Japanese Challenge. Reading, Mass.: Addison-Wesley.

Prigogine, I. (1980) From Being to Becoming. San Francisco: Freeman.

Rall, W. & Hagemann, H. (1984) Organisation. In: Management Enzyklopädie. Lexikon der modernen Wirtschaftspraxis. Vol. 7. Weinheim: Zweiburgen Verlag, 321-337.

Schein, E. H. (1985) Organizational Culture and Leadership. San Francisco: Jossey-Bass.

Schneider, D. (1987) Allgemeine Betriebswirtschaftslehre. München: Oldenbourg. (3rd ed.)

Schiepek, G. & Tschacher, W. (1992) Application of Synergetics to Clinical Psychology. In: Tschacher, W., Schiepek, G. & Brunner, E.J. (Eds.), Self-Organization and Clinical Psychology. Berlin: Springer.

Sprüngli, K. (1981) Evolution und Management. Ansätze zu einer evolutionistischen Betrachtung sozialer Systeme. Bern: Haupt.

Staehle, W.H. (1989) Management. Eine verhaltenswissenschaftliche Perspektive. München: Vahlen. (4th ed.)

Tschacher, W. (1990) Interaktion in Selbstorganisierten Systemen. Grundlegung eines dynamisch-synergetischen Forschungsprogramms in der Psychologie. Heidelberg: Asanger.

Tschacher, W., Brunner, E.J., & Schiepek, G. (1992) Self-Organization in Social Groups. In: Tschacher, W., Schiepek, G., & Brunner, E.J. (Eds.), Self-Organization and Clinical Psychology. Berlin: Springer.

Ulrich, H. (1984) Management. Bern: Haupt.

Ulrich, H. (1985) Organisation und Organisieren in der Sicht der systemorientierten Managementlehre. In: Zeitschrift Führung und Organisation, **54**, p. 7ff.

Ulrich, H., & Probst, G.J.B. (1984) Self-Organization and Management of Social Systems. Berlin: Springer.

Ulrich, H., & Probst, G.J.B. (1990) Anleitung zum ganzheitlichen Denken und Handeln (Ein Brevier für Führungskräfte). Bern: Haupt. (2nd ed.)

Ulrich, P. (1986) Transformation der ökonomischen Vernunft. Fortschrittsperspektiven der modernen Industriegesellschaft. Bern: Haupt.

Watzlawick, P., Beavin, J.H., & Jackson, D.D. (1967) Pragmatics of Human Communication. New York: Norton.

Wiedl, K.H. & Greif, S. (1991) Störungen betrieblicher Organisationen: Intervention. In: M. Perrez & U. Baumann (Eds.), Lehrbuch Klinische Psychologie. Vol. 2: Intervention. Bern: Huber, 395-407.

Part VII

Epistemology

Schelling's Concept of Self-Organization

M.-L. Heuser-Keßler

Philosophisches Institut der Heinrich-Heine-Universität Düsseldorf, Universitätsstr. 1, W-4000 Düsseldorf 1, Fed. Rep. of Germany

Abstract: In the early 19th century the classical "Naturphilosophie" of F.W.J. Schelling (1775-1854) had many repercussions in the sphere of science. The term "self-organization" was originally introduced by I. Kant to characterize processes of organic nature. Schelling was the first to extend the concept to inorganic nature, for he applied it to the complete evolution of the universe, from the primordial beginnings of matter up to the origin of life and human mind. Schelling attempted to go beyond the contemporary mechanistic theory of nature by designing a program for science that would focus on the self-creation of "organizations". Of course he could not fall back upon elaborate scientific theories of self-organization, and at his time even empirical phenomena of self-organization were scarcely known. So Schelling created a "speculative physics" which was based on philosophical principles and intended to yield a systematic explanation of the history of nature. This paper will deal especially with the origin of Schelling's concept of self-organization, the specific meaning of this concept in the context of his work, and his method of construction. Apart from that it will be shown how the reception of Schelling's ideas in 19th century science led to first attempts at mathematical theories for a non-mechanistic concept of nature, especially concentrating on the works of Bernhard Riemann.

1. Introduction

Schelling's works on "Naturphilosophie" - especially those of the early phase: "Ideen zu einer Philosophie der Natur" (1797), "Von der Weltseele, eine Hypothese der hoeheren Physik zur Erklaerung des allgemeinen Organismus" (1798), "Erster Entwurf eines Systems der Naturphilosophie" (1799), "Allgemeine Deduktion des dynamischen Prozesses oder der Kategorien der Physik" (1800), but also those of the later period: "Die Weltalter" (1811/1813), "Darstellung des Naturprozesses" (1843/44) - represent a first extensive attempt to establish a theoretical basis for a scientific treatment of the history of nature and of the self-organizing character of the inorganic, organic, and cognitive spheres. His objective is to find the universal principles of the "productivity of nature", which can explain on an evolutionary basis the underlying unity of organic and inorganic nature, but also the unity of the external, visible nature with our internal, invisible nature.

Springer Proceedings in Physics, Vol. 69
Evolution of Dynamical Structures in Complex Systems
Editors: R. Friedrich · A. Wunderlin

Schelling believed that the creative productivity of man - which is responsible for the fact that human beings have a history - is intimately connected with the process of nature as a whole. As a consequence the innovative faculties peculiar to human beings do not separate or distinguish them from nature. On the contrary, it is the core of nature which is reflected in human creativity. In contrast to the transcendental philosophies of Kant and Fichte, which are concerned with the productive powers of the spirit, Schelling proceeds from the presumption that - apart from man's faculty to build up a world in theory - nature itself must have created the diversity of organizations in reality. In his "Erster Entwurf eines Systems der Naturphilosophie" of 1799 Schelling stated the following as a first principle: "... nature takes its reality out of itself - it is its own product - a self-organizing and self-organized whole." /23, III p.17/ For Schelling this means that nature has "unconditional reality", that it does not depend on external conditions, but is self-creative, and he assumes that there must be also empirical evidence for this process of self-organization. The "Unconditional" - a term that has often been misconstrued - is nothing other than the original productivity which produces each level of development in a non-mechanistic manner, i.e. without external determination.

On the basis of this fundamental principle of self-organization Schelling criticizes the empiricism as well as the mechanicism of his age. Empiricism - so he argues - takes the objects of nature as given and then proceeds to observation, analysis and description of the relations pertaining between them. The reconstruction of the origin and genesis of natural objects - according to him the primary task of a future science - is left out of account. Schelling writes: "The difference between empiricism and science is based on the fact that empiricists consider their object as something complete and as a product, whereas science observes the object in the process of emerging and as something that has yet to be brought about." /23, III p.283/ Schelling's intention was not to advocate a science without empirical methods. But it was important for him to construe the objects of experience as products (natura naturata) of an original generative process (natura naturans). This process can only be understood by an "unbedingter Empirismus", i.e an empiricism with respect to the self-construction of nature. The mechanistic theories of nature were criticized by Schelling for limiting their explanation to external causation, an approach which excludes a self-organizing process of nature right from the start.

I. Kant had shown that there are at least two areas which cannot be adequately described in mechanistic terms: the area of living things and the area of practical reason that is based on freedom. In his "Critique of Judgment" he had analysed in detail that organisms are not susceptible to mechanistic explanation. He was the first to introduce the term "self-organization" in order to characterize the dynamic process of the organic sphere as opposed to the mechanic sphere of physics. Kant puts it this way: "An organized being is, therefore, not a mere machine. For a machine has solely **motive power**, whereas an organized being possesses inherent **formative** power, and such, moreover, as it can impart to material devoid of it - material which it organizes. This, therefore, is a self-propagating formative power, which cannot be explained by the capacitiy of movement alone, that is to say, by mechanism. - We do not say half enough of nature and her capacity in organized products when we speak of this capacity as being the **analogue of art**. For what is here present to our minds is an artist - a rational being - working from without. But nature, on the contrary, organizes itself, and does so in each species of its organized products - following a single pattern, certainly, as to general features, but nevertheless admitting deviations calculated to secure self-preservation under particular circumstances."/16, p.557/ In an organized being every part is thought of as owing its presence to the agency of all the remaining parts, and also as existing for the sake of the others and of the whole. Every part is an organ producing the other parts. Therefore Kant writes: "Only under these conditions and upon these terms can such a product be an **organized** and **self-organized being**, and as such, be called a **physical end.**"/16, p.557/

One can in particular deduce from Kant's choice of examples that his interpretation of the term self-organization differs from modern usage. He does not refer to the origin of organization, but to the **reproduction** of existing organized beings, especially those of organisms. Kant excluded the evolutionary perspective as a "daring venture on the part of reason" /16, p.579/ and refrained from taking up the topic of the emergence of life from the inorganic sphere, which would have led him to questioning his mechanistic concept of physics. He declared Newton's physics to have unrestrictedly validity. To him a physics of self-organization was unthinkable. Therefore he only accepted the concept of self-organization in biology as a merely regulatory but not as a constitutive principle. The limitations of Kant's concept were fully realized by Schelling. He extended the concept of self-organization as an emergent process to the whole of nature and to the coming about of new forms of organization as opposed to the reproduction of existing ones. Schelling writes: "The **whole** of nature, not only a **part** of it, should be like an ever-**becoming** product. All of nature must be thus in perpetual formation, and everything must be included in this process." /23, III p.33/ The history of nature is thus for Schelling a progressive "Potenzierungsprozeß" which goes "from the simplest to the highest and the most complex." /23, IV s.89/

Before one can draw a parallel between Schelling's idea of self-organization and the modern one /8/, we have to take into account that he developed this idea as a philosopher and not as a scientist. As a consequence of his interest in "Naturphilosophie" he succeeded, however, in making himself familiar with the state of the art in contemporary science, and his philosophical reflections are certainly on the level of the scientific knowledge of his period. At that time there was no such thing as a physics of self-organization, not even an accepted theory of evolution in biology. Lamarck published his work "Philosophie Zoologique" only in 1809, and Darwin his "Origin of Species" in 1859. Not even empirical phenomena of self-organization in the sphere of physics were known. The Laser, Bénard cells, the Belousov-Zhabotinsky-reaction were all completely unknown at the time. The only processes in physics which Schelling could point to were hydrodynamic whirls, crystallization, chemical and physiological processes. Nevertheless Schelling adopted the far-reaching hypothesis of a universal self-organization of nature as the central principle of his "Naturphilosophie" and furthermore attempted to establish a preliminary theoretical basis for a future research program of a science of self-organization. There was no chance for this program in the 19th century, because neither the necessary instruments of empirical research nor the mathematical theories were available. So the advocates of a mechanistic theory of nature won their case as opponents of the Schelling school in a polemic disqualifying dynamic "Naturphilosophie" as "speculative pipe dreams". As a consequence "Naturphilosophie" as a discipline almost fell into oblivion.

Schelling himself was quite aware of the fact that his project could not be more than a preliminary stage. He was primarily concerned with "stating the fact" that nature is not a static and mechanistic system, but a "progressive genesis" proceeding from the origin of inorganic systems up to the emergence of human consciousness /23, X p.229/. He compares the intention of philosophy to explain the fact of the world with the role of experiment in science and writes: "Nothing seems easier than naming the fact that philosophy has to explain. But think of all the work and efforts which are necessary in science to arrive at a single true fact concerning a simple phenomenon. People will say: philosophy is supposed to explain the fact of the world. But what about this world is the fact?" /23, X p.228/ The genesis of nature could be considered as a hypothesis and submitted to experimental treatment only after certain experiences with respect to the self-organization and history of mankind were made resulting in a new self-image of mankind, which were formulated by the philosophers of the "Enlightenment" (more about this in the next section).

Schelling regretted greatly the limitations of contemporary experimental science: "The more we describe the sphere of investigation, the more we see how scanty and defective the relevant experience is, and few will feel this imperfection more deeply and vividly than the investigator himself." /23, II p.351/ Schelling thought that still he deserved credit for at least "having dared anything at all in this science, so that others may be in a position to practice their wit by detecting and refuting the errors."/23, II s.351/

His courage in formulating new hypotheses was rewarded. Though Schelling's "Naturphilosophie" was criticized by scientists and philosophers of empiricist and mechanistic standing, it was received favourably in mathematical physics, and it seems to have survived there in covert and transformed shape the attacks of 19th century science. Schelling's "Naturphilosophie" can be shown to have served as an orientation and guideline for a whole generation of theoretical physicists, who created an "abstract" science of physics working with mathematical-theoretical models besides the "concrete" science of physics with its strong empirical bias. (viz. Caneva who shows this for the context of electromagnetism, /1/). Some examples of this can be also drawn from the work of a research project that is under way at Duesseldorf University. Unfortunately this project is still in its early stages so that results cannot be fully stated as yet. One of its objectives is to explore the historical continuity of Schelling's philosophy of self-organization into present-day physics of self-organization. Apart from that the project intends to examine the chances to maintain Schelling's concepts under present conditions in the light of certain mathematical considerations which were made in the 19th century. This would require a detailed analysis of the particular mechanisms conceived by Schelling in his "speculative physics" and an analysis of the works of those mathematicians, who were influenced by Schelling's philosophy, but this analysis is still under work, and so only the general ideas can be presented.

One of the mathematicians influenced by the philosophy of Schelling is Hermann **Grassmann**, who in his "Ausdehnungslehre" of 1844 developed a vector calculus for multidimensional space in order to formulate better explanations for morphogenetic processes. He applied his new mathematics especially to the process of crystallization and tried to find a more adequate mathematical basis for the dynamic principle of Schelling's philosophy. It is interesting to see how closely his abstract conception of multidimensional "power space" is connected with Schelling's concept of space as developed in "Allgemeine Deduktion des dynamischen Prozesses oder der Kategorien der Physik"(1800). According to Schelling, space is not the empty Euclidean space or a Kantian form of pure intuition, but constituted by means of magnetism, electricity, and chemical process. The three dimensions of space are not an absolute fact but attributable to particular physical conditions /also see 17/.

Schelling's ideas had a great impact on those theories of crystallization in the early 19th century which were based on dynamic principles. In the visible shape of crystals they saw the expression of an invisible system of production principles, and they were the first to introduce point symmetries and description in terms of vectors to the theory of crystals. Prominent in this sphere were Christian Samuel **Weiss** and Justus Guenther **Grassmann**, the father of Hermann Grassmann. /27/

Bernhard **Riemann** was also familiar - by way of Herbart and Fechner - with Schelling's "Naturphilosophie", as we can learn from his posthumous writings, which have not all been published yet. He laid the foundations for "mehrfach ausgedehnte Mannigfaltigkeiten" (multiply extended manifolds) with different metrical systems and for non-Euclidean geometry, which at first sight seemed to contradict empirical evidence. Before he wrote his *thesis for habilitation* "Über die Hypothesen, welche der Geometrie zugrundeliegen" (1854) he was - to follow his own words - almost exclusively occupied with "Naturphilosophie" with the intention to establish a "novel conception of nature" that would go beyond mechanistic theories and be based upon an integration of magnetism, electricity, light and gravity. The relation between Schelling and Riemann will be elaborated later. What concerns us here is that Schelling's "speculative method" was adopted especially by those mathematical physicists who undertook a (field-theoretical) construction of material phenomena on the basis of abstract mathematical considerations and whose overall objective was to find the dynamical organizing principles of nature. These principles cannot be extracted from the data, but are supposed to be the synthetic-constructive generation of the phenomena.

As historians of science have very well pointed out the pioneers of field theory in physics were mostly advocates of Schelling's philosophy (for further literature see /8/): i.e. **Oersted**, who was encouraged by Schelling's ideas to look for experimental evidence of the connection between electricity and magnetism. Thanks to "Naturphilosophie" he was in a position to give a coherent interpretation to his observation of the fact that a magnetic needle is deflected by electricity. Without the particular theoretical guideline the special significance of this observation would have escaped him. **Faraday**, whose mentor Humphry Davy was an admirer of the ideas of Coleridge, an intimate connoisseur of Schelling's works, discovered that the relation works the other way as well: magnets can induce electric current. (The macroscopic standpoint of coherence in Haken's theories was likewise induced by field physics. The enslavement of the particular motions of electrons is due to the amplitudes of the light field, even if conversely the amplitudes are generated by motions of the electrons.)

Last but not least William Rowan **Hamilton**, to whom we owe an abstract formulation of mechanics, was a follower - via Coleridge - of the dynamical concept of nature created by Schelling and Kant. /4/

2. How Did Schelling Conceive the Idea of Self-Organization?

In the introduction of his first work concerning "Naturphilosophie", i.e. "Ideen zu einer Philosophie der Natur", Schelling remarks: "We are still pressed by the same failure to understand how mind and matter can be connected. One can try to conceal the harshness of the contrast by all kinds of illusions, we can insert more and finer intermediate stages between matter and mind, but still we finally get to the point, where matter and mind must either unite or the great leap we wanted to avoid all the time becomes inevitable, and this is the same in all theories (...) We leave man behind as the visible and walking problem of philosophy." /23, II 53sqq./ In order to solve this problem, Schelling formulated as a preliminary integrative objective the following postulate: "Nature shall be visible mind and mind shall be invisible nature." /23, II p.54/

How are these formulae of "identity philosophy" to be interpreted? At first sight we could be persuaded to assume that Schelling is expressing an epistemological relation, because human mind recognizes nature or at least tries to do so, and by the process of cognition pervades natural mechanisms and thus transforms them into something spiritual. Of course the sciences are not themselves nature, but at least they are implicitly based on the assumption that nature is susceptible to cognition. It can be expressed theoretically and thus spiritualized. On the other hand our mind must be in some way related to nature, because otherwise the mind wouldn't be able to recognize the nature.

But Schelling wants more than this epistemological relation between mind and nature. According to Schelling we not only expose an epistemic attitude towards an existing world. We are creative beings, and to that extent we have a history. Human beings are not inherently limited to a particular biological program of behaviour. This recognition was the deciding factor of the "Enlightenment", which was taken up by Schelling. Accordingly we differ from animals in our faculty to liberate ourselves from existing conditions and circumstances. In a way we can separate ourselves from immediate experience and abstract from empirical data. This faculty of **abstraction** is the prerequisite for our ability to set aside all particulars and situational elements in order to reflect the whole fundamentally and with respect to its principles (=philosophy). With our critical assessment of the fundamentals of a given situation we can invent alternatives, improvements and new circumstances, in the sphere of science as well as in society.

The invention of new systems is an expression of the creative nature of mankind, which is not exhausted in one particular system or satisfied with one product. We find in ourselves an infinite tendency to self-organization, that pervades all formations without ever completely realizing itself. Schelling's early hypothesis (1796/1797) is the following: "In our mind we find an infinite striving for self-organization, and something like this must also reveal itself in the external world as a general tendency towards organization. And it is really so. The system of the world is a kind of organization extending from a common centre. The powers of chemical matter are already beyond the limitations of the merely mechanical. Even raw matters settling out of their medium form regular figures (i.e. crystallization M.H.)." /23, I p.386/

This is the background for Schelling's postulate that "nature be visible mind and mind invisible nature". It means that the productivity of human mind, which is responsible for our having a history, is not a point of separation or distinction between man and nature. On the contrary, productivity is at the core of nature. Nature produces out of itself the whole diversity of natural products, including ourselves. This unconscious activity of nature is basically identical with what we do consciously. Thus productive nature is visible mind with respect to reality, and productive mind is invisible nature with respect to ideality. Schelling's philosophy is essentially a philosophy of productivity. The transcendental part of Schellings's philosophy takes care of epistemology. It explains the reality of visible nature ideally, i.e. theoretically, whereas philosophy of nature explains the ideal on the basis of the real. With his "Naturphilosophie" Schelling is seeking a "physical explanation of idealism", i.e. an explanation for the question how we could erupt at the borderline of nature the way we did. If self-organizing mind is a product of nature, then nature cannot be a merely mechanical system. It must at least have made freedom possible.

3. Schelling's Method of Construction and Mathematics

Schelling derived his concept of construction from Kant's philosophy of mathematics, as described in "The Critique of Pure Reason". In Kant's view, only mathematics is able to construct objects by "non-empirical intuition" and to make a priori synthetic judgements, so that mathematics is the only science which presents "the most brilliant example of the extension of the sphere of pure reason without the aid of experience" /15, p.211/. For Kant, creativity which goes beyond experience and constructs new objects not learned from experience, is only possible in mathematics. Kant argues as follows: When for example the geometrician conceives a triangle, he makes not a specific, empirical triangle, but the general, invariant triangle form, which includes all possible triangles, whatever their size, position or special shape. He constructs the "general in the particular", meaning the general conception of the triangle form, which is still a specific general conception in contrast to other geometrical figures, such as the circle. According to Kant: "The **construction** of a conception is the presentation **a priori** of the intuition which corresponds to the conception. For this purpose a **non-empirical** intuition is requisite, which, as an intuition, is an **individual** object; while, as the construction of a conception (a general representation), it must be seen to be universally valid for all the possible intuitions which rank under that conception." /15, p.211/

According to Kant, only the mathematician can construct objects independent of sensory experience and develop new insights because he constructs in the form of pure intuition of space. The geometrical figures are each specific limitations of space and concern the form of reality, not the matter. Accordingly, only sizes can be constructed, i.e. shown a priori in the intuition, whilst qualities can only be registered a posteriori. Therefore natural scientists, like philosophers, can only learn things by experience. Natural scientists structure these experiences under the category of causality; philosophers sub-classify individual experiences under general conceptions discursively.

Schelling disputes Kant's view. According to him, creativity is also possible in other fields besides mathematics. In particular, it must be possible for "speculative physics" to construct natural objects in terms of both form and matter if one wishes to go beyond the mere surface of phenomena and reveal the "internal drive system" of nature's self-construction (also see /12/, /13/). As nature itself produces things by means of an original process of generation, it must also be possible to apprehend nature's self-construction in theoretical terms if there is to be a science of nature that not only describes these manifestations phenomenologically, but also aims to explain them. The natural objects and phenomena we perceive can only be a provisional starting point and must be the starting point, for, according to Schelling: "**We do not know only this or that through experience; rather, we originally know nothing at all except through experience and by means of experience**, and, in this respect, our entire knowledge consists of sets of experiences." /23, III p.278/ But - and, according to Schelling, this is the starting point for science - these sets of experiences must be transformed into sets a priori by recognizing their inherent necessity. We gain such insight into their inner necessity, and thereby certainty about nature, when we penetrate the inner constructional and evolutionary principles of nature itself, from which individual natural phenomena can then be explained as products.

To illustrate his notion of construction, Schelling cites the example of an engineer. The person who really knows a machine is not the one who perceives it through his senses and merely describes its outward appearance and names the parts, and it is also not the one who analytically dismantles it, then reassembles the parts, and, in this way seeks to unravel its functions (although this method will reveal considerably more than a mere description), but it is the engineer who invented and built it. Not only does the inventor know how each part relates to the other parts, but also how the machine itself and above all its total complexity and coherence originally came into being. He knows the idea of the machine by having conceptualized the first **prototype**. Schelling's method of construction is essentially orientated to the paradigm of the inventor and artist, not to the aristotelean paradigm of the craftman, who normally works from the traditional prototypes and does not create his own.

Schelling's idea, then, is that it is not just the inventor who creates qualitatively new entities in the form of prototypes, but that nature does this as well by way of a process of self-organization - but in this instance not ideally, but really by means of natural powers. For example, with the construction of the first living creature, a prototype of organismicity is produced that is a specific general form insofar as it is a new type of natural object that essentially differs from inorganic objects. Thus, the first living creature is an identity of the particular and the general, the "lasting" invariant of which - in contrast to Kant's example of the triangle - is not to be found in the mere outer form, but in a specific production dynamism. By deducing this prototype of production dynamism from principles of self-organization, the natural scientist proceeds just as constructively as a mathematician. Unlike Kant's geometrician, who constructs invariants of specific geometrical **objects**, according to Schelling, the natural scientist should seek to construct invariants of **process types**, from which the entire multiplicity of a specific "sphere" can be derived.

Another important difference to Kant's example is that the geometrician constructs his objects as limitations of space, while in Schelling's example, the physicist constructs the specific spheres as limitations of natural activity by applying a principle of inhibition. Like space, which is not a single object and therefore cannot be grasped through empirical but only through non-empirical intuition, the original activity of nature also can only be perceived by non-empirical intuition. Only the objects of nature and their changes can be empirically experienced, but not their original construction. (Note 1) Because of the possibility of conceiving the original productivity of nature by means of non-empirical intuition, it follows that speculative physics also can apply the constructive method.

Schelling endows the basic invariants of all Being with a certain dynamism. Not the atoms in absolute space and absolute time and the forces interacting with them are the basic invariants, but the process of self-construction in its unlimited definition. For Schelling "Being in itself = activity" (more about this in the next section). In his "Allgemeine Deduktion des dynamischen Prozesses oder der Kategorien der Physik" Schelling tried to demonstrate this field-theoretically in terms of magnetic, electrical, and chemical processes. Further research would be needed to determine whether or not he succeeded. Schelling also postulated that the process of self-construction is in some way linked to a unification of the geometrical continuum and arithmetical discontinuum. /23, V 125 sqq./

The limiting conditions, i.e. inhibitions, of unlimited natural activity each determine certain spheres (transformation groups) of nature, which can be characterised by specific forms of production. For Schelling evolution is not a chain of product stages but a chain of productivity. According to Schelling: "To date it has been claimed that every organisation corresponds to a certain development stage. I can now claim the reverse: the diversity of the stages alone accounts for the diversity of organisations." /23, III p.54/ The different stages of development are not designated by objects and their invariants, but by production and their invariants. Invariants of specific types of production result from restrictions of unrestricted activity by the inhibition principle. Schelling writes: "Each stage of development thus has its own character. **At each stage of development, nature is restricted to a certain - only one possible - Gestalt**, in respect of this Gestalt she is completely tied, in the production of this Gestalt she shows no freedom." /23, III p.43/ The fixation of development at a certain development stage is not the same as nature being entrenched at this stage. For in Schelling's words on dynamic systems: "That the product is inhibited at a certain development stage does not mean that it ceases to be active, but that it is restricted in its productions, i.e. it can reproduce ad infinitum nothing but itself". /23, III p.59/

If, like Schelling, one postulates a history of nature that has successively produced different evolutionary stages and successively created the levels of inorganic, organic and cognitive nature as various "Potenzierungsstufen", the construction of inorganic matter must then be the most basic work of the natural sciences, from which all other stages must derive. Schelling does in fact represent a certain **physicalistic standpoint**. He is of the opinion that the theory of the self-construction of matter is the basis for the theory of the origin of life and the biological evolutionary process. Schelling rejects the vitalistic point of view. According to him, life must be explainable in physical terms, otherwise it is impossible to comprehend its origin from inorganic matter. /9/ For Schelling, evolution is selfsimiliar because of "the recurrent self-construction of matter at various stages". The theory of the history of nature is therefore equivalent to the theory of the self-construction of matter. /23, VI p.4/

According to Schelling: "The only task of the natural sciences is: **to construct matter**. This exercise can be solved, although the application derived from this general solution is one that is never complete. If it were the intention of a general theory of nature to reach a consciousness of the infinite multiplicity and depth of phenomena, which are unconsciously produced in nature, it would have to be considered an impossibility." /23, IV p.3/ To penetrate the endless variations of nature's products in detail would be an exercise far beyond all our finite powers. Schelling therefore takes the view: "Our entire effort can only be dedicated to researching the **general** principles of all nature's productions - their application, however, which extends in all dimensions into infinity, must be regarded as an unending exercise." /23, IV 3 sq./

It was Schelling's basic idea that it must be possible to construct the objects of nature from a few underlying, physical production principles. His aim was a unified theory of the history of nature. Given the limited knowledge of the natural sciences and mathematics at that time, this demanding research programme could not be carried out in Schelling's day. However, it encouraged a number of scientists to look for nature's basic process invariants, for example Robert **Mayer**, an admirer of Schelling, who formulated the principle of energy conservation.

By attempting to establish the construction method which Kant wanted to reserve for mathematics in the realm of physics as well, Schelling philosophically upgraded theoretical physics. Schelling was thus fascinated by Le Sage's theoretical mechanics because they deduced the whole structure of this science from a few basic principles. Schelling merely thought that the principles of mechanics should be supplemented by dynamic self-organisation principles. By the same token, there were a number of theoretical physicists who appreciated Schelling's methodical approach. **Hamilton** can be cited as representative of many: "There are, or may be imagined, two dynamical sciences: one subjective, **a priori**, metaphysical, deducible from mediation on our ideas of power, space, time; the other objective, **a posteriori**, physical, discoverable by observation as a generalization of facts and phenomena." /4, p.179/

It was also Schelling who realized early on that a theory of natural history in strict terms of a priori-constructions is problematic. In an interesting work entitled "Is a Philosophy of History possible?" (1797/1798) he stated: "whatever has to be calculated a priori or happens according to necessary laws, is not the object of history; and in reverse, whatever the object of history is, does not need to be calculated." /23, I p.467/ Calculable processes, according to Schelling, include all periodically recurring events. Consequently, history cannot happen in periodical cycles, but must be "progressive". However, this progress, if historical, can only be based on freedom which enables specified boundaries, or spheres, to be exceeded. In this essay, Schelling therefore concludes that an a priori science of history is not possible, for such a science would have to show the necessity of events which would effectively exclude freedom. In his subsequent writings he attempts to solve this problem by assuming a "law of freedom", in which necessity and freedom are not opposed and which occurs in the transition from one sphere to another. Schelling's definition of freedom should not be confused with coincidence or mere arbitrariness. In the case of mere arbitrariness we would only be infinitesimally different from Buridan's mule, who could not decide between two haystacks of the same size and composition at the same distance from each other. Truly free are those who can transcend causal connections and shape them autonomously. Here, again, as already mentioned at the outset, abstract thinking is required to lead us to a higher plane. According to Schelling, nature's freedom is not demonstrated so much in incidental fluctuations, but in the production of higher levels of multiplicity or "Potenzierungsstufen", which each determine a higher hierarchy level.

4. Schelling's "Speculative Physics" of Self-Organization

Schelling published several "drafts" on the physics of self-organization. The "Ideen einer Philosophie der Natur" (1797) is still strongly influenced by Kant's problem of self-reproduction of the biological sphere. In his subsequent work "Die Weltseele" (1798) he explores for the first time in detail the question of the origin of life from the inorganic sphere and, since he considers both physiological chemism and the vitalistic position as inadequate, he is driven to the problem of outlining a physics of self-organization. In his paper "Erster Entwurf eines Systems der Naturphilosophie" (1799) he explores the idea of a systematic natural history which attempts to deduce all evolution from basic principles of self-construction of matter. This idea, that only a theory of the original genesis of matter and its organization can provide a basis for an evolution theory, is taken a stage further in the "Allgemeine Deduktion des dynamischen Prozesses oder der Kategorien der Physik" (1800), in which Schelling seeks to derive spatial dimensions and matter from magnetism, electricity and chemism. For the first time here, Schelling no longer regarded magnetism and electricity as specific characteristics of certain types of matter, but defines these as constituents of all matter and as such as "categories" of physics, because they supply the schemata for their "self-construction". This was a very courageous hypothesis, for in Schelling's day, most physicists regarded magnetic and electrical phenomena as the effects of certain fluids, but not as a universal constituent of matter itself. In the following, I should like to focus on Schelling's "Erster Entwurf eines Systems der Naturphilosophie".

As Schelling's question concerns the origin and genesis of nature which refers to the entirety of real and possible objects, he cannot begin his construction with a single given object or a set of objects. This seems to be one of the major differences between the philosophy of self-organization and the natural science of self-organization. Schelling's philosophy explores the universal, basic conditions for objectivization, i.e. it asks how it is possible for isolated, limited forms to emerge at all. In contrast, the self-organization theories of the various branches of science begin with a prescribed system of parts. Furthermore, the external boundaries of the systems to be investigated are demarcated by means of an experimental arrangement. In the self-organization theories of natural sciences, "open systems" with limitations are prerequisites which allow the flow of energy or material induced from outside. /10/ Schelling's approach is different. He writes: "Originally there is for us in nature no one **single Being** (as one that has emerged), for otherwise our approach is not philosophical, but empirical. - We must see the **object** in its **first origin**. Initially, everything in nature, and nature as the quintessence of **Being**, does not exist for us. To philosophize about nature is to create nature." /23, III p.13/ Self-organization in this case is therefore not taken as an attribute of prescribed systems, but ontologically as their genetic ground.

But how should we make a start with our construction if not with an object or a number of objects? In the "Erster Entwurf" Schelling begins with "pure constructing as such" or "purely productive activity" which, as a basic principle of everything that is objective, is itself not an object and which has no corporeal-material substrate as its carrier. Individual things are only isolated restrictions of this activity; Being itself is meant as productivity in its unrestricted sense. (Note 2) - How then from unrestricted, constructive activity can individual, final products now come into being? To explain this, Schelling lists a series of conditions or restrictions.

Firstly he states that unrestricted activity is manifested not by a finite but only by an infinite product, namely by infinite becoming ("das unendlich Werdende"). This statement expresses the fact that universal productivity is not exhausted in a single product, for then evolution would already have come to an end. Infinite becoming would, however, manifest itself empirically as a becoming with infinite speed that produces nothing determinable if, apart from the production principle, there were no inhibiting principle, which, for its part, would collapse everything into a point if able to manifest itself unrestrictedly. Pure productivity only becomes empirical becoming through the inhibiting principle: "The true concept of empirical infinity is the concept of an action which is inhibited into infinity." /23, III p.16/ But at each point of the line it is still infinite. In this context Schelling takes recourse to mathematics: "The original, infinite series, of which all specific series (in mathematics) are only imitations, does not come into being by addition, but by **evolution**, by the evolution of **One size already infinite in its origin**, which flows through the entire series; in this one size is originally concentrated the entire infinity, the successions in the series are but the manifestations of the specific inhibitions (through reflection, i.e. self-reflexivity M. H.), which continually set limits to the expansion of that size in an infinite series (infinite space), which otherwise would occur with infinite speed allowing no real intuition." /23, III p.15/

From this construction of infinite becoming, Schelling concludes that each product of nature only appears fixed to our intuition, whilst in reality it is constantly undergoing change. The apparent fixation is brought about by natural products being reproduced, like a whirl in currents which, by constant reproduction, appears constant, whilst in reality it is in constant flux. The process, however, does not stop at reproduction, but contains further development potential.

The production and inhibiting principle can elucidate an evolution with finite speed, but not the emergence of different qualities. Thus Schelling adds an "infinite multiplicity" of original inhibiting, selfreflexive points which serve as a device for conceptualizing a "**dynamic atomic theory**". The evolutionary continuum is to a certain degree broken down into singularities, which are conceived as "original productivities" (now already in the inhibited sense). In this he observes that the most difficult problem of his Naturphilosophie is to explain how this infinite multiplicity of original actions could come about. (In his subsequent paper he explains this as points of indifference at which the duality of the production and inhibiting principle is overcome, thereby producing new differences.) These original actions, which Schelling also compares with Leibnitz' "Monads", demonstrate a common tendency to fill space from the inside out. Filled space is only the phenomenon of an aspiration, the principle of which is not in space itself. Schelling first considers this common filling of space as a completely **chaotic mix** in which, down to the smallest part, all "tendencies" are to be found. As a result, matter is not only infinitely separable like homogeneous space, but is really infinitely separated, differentiated and heterogenous. The first product of the chaotic mix is the "amorphous" in which each action is restricted and disturbed by all other actions. This dynamic equilibrium of actions can be disturbed by the "slightest change". How can order arise from this chaos?

As an answer, Schelling introduces the "**principle of subordination**", which restricts the degree of freedom of each action, thereby imposing an organization and design: accordingly, no matter can leave the amorphous state "without some action becoming predominant. However, no action can become predominant without suppressing or destroying another." /23, III p.37/ He explains this as follows: "Since each action is highly individual, and as each strives to produce what it has to produce according to its nature, this generates the drama of conflict in which no force wins or loses completely. The egoism of each individual action will have to bow to that of all the others; whatever happens is the product of subordination of all to one and one to all, i.e. completely mutual subordination." /23, III p.41/

Should a certain macroscopic organization of elementary actions come about by subordination, a certain **proportion** of actions to one another can be observed. The difference in the various organizations is then the result of each of the different proportions of the actions to one another. "As many variations in the proportions of these actions as are possible, so many variations of different forms and development stages." /23, III p.43/

The transition from one form to another is enabled by cancelling the equilibrium of the actions, which manifests itself by a certain proportionality. The more independent actions become from each other, the more "the equilibrium within a certain natural sphere, which they define, will be disturbed. If they arrive at the apex of mutual independence, they reach the highest moment of disturbed equilibrium. But in Nature the highest moment of disturbed equilibrium is one and the same as the restoration of (a new, M.H.) equilibrium. Between the two, no time passes." /23, III p.50/

Schelling also defines the term organization as the common expression for a multiplicity of actions, which mutually confine one another to a certain sphere. The sphere is "that which has emerged in conflict, so to speak the monument of those intermeshing activities" or "the lawfulness of the product itself" /23, III p.65/. This strange term "**sphere**", which appears to be Schelling's own creation, does not mean the same as the mere sum of single activities, but contains a synthetic, macroscopic quality: and that is the lawfulness of the product itself. In his work "Weltseele", Schelling had already stated: "The perennial would then not be the phenomena within this sphere (for they would in this case also emerge and disappear, disappear and emerge again), but the perennial would be the sphere itself, within which each phenomenon is contained: the sphere itself could not be just a mere phenomenon, for it would be that which has emerged in the conflict of those phenomena, the product, so to speak, the concept (the enduring) of those phenomena." /23, II p.516/ Accordingly, the sphere does indeed consist of components and individual processes, but only the integration of these factors into an "indivisible", "coherent" whole gives rise to the sphere itself. This is now governed by laws which can only be explained in terms of this cooperative whole, the product itself, but not in terms of individual components and their characteristics.

5. Schelling and Riemann

Riemann came into contact with Schelling's philosophy by means of Herbart. Herbart was Riemann's colleague at the university of Goettingen, where he taught philosophy. From the unpublished posthumous writings of Riemann in the manuscript collection at Goettingen university we can conclude that Riemann had acquainted himself with Schelling's philosophy. Moreover Riemann very much appreciated Fechner's philosophy of nature that included many of Schelling's points, especially the idea of nature as an organized whole. The notes Riemann made of Herbart's writings show that Riemann was primarily interested in questions of genesis and the development of nature. Here we can only mention a few instances of agreement between Schelling and Riemann's concept of nature and epistemology. A detailed comparison is still in preparation:

1) Riemann's method differed from that of the empirically oriented scientists of the 19th century. But at the same time he avoids an a priori mathematical approach. According to Riemann science proceeds by improving the traditional conceptual systems. By virtue of new concepts "we do not only add new perceptions each moment, but also classify future perceptions as necessary or - if the conceptual system is incomplete - as probable." /22, p.521/ On the other hand, unexpected perceptions can lead to changes in the conceptual system. "By this process our understanding of nature is improving continously and at the same time going further **back behind the surface of phenomena**.(my emphasis, M.H.)" /loc.cit./ Science should not content itself with the observation and calculation of the outside of nature, but aim at the "**inside of nature**". In his own theory-building work Riemann relies on an "**inner perception**", a non-empirical intuition /22, p.528/. The difference between the external perceptible objective side of nature and its internal workings was Schelling's main argument against the empirically oriented scientists of his time, and he intended to establish a science of nature which seeks to decipher the internal productive processes of nature.

2) Riemann shared Schelling's persuasion that the internal workings of nature cannot be purely mechanical. Therefore he aimed at a **new concept of nature**: "My main efforts concern a new conception of the well-known laws of nature - an expression in different basic terms -, and this enabled me to use experimental data on the interaction between heat, light, magnetism and electricity for an investigation of the connection between them." /22, p.507/ Riemann intended to develop a concept of nature "beyond the foundations of astronomy and physics laid

by **Galilei** and **Newton**" /22, p.528/ that would be in accordance with continuum theory. The atomistic and mechanistic hypothesis, according to which natural phenomena can be explained by reduction to elementary particles and their mechanistic interaction in an empty Euclidean space, seemed insufficient to him. Riemann's ideas aimed at a general field theory, although he assumed - like traditional continuum theory - the existence of ether /Scholz, 24, p.99/. He imagined cosmic space as filled with an incompressible, homogeneous fluid without inertia, i.e. he did not support a mechanistic hypothesis of ether. Riemann, quite like Einstein, speaks of a "physical space" whose "forms of motion" are supposed to explain the forms of motion of light and gravitation. Riemann's work on non-Euclidean geometry originates from a period, during which he was almost exclusively occupied with "Naturphilosophie" /18, p.547/ and much concerned with finding a unifying concept for "electricity, galvanism, light, and gravitation" (loc.cit.). The same idea was central for Schelling, and had inspired Oersted certainly, and perhaps Faraday, in their attempts at unification /8/. Furthermore, and this is the really surprising point with respect to Riemann, Schelling had already arrived at a concept of space quite beyond the Kantian. For Schelling space was not an absolute intuition, given a priori by the organization of our mind, in which material interaction can be localized. He considered the dimensions of space as generated by the physical processes of magnetism, electricity and the chemical processes /23, IV p.1-78/. For Schelling space was not a Euclidean container for physical processes, but - like Riemann - he saw its structure as generated by "the self-construction of matter" /23, IV p.4/.

3) Furthermore Riemann tried to combine his continuum concept with particle physics. The fact that he considered the continuum to be a kind of fluid (without properties of inertia) suggested a treatment of the problem in the context of hydrodynamics. In his 1860 paper "Ueber die Fortpflanzung ebener Luftwellen von endlicher Schwingungsweite" Riemann developed a process-theoretical approach to the problem of continuity and discontinuity. In this paper he dealt with the emergence of shock waves, i.e. stable, definite entities with particle properties. Riemann found out that differential equations of plane air waves expose a discontinuity, a singular instability at exactly the point where "sudden compression" of waves produce "shocks of compression" /21, 164 ff./. Riemann sees the reason for this discontinuity in a "salient" change of the physical parameters (essentially in density and speed) which make a linear description impossible. According to Riemann the differential equations lose their validity at the point where the shock waves originate. They cannot serve to calculate the process of emergence, because some of the terms assume the value of infinite at the singulary point. Quite in accordance with contemporary bifurcation theory Riemann comes to the conclusion that spontaneous, dynamic structure-building processes exhibit a non-differentiable quality in a critical point, a fact which points out a **non-deterministic** and **non-mechanistic** process of physical nature. Instead of introducing - like modern theories of self-organization - accidental "fluctuations" by way of stochastic methods, Riemann looks out for characteristic **invariants** in order to decipher the dynamics of shock waves in the

point of origin. He is looking for non-mechanistic laws of nature, or - in terms of Schelling - for a law of freedom, for Riemann was convinced that "freedom is very well compatible with strict lawfulness of the course of nature."/22, p.519/

Schelling also tried to deduce discontinuity from the continuum, but in contrast to Riemann he interpreted the continuum as an "absolute productivity" which become discontinous by the agency of the selfreflexive inhibitions. In this context Schelling was also interested in hydrodynamic processes. As he did not know shock waves, he often used the formation of whirls as an example. /e.g. 23, III p.289/.

4) The fact that Riemann was influenced in his reflections on the self-organization of nature by Schelling's philosophy is supported by a remark he makes on the origin of life /22, p.514/. There he assumes - using the same words as Schelling - an "**organizing principle**" in order to explain the emergence of the first organisms on earth out of the inorganic. According to Riemann this organizing principle has to be interpreted as analogous to the process of thinking. Schelling was one of the first to reflect upon the origin of life from inorganic matter, and in his reflections was led to an "organizing principle" that is very similar to present-day ideas of self-organization /8/, /9/. Furthermore Schelling saw the self-organizing principle as a principle that is at work in human mind as well as in nature. Obviously Riemann agrees with this "identity philosophy", because he identifies the organizing principle with mental activity. Mental process takes place not only in us, but is analogous at the same time to a "process on the inside of ponderable matter", for "the impossibility to explain mental activities by way of material motions in space can easily be affirmed by anyone who tries to analyse his own inner perceptions; but still we can concede the abstract possibility here." /22, p.516/ Thus Riemann - like Schelling - assumed an identity of the internal principle of matter and the principle of mental activity.

5) Like Schelling Riemann is not merely an advocate of biosophy. For him the primary ontological reality is not the organism, but the organizing principle, which is the basis of the emergence of life from the inorganic sphere and of its evolution and therefore cannot be construed as an inherent principle of the organisms itself. The organizing principle always leads from existing formations to more perfect formations: "In the course of its life (the earth's, M.H.) always new and more perfect organisms appear. So we must look for the reason in a mental process which is proceeding to higher stages."/22, p.517/ Riemann was obviously interested in the **progressive moment of development towards higher levels**, the emergence of new stages of diversity ("Mannigfaltigkeitsstufen"), which he saw in analogy to the progessive, creative process of thinking. This seems to me to be the strongest evidence for Schelling's influence, for the later so-called "Romantische Naturphilosophie" placed much more emphasis on the metaphor of the organism, a metaphor that belongs to the context of hylozoism, i.e. a theory that Schelling rejected. /11/ This has to be taken into account especially for an assessment of the influence of Fechner, to whom Riemann refers repeatedly.

(6) Riemann refers to **Herbart** and **Fechner** as philosophical sources. /22, 507 ff./, /Scholz 25/ Riemann could follow Herbart in his epistemology and psychology, but not in his philosophy of nature, which was conceived as a programm opposing the Schelling School. Herbart was one of the acutest critics of the well-known doctrine of "Naturphilosophie", namely absolute identity, which "is contrary to experience which shows us a multiplicity of things in accidental connection and separation." /7, VII p.130/ Instead of conceiving the invariants of productivity a future philosophy of nature should rely on the chemical elements as invariants. "Our chemistry leads us to the opposite conclusion, namely to elements which maintain their inherent nature through all their changes of state in **accidental** mixtures..."/7, V p.134/

Riemann obviously could not agree with this chemical atomism. At this point he favoured Fechner /2/, /3/. Fechner - via Oken - was influenced by Schelling's philosophy /5/, /6/ but deviates from Schelling in certain points. As a matter of fact, Riemann returns to Schelling's position in exactly these points. In agreement with Fechner Riemann was convinced "that earth is an animated being", for the well-organized expediency of the biosphere suggests an animated nature. In contrast to Fechner Riemann did not consider inorganic matter a later product of evolution, but agrees with Schelling in the conviction that organic matter emerges from inorganic by virtue of an organizing principle.

6. Final Remarks

In 1832 Schelling held a public lecture before the Bayerische Akademie der Wissenschaften about "Faradays neueste Entdeckung" /23, IX p.439-452/, an invention which, although scarcely publicly known, he praised enthusiastically. This speech testifies to the keen interest Schelling the philosopher took in major research advances of the natural sciences. If Schelling were able to experience the discoveries being made today about the self-organization of nature, he would certainly be very pleased, not only because they confirm his fundamental hypothesis that self-organization also extends to physical nature.

I am glad to have had the opportunity, on the occasion of Hermann Haken's 65th birthday, to give you an account of some of Schelling's ideas, albeit allowing for the fact that Schelling's thinking was far more profound than I am able to convey here. Today of course we can only look back on Schelling's work and read that which our minds comprehend today. But nevertheless it is astonishing to find so much in this work, written some 200 years ago, that is relevant to us today. In fact, for many it was an amazing revelation to find that the idea of nature's self-organization, which has revolutionized our view of nature since the 1960's and 1970's, had already been elevated to a comprehensive research programme by a Naturphilosoph in 1800, and had even inspired a number of scientists and mathematicians in the 19th century in their attempts to develop a non-mechanistic, dynamic theory of nature.

Acknowledgments:

For reading the paper I thank Wolfram Hogrebe, Jochen Lechner, Hans Nellißen and Rainer Scharf.

Notes:

(1) Non-empirical intuition, i.e. "productive intuition", refers to the creative process or, on an even more basic level, to the original creation of the ego, which is decribed by Schelling in his transcendental philosophy, and cannot be discussed here.

(2) In his later works, e.g. his "Darstellung des Naturprozesses" of 1843/44 on the other hand, Schelling assumes an as yet intangible potentiality, a pure "Seinkönnendes", as he calls it, which is not yet, but just could be. This "Seinkönnendes" does not manifest itself at once in the form of objects with a material body, but initially to a certain extent as an energetic field from which the matter then comes into being, namely by a inversion (= "uni-versio") of pure energy back into latency. It is a principle, which though it is in space, still does not actually "fill it, i.e. make it impenetrable, just as light is in space without filling it". /21, X p.352/

References

(1) Caneva, Kenneth L., Conceptual and Generational Change in German Physics: The Case of Electricity 1800-1846, Diss. Princeton 1975

(2) Fechner, Gustav Theodor, Nanna oder über das Seelenleben der Pflanzen, Leipzig 1848

(3) Fechner, Gustav Theodor, Zend=Avista oder über die Dinge des Himmels und des Jenseits. Vom Standpunkt der Naturbetrachtung, 3 Teile, Leipzig 1851

(4) Hankins, Thomas L. Triplets and Triads: Sir William Rowan Hamilton on the Metaphysics of Mathematics, in: Isis (1977) 68, pp. 175-193

(5) Hartung, Walter, Die Bedeutung der Schelling-Okenschen Lehre für die Entwicklung der Fechnerschen Metaphysik, Diss. Bonn, Leipzig 1912

(6) Heidelberger, Michael (1990), Selbstorganisation im 19. Jahrhundert, in: Wolfgang Krohn, Günter Küppers, Selbstorganisation. Aspekte einer wissenschaftlichen Revolution, Vieweg-Verlag, Braunschweig/Wiesbaden 1990, pp. 67-104.

(7) Herbart, Johann Friedrich, Sämtliche Werke, hg. v. Karl Kehrbach, Langensalza 1887 ff.

(8) Heuser-Keßler, Marie-Luise, Die Produktivität der Natur. Schellings Naturphilosophie und das neue Paradigma der Selbstorganisation in den Naturwissenschaften, Duncker&Humblot-Verlag, Berlin 1986

(9) Heuser, Marie-Luise, Schellings Organismusbegriff und seine Kritik des Mechanismus und Vitalismus, in: Allgemeine Zeitschrift für Philosophie, 14 (1989) 2, pp. 17-36

(10) Heuser, Marie-Luise, Zur Kritik gegenwärtiger Selbstorganisationstheorien, in: Philosophy of the Natural Sciences, Proceedings of the 13. International Wittgenstein-Symposium, Kirchberg am Wechsel, Austria 1988, Wien, pp. 246-249

(11) Heuser, Marie-Luise, Wissenschaft und Metaphysik. Überlegungen zu einer allgemeinen Selbstorganisationstheorie, in: Wolfgang Krohn, Günter Küppers (ed.), Selbstorganisation. Aspekte einer wissenschaftlichen Revolution, Vieweg-Verlag, Baunschweig/Wiesbaden 1990, pp. 39-66.

(12) Heuser-Keßler, Marie-Luise, Maximum und Minimum. Zu Brunos Grundlegung der Geometrie in den Articuli adversus mathematicos und ihrer weiterführenden Anwendung in Keplers Schrift Neujahrsgabe oder Vom sechseckigen Schnee, in: Klaus Heipcke, Wolfgang Neuser, Erhard Wicke (ed.), Die Frankfurter Schriften Giordano Brunos und ihre Voraussetzungen, Weinheim 1991, pp. 181-197.

(13) Heuser-Keßler, Marie-Luise, Keplers Theorie der Selbststrukturierung von Schneeflocken vor dem Hintergrund neuplatonischer Philosophie der Mathematik, in: Selbstorganisation. Jahrbuch für Komplexität in den Natur-, Sozial- und Geisteswissenschaften, Berlin 1991, S. 43-62.

(14) Hogrebe, Wolfram, Prädikation und Genesis. Metaphysik als Fundamentalheuristik im Ausgang von Schellings "Die Weltalter", Suhrkamp-Verlag, Frankfurt a. M. 1989

(15) Kant, Immanuel, The Critique of Pure Reason (1781), translated by J. M. D. Meiklejohn, in: Great Books of the Western World, ed. by Robert Maynard Hutchins, Bd. 42: Kant, Chicago/London/Toronto/Geneva 1952, pp. 1-252

(16) Kant, Immanuel, The Critique of Judgement (1790), translated by James Creed Meredith, in: Great books of the Western World, ed. by Robert Maynard Hutchins, Bd. 42: Kant, Chicago/London/Toronto/Geneva 1952, pp. 461-613

(17) Poincaré Henri, Why Space has Three Dimensions?, in: Mathematics and Science: Last Essays, Dover Publications, New York 1963

(18) Noether, M./ W. Wirtinger, Bernhard Riemann's Lebenslauf, in: B. Riemann, Gesammelte Werke, hg. v. e. Zermelo, Berlin 1932, pp. 539-538.

(19) Otte, Michael, the Ideas of Hermann Grassmann in the Context of the Mathematical and Philosophical Tradition since Leibniz, in: Historia Mathematica 16 (1989), pp. 1-35

(20) Riemann, Bernhard, Gesammelte Werke, hg. v. E. Zermelo, Berlin 1932

(21) Riemann, Bernhard (1860), Über die Fortpflanzung ebener Luftwellen von endlicher Schwingungsweite, in: B. Riemann, Gesammelte Werke, hg. v. E. Zermelo, Berlin 1932, pp. 157-175.

(22) Riemann, Bernhard, Fragmente philosophischen Inhalts, in: B. Riemann, Gesammelte Werke, hg. v. E. Zermelo, Berlin 1932, pp. 507-538.

(23) Schelling, Friedrich Wilhelm Joseph, Sämmtliche Werke, hg. v. K.F.A. Schelling, Stuttgart 1856-1861.

(24) Scholz, Erhard, Geschichte des Mannigfaltigkeitsbegriffs von Riemann bis Poincaré, Birkhäuser-Verlag, Boston/Basel/Stuttgart 1980

(25) Scholz, Erhard, Herbarts Influence on Bernhard Riemann, in: Historia Mathematica, 9 (1982), pp. 413-440

(26) Scholz, Erhard, Riemanns frühe Notizen zum Mannigfaltigkeitsbegriff und zu den Grundlagen der Geometrie, in: Archive for History of Exact Sciences, 27 (1982), pp. 213-232

(27) Scholz, Erhard, Symmetrie, Gruppe, Dualität, Birkhäuser-Verlag, Basel/Boston/Berlin 1989

Index of Contributors